MW00615743

# Fluvial Processes in River Engineering

# Fluvial Processes in River Engineering

**Howard H. Chang**
San Diego State University

A WILEY-INTERSCIENCE PUBLICATION
**John Wiley & Sons**
New York • Chichester • Brisbane • Toronto • Singapore

***Library of Congress Cataloging in Publication Data:***

Chang, Howard H., 1939–
Fluvial processes in river engineering.

"A Wiley-Interscience publication."
Includes bibliographies and indexes.
1. River engineering. I. Title.

TV405.C47 1987 627′.12 87-21591
ISBN 0-471-63139-6

Printed in the United States of America

10 9 8 7 6 5 4 3 2 1

# PREFACE

Rivers have been a focus of human activities throughout ancient and modern times. So important to mankind are the benefits obtained from rivers, and so necessary is the protection against floods and other river disasters. While engineers are interested in water supply, channel design, flood control, river regulation, navigation improvement, and so on, it has been clear that rivers, as a part of nature, can be mastered not by force but by understanding. Rivers have been a subject of study by engineers and scientists who have been fascinated by the self-formed geometric shapes and their responses to changes in nature and human interferences. In addition to engineering, understanding river behavior is also necessary to environmental enhancement. Few other subjects have been studied more extensively than rivers, yet some major aspects of the hydraulics, sedimentation, and fluvial processes have only become clear to us in recent years. Of course, many more aspects are yet to be understood.

River flow is a type of open-channel flow because of the free surface. Considerable knowledge is required to make a determination of the free surface for rigid channels. But all boundaries of a river are free surfaces, a fact well described by J. F. Kennedy in the *Journal of Hydraulic Engineering,* 1982. The hydraulics and fluvial processes of rivers are far more complex than rigid channels. For example, analytical determination of the self-formed width and its adjustments has only become possible in recent years. Width formation is one aspect of fluvial processes to which other aspects, such as depth and meandering pattern, are closely related.

Recent development in river research has substantially extended beyond the hydraulics of sediment transport, stimulated by the abundance of erosion and sedimentation problems. Much progress has been made to provide analytical methods for alluvial channel design, river morphology, and mathematical simulation of river channel changes. My motivation to write this book stems from recent advances, which need to be assembled in book form.

Fundamental principles and applications are presented in this book. Because of the close interrelationship between river engineering and fluvial geomorphology, geomorphic approaches are integrated with the engineering principles. The river as a part of the fluvial system is also analyzed from the system point of view. The selection of key references and the bias of this book are dictated by my personal experience and interest, which may be different from those of others.

Through teaching, research, and consulting, I have come across different problems that require engineering analysis and solution. Approaches to solving these problems have been developed by engineers and scientists, including myself. Underlying such approaches are the common principles for river hydraulics and fluvial processes. The large variety of problems covered in this book are actually governed by common principles of fluvial processes.

This book is intended primarily as a textbook for civil engineering students at the senior or graduate level, to be used in a course on river and sedimentation engineering. The materials in this book may be covered in three to five units for college credit. Suggested prerequisites for this book include basic training in mathematics (calculus and differential equations), computer programming, and basic hydraulics. This book is also a reference book for professionals directly involved in flood control, sedimentation, fluvial processes, bridge design, waterways, irrigation, and so on. It is also useful for researchers in hydraulics, agricultural engineering, geomorphology, environmental sciences, and geography.

My initial fascination with regard to rivers was attributed to the stimulation by Daryl B. Simons and Stanley A. Schumm, to whom I am greatly indebted. I am also very grateful for the ideas obtained through professional association with Joseph C. Hill, V. Miguel Ponce, Vito A. Vanoni, and C. Ted Yang.

## BOOK COVERAGE

The subject matter of this book includes river flow, river channel formation, the physical characteristics of rivers, responses of rivers to natural and human-made changes, and analytical methods of design and evaluation. A comprehensive presentation of river processes and engineering must be built upon the foundations of fluvial geomorphology, hydraulics of river flow, and sediment transport. On the basis of this logic, this book is organized into the following five principal parts:

Part I. Fluvial Geomorphology
Part II. Foundations of Fluvial Processes
Part III. Regime Rivers and Responses
Part IV. Mathematical Modeling of River Channel Changes
Part V. River Engineering

Part I, which is on fluvial geomorphology, presents an overview of the fluvial system, including the river and its drainage basin. This part covers fundamental materials related to rivers (developed following essentially the geomorphic approach) such as the variables for alluvial rivers, the regime concept, channel-forming discharge, river classifications, hydraulic geometry, meander planform, thresholds in river morphology, and geomorphic analysis of river responses. These topics provide the general framework and set the stage for more in-depth and analytical treatment of river processes. Geomorphic principles described in this part will be extended by engineering approaches in the remainder of the book.

River channel formation and responses to change are direct results of the complex interaction of the flow and its boundary. Physical and analytical foundations of such fluvial processes are presented in Part II. Major topics include open-channel hydraulics, physical properties of sediment, scour criteria and scour-related problems, alluvial bed resistance, sediment transport in rivers, and flow and sediment processes in curved channels. These subjects provide the foundation for the comprehensive study of rivers. Such materials are typically covered in a standard textbook on sedimentation. This book incorporates recent advances in these subjects; it also includes the subject of flow in curved river channels, which is usually not included in a book on sedimentation. While materials presented in Part II serve as the analytical bases for the remainder of the book, they are also useful in engineering analysis as they stand alone.

Part III is on regime rivers and responses. While important empirical methods are included in the book, the thrust is on the analytical approach to determine the basic parameters for the hydraulic geometry of alluvial rivers under the dynamic equilibrium and to quantify the fluvial processes of river channel formation and adjustments of equilibrium. In this coverage, basic physical relationships for river channel formation are first outlined. They are then applied to develop design methods for stable alluvial channels and to establish hydraulic geometry for natural rivers. Complete analyses are provided for the important features of alluvial rivers, such as channel geometry, river meanders, channel patterns, and the thresholds with regard to distinct characteristics of river channels. A unique feature is the quantitative prediction of channel response to change, illustrated by case histories.

The planform of river meanders and the processes of migration are also covered in Part III. The process that characterizes the flow through meanders is the streamwise variation of the helical motion or secondary currents, to which many features of river meanders are related. Analytical determination of the self-formed meander planform is based on the streamwise variation in secondary currents.

Part IV is devoted to mathematical modeling of alluvial channels, in which modeling techniques and applications are described and illustrated by case studies. The scope is on quasi-two-dimensional modeling to simulate fluvial processes in river channels with a changing boundary. River channel changes so simulated include channel-bed scour and fill (or degradation and aggradation), width variation, and changes in bed topography induced by curvature effects. The mathematical modeling has its physical foundation in the fluvial process-response, which is characterized by the river's constant adjustment toward dynamic equilibrium subject to the physical constraints. In response to a natural or human-made change, the transient behavior of an alluvial river is reflected in its adjustment toward dynamic equilibrium, although the dynamic equilibrium may never be attained in nature because of the changing discharge.

Mathematical modeling is presented in modular form, with major components of water routing, sediment routing, and simulation of river channel changes. The physical foundation and modeling techniques for each component are described in detail.

A variety of transient problems for erodible channels are elucidated using the approach of computer-aided study, including general scour at bridge crossings, gradual breach morphology, erosion and deposition induced by instream sand and gravel mining, tidal responses of the river–delta system, water and sediment routing through curved channels, fluvial design of river bank protection, and stream gaging of fluvial sediment. Several field tests of modeling studies are also presented.

Part V, which is on river engineering, is devoted to engineering measures for achieving river training, including bank protection, dikes, grade-control structures, and so on. Each measure is illustrated by examples.

## SPECIAL FEATURES

This book is one of the first to present a complete analytical treatment of river morphology and its responses to environmental and human-made changes from the engineering point of view. This comes about because of recent advances on the analytical determination of the regime width of rivers and transient adjustments in width. Basic principles underlying fluvial processes are summarized.

Computer modeling of alluvial rivers and its applications are within the theme of this book. This approach has been a focus of recent research and development in this evolving field. From a sound physical foundation, mathematical techniques are employed in this book to integrate those convoluted physical relationships into a model for simulating river channel changes. Computer-aided analysis and design for river projects are illustrated by abundant examples. Such applications are especially useful for rivers, whose natural geometry would prohibit the traditional approach.

In addition to engineering application, mathematical modeling is also valuable for researchers who have been helped by the computer to understand physical phenomena and to test hypotheses. My research has been greatly benefited by the countless trials made on the computer.

This book presents considerable discussion on the flow and sediment processes in curved channels. In addition to the mechanics of flow in the channel, it also associates the channel morphology, such as meander planform and bed topography, to the secondary currents inherent in curved channels.

This book presents materials that are related to engineering design. Such design-related topics are included in many chapters, from traditional design methods to computer-aided channel design.

Predicting channel response to change, whether caused by nature or human interference, is important in river engineering and environmental studies. This book is unique in presenting quantitative methods for predicting river's long-term adjustments in regime as well as short-term changes during a flood event.

The basic topics of alluvial flow resistance and sediment transport are generally included in existing books on sediment transport. This book incorporates recent development and thus brings up to date such coverage. The subjects of sediment diffusion and sorting in nonequilibrium sediment transport are among the recent development in sedimentation covered in this book.

*San Diego, California*
*October, 1987*

HOWARD H. CHANG

# CONTENTS

# PART I

# FLUVIAL GEOMORPHOLOGY

# 1

# INTRODUCTION

River engineering involves the control and utilization of rivers for the benefit of mankind. Its scope, in the broad sense, may include river training, channel design, flood control, water supply, navigation improvement, hydraulic structure design, hazard mitigation, and environmental enhancement. River engineering is somewhat different from other aspects of civil engineering because the emphasis is often on the river responses, long term and short term, to changes in nature and to control and utilization, such as damming, channelization, diversion, bridge construction, and sand and gravel mining. Evaluation of river responses is essential at the conceptual, planning, and design phases of a project. The use of fundamental principles of river and sedimentation engineering is required in analysis for the special conditions of each river project.

Although the domain of river engineering is limited to the river channel, and the time scale is for the engineering interest, river behavior is affected by the fluvial system as a whole, thus involving the geologic time scale. River engineering must therefore be based on proper understanding of the fluvial system and geomorphic concepts.

## 1.1 THE FLUVIAL SYSTEM

A river is within the domain of the fluvial system, which also consists of the drainage basin (or watershed) and the downstream reservoir, lake, or ocean. Schumm (1977) divided the fluvial system into three parts, as shown in Fig. 1.1. The upper part, or Zone 1, is the watershed, where most of the water and sediment for the river originate. Small streams in the area are characterized by unstable, and often braided, channels. Because of the unstable channel pattern, the study of stream morphology can only be accomplished on certain gross features but not on the details.

The middle part, or Zone 2, is the reach where the river channel is the most stable and where its configuration is the best defined. Large rivers have long reaches of Zone 2, but this zone may be missing in small streams. This is the reach for which extensive river studies, modeling, and control have been made. Despite the

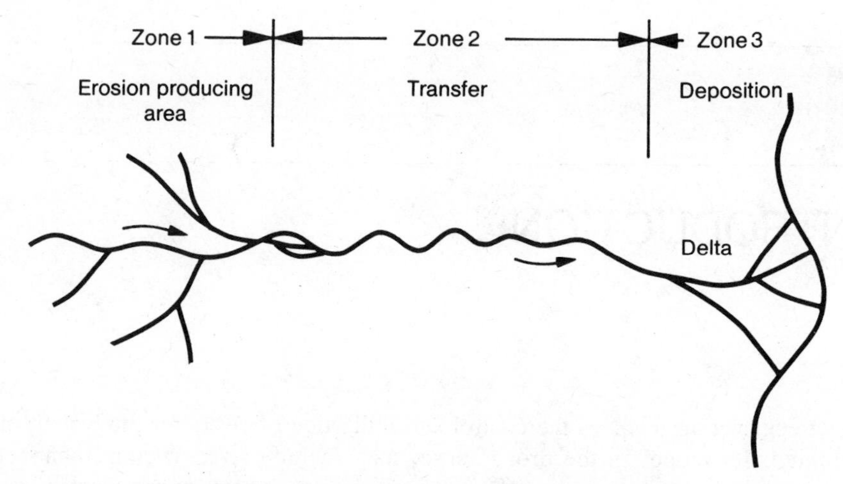


**Figure 1.1** The fluvial system (after Schumm, 1977).

relative stability, a river channel, as a dynamic system, still undergoes changes that can be rapid and significant at times.

Zone 3 is near the river mouth, where the alluvial river is also under the influence of the tidal, or base level, variation. Because of continuous delta growth, it requires a rise in grade (aggradation) to maintain the channel slope and transport capacity. Rivers in this zone are often braided.

## 1.2 VARIABLES FOR ALLUVIAL RIVERS

River flow is an open-channel flow because of the free surface. A rigid channel has only one free surface, but an alluvial river that is unconstrained in developing its own geometry has all its boundaries as free surfaces. The self-formed geometry of the alluvial river encompasses many variables. The abundance of variables has made river morphology and river mechanics a complex subject that has attracted the attention of many engineers and scientists. As an initial step in searching for an adequate understanding of river channel formation and changes, the variables and their interlocking relationships involved in the fluvial processes need to be identified.

Variables for the alluvial river are classified as independent variables and dependent variables, that is, cause and effect. Those that are imposed upon the river by any source are independent variables or controlling variables, whereas those that result are dependent variables (Kennedy and Brooks, 1963). The river has no control of the independent variables but instead is controlled by such variables. Variables included in the present discussion are the fluid property, sediment property, and characteristics of the flow system, including water discharge ($Q$), sediment discharge ($Q_s$), channel width ($B$), flow depth ($D$), mean velocity ($U$), hydraulic radius ($R$), channel slope ($S$), and friction factor ($f$).

Choices of independent and dependent variables for open channel systems in quasi-equilibrium have been summarized by Langbein (see Leopold et al., 1964), Kennedy and Brooks (1963), Leopold et al. (1964), and Schumm (1971). The flume and field cases are distinguished because, in the former, certain variables can be controlled at will by the experimenter. Any variable so selected is an independent variable. In the field case for nontidal rivers, it is necessary to differentiate between long-term and short-term points of view. Schumm (1971) divided the time scale into the following classifications: steady time, graded time, and geologic time, which correspond to the short-term, long-term, and very long-term time spans, respectively. The steady time span is measured perhaps in days, the graded time is in the order of a few hundred years, and the geologic time may be in millions of years. The length of the time span should also vary directly with the size of the drainage basin. The absolute length of these time spans is not important. What is significant is the concept that a variable can be a cause or an effect, depending on the time scale.

River variables and their status as independent or dependent variables during designated time spans are summarized in Table 1.1. During the short-term, or steady, time span, sediment discharge is a dependent variable that may be expressed as a function of the velocity. Such a functional relationship implies that velocity is the cause and that sediment discharge is the effect. This is usually true for the short-term situation but not so for the long-term (graded-time) equilibrium under which the water discharge and sediment inflow that are determined in the drainage basin are independent variables for the river. These quantities are imposed upon the river reach from upstream and tributaries and are introduced at the upstream and tributary entrances to the reach. In the long term, a river must establish its velocity, geometry, slope, and other characteristics to maintain a balance between its transport capacity and the inflow quantities imposed. Therefore, the velocity and channel characteristics (width, depth, and slope) so established

**TABLE 1.1 River Variables for Different Time Spans (After Schumm, 1971)**

| | Status of Variable[a] | | |
|---|---|---|---|
| Variable | Steady (Short-Term) | Graded (Long-Term) | Geologic (Very Long-Term) |
| Geology (lithology, structure) | I | I | I |
| Paleoclimate | I | I | I |
| Paleohydrology | I | I | D |
| Valley slope, width, and depth | I | I | D |
| Climate | I | I | X |
| Vegetation (type and density) | I | I | X |
| Mean water discharge | I | I | X |
| Mean sediment inflow rate | I | I | X |
| Channel morphology | I | D | X |
| Observed discharge and load | D | X | X |
| Hydraulics of flow | D | X | X |

[a]I, independent variable; D, dependent variable; X, indeterminate.

are dependent variables. The velocity in a channel as agent of transportation is determined by, or adjusted to, the sediment load. Velocity is an effect, and load is the cause: This relationship is not transposable (Mackin, 1948).

In the evolution of a river channel, sediment transport rate is considered as a dependent variable for the short term. Any departure of this rate from the inflow rate to the reach is absorbed in temporary storage, which ceases at the stage of equilibrium when the transport rate and inflow rate are equalized.

For the graded-time span, the gradient of the alluvially formed valley floor, or valley slope, is considered as an independent variable for the river because the time scale for river valley formation is much longer as compared with that for river channel formation. The interest of engineers is more or less confined to the steady-time span or graded-time span. Therefore, the valley slope is generally accepted as a constant for practical engineering purposes. Among the various possible choices of independent and dependent variables, certain relations are unique and other are nonunique. The multiplicity in functional values is attributed to the nonuniqueness in channel roughness, which will be described in Chapter 6.

For the very long-term, or geologic, time span, the climate and geology are independent variables; the discharge, load, river morphology, and valley slope are dependent variables. A fluvial system, when viewed from this perspective, is undergoing continuous changes.

# 2

# OVERVIEW OF RIVER MORPHOLOGY

River morphology has been a subject of great challenge to scientists and engineers who recognize that any effort with regard to river engineering must be based on a proper understanding of (1) the morphological features involved and (2) the responses to the imposed changes. In this chapter, an overview of river morphology is presented from the geomorphic viewpoint. Included in the scope are the regime concept, channel-forming discharge, longitudinal stream profile, river channel classifications, thresholds in river morphology, hydraulic geometry, meander planform, and geomorphic analysis of river responses. Analytical approach to river morphology relies on adequate techniques for the hydraulics of flow and sediment transport processes; it is presented in Parts III and IV, after the foundation of fluvial processes.

## 2.1 REGIME CONCEPT

The regime concept is generally considered synonymous with that of equilibrium. This concept originated from the study of stable alluvial canals, which, with a mobile bed and earth banks, are nonscouring and nonsilting over an operating cycle. An alluvial canal used for irrigation is usually operated under a fairly constant discharge. Because of natural discharge variation, the true regime or dynamic equilibrium of a natural river may never be attained, although each river is constantly adjusting itself toward that direction. Mackin (1948) defined grade as a condition of equilibrium in streams acting as agents of transportation; he defined *graded stream* as one in which, over a period of years, slope is delicately adjusted to provide, with available discharge and the prevailing channel geometry, just the velocity required to transport the load supplied from the drainage basin. A graded stream (i.e., a regime river) is a system in dynamic equilibrium, or, to be more precise, a system in quasi-equilibrium. The regime concept has been reaffirmed by Ackers and Charlton (1970) on the basis that the channel geometry does not adjust with short-term variation in discharge.

## 2.2 CHANNEL-FORMING DISCHARGE

River channel formation is a result of the constantly changing discharge, and the bankfull discharge is usually used as the channel-forming discharge for downstream changes in channel geometry. This simplified approach is justified in view of the fact that lower discharges, which move less sediment, contribute less to the channel formation. Also, the rise in discharge above the bankfull stage is largely absorbed by the broad flood plain and therefore it generally has less effect on the channel shape. From the 13 gaging stations in the eastern half of the United States, Leopold et al. (1964) found that the bankfull stage has a return period averaging 1.5 yr. However, the bankfull discharge among the rivers studied by Williams (1978) does not have a common recurrence frequency. Using 233 sets of data, Williams obtained the following regression equation for the bankfull discharge

$$Q = 4.0A_f^{1.21}S^{0.28} \tag{2.1}$$

where $Q$ is the bankfull discharge (in cubic meters per second), $A_f$ is the bankfull flow area (in square meters), and $S$ is the water-surface slope. For a given river channel, the bankfull discharge may also be determined using a flow resistance equation, one of the simplest being the Manning equation (see Sec. 3.2).

The bankfull discharge is usually greater than the mean annual discharge. Based on published data by Schumm (1968) and Carlston (1965), Chang (1979) obtained a relationship between bankfull discharge and mean annual discharge, as shown in Fig. 2.1.

## 2.3 LONGITUDINAL STREAM PROFILE

The slope of a stream is determined by conditions imposed from upstream, but the elevation and location of each point of the profile are also determined by the downstream base level. Major variables controlling the slope are the discharge, sediment load, and caliber. Longitudinal stream profiles in Virginia and Maryland were studied by Hack (1957). An empirical equation for the slope was obtained as follows

$$S = 18\left(\frac{d}{A_d}\right)^{0.6} \tag{2.2}$$

where $S$ is channel slope (in feet per mile), $d$ is median particle size of bed material (in millimeters), and $A_d$ is drainage area (in square miles). According to this relationship, channel slope of rivers with the same sediment size is inversely related to the drainage area, namely, discharge. Where drainage area is the same, the slope is directly proportional to $d^{0.6}$. In Hack's studies, the drainage area covered the range from 0.12 to 370 square miles and the bed material was in the

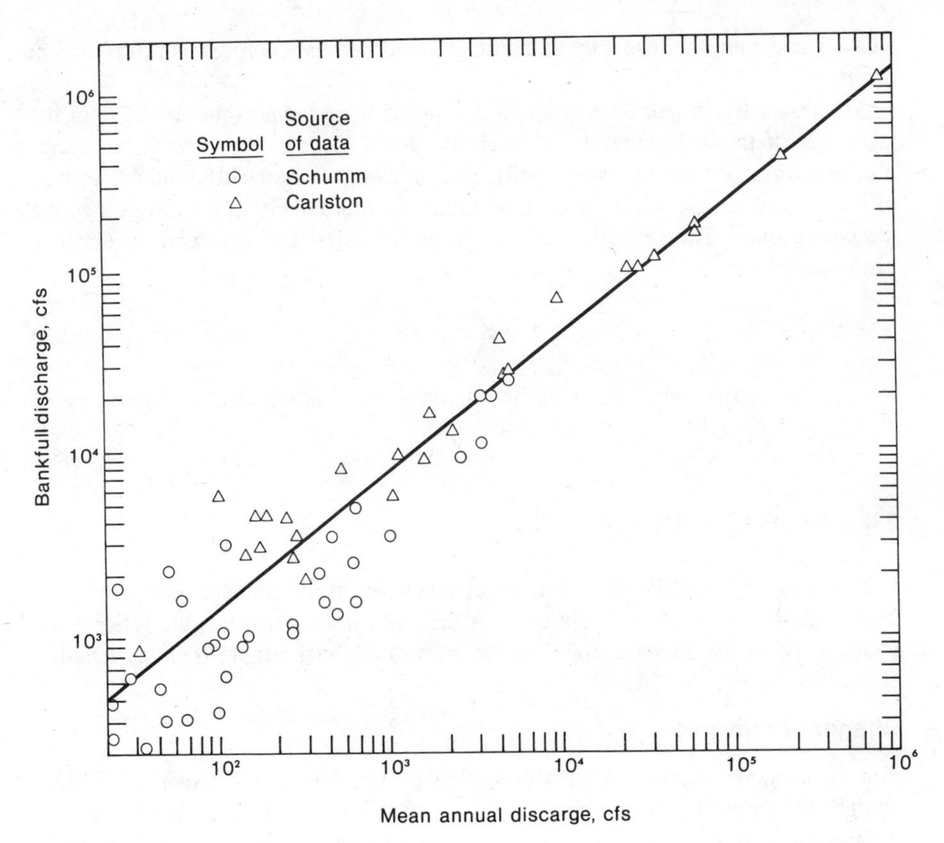


**Figure 2.1** Relationship between bankfull discharge and mean annual discharge (Chang, 1979).

range of 5–600 mm. One of the most significant results of the data analysis is the finding that rivers with the same geology and drainage area are adjusted to load, slope, and channel cross section in the same way.

The longitudinal slope profile was fitted by Shulits (1941) as an exponential decay function:

$$S = S_0 e^{-\alpha x} \tag{2.3}$$

where $S$ is the slope at a distance $x$ downstream of a reference section where the slope is $S_o$; $\alpha$ is a coefficient of slope reduction. Replacing $S$ by $dz/dx$ in Eq. 2.3 and integrating gives the equation for stream profile

$$z = \frac{S_0}{\alpha}(e^{\alpha x} - 1) \tag{2.4}$$

where $x$ and $z$ are the respective longitudinal and vertical coordinates of the stream profile.

The downstream decrease in slope is attributed, in part, to the decrease in the grain size of the bed material due to abrasion and sorting. *Abrasion* means wearing, grinding, or rubbing away by friction. *Sorting* refers to differential transport of particles of various sizes, since fine grains are more likely to be moved than are the coarse ones. The reduction in median particle size with downstream distance can be expressed as:

$$d = d_0 e^{-\beta x} \tag{2.5}$$

where $d_o$ is median particle size at the reference section, and $\beta$ is the coefficient of particle-size reduction.

## 2.4 RIVER CLASSIFICATIONS

Rivers may be classified according to channel pattern or channel type. An index used to describe the channel planform is the *sinuosity*, defined as the ratio of the valley slope to the channel slope, or the ratio of channel length to valley length.

### Channel Patterns

The three major channel patterns classified by Leopold and Wolman (1957) are straight (or sinuous), meandering, and braided. Examples for sinuous and meandering rivers are given in Fig. 2.2, and a braided river is shown in Fig. 2.9. Truly straight rivers are rare in nature. A straight river in this classification refers to one that does not have a distinct meandering pattern; that is, its sinuosity is less than about 1.5. Although a river may have a relatively straight alignment, its thalweg, or the flow path, usually wanders back and forth from one bank to the other.

A meandering river has a sinuosity greater than about 1.5, and it consists of alternating bends and a distinct sinuous planform, as shown by examples in Fig. 2.2. Although the sinuosity varies among meandering rivers, there exists a marked similarity in the ratio of radius of curvature to channel width $r_c/B$ (see Fig. 2.14), common to bends in rivers of different sizes and in different physical provinces. In a sample of 50 different meandering rivers, Leopold et al. (1964) found that two-thirds of the ratios were in the range 1.5–4.3, with a median value of 2.7. This geometric similarity often makes it difficult to tell one meandering river from another, as viewed from a plane or observed on an aerial photograph. In view of this striking regularity of winding rivers, Leopold and Langbein (1966) suggested that meanders are no accident and that they appear to be in the form in which a river does the least work in turning. An analysis is presented in Sec. 11.1 to explain this characteristic geometric relation common to meandering rivers. This nearly constant ratio of $r_c/B$ will be analytically established as the maximum curvature for which a river does the least work in turning.

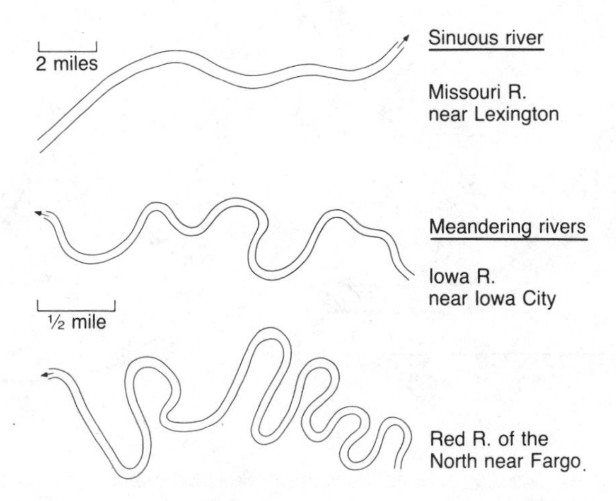


**Figure 2.2** Examples of sinuous and meandering rivers .

The maximum curvature certainly limits the sinuosity of regular meanders. However, the river usually also meanders partly because of topographical effects. The total sinuosity (TS) is the channel length divided by air distance, and the valley sinuosity (VS) is valley length divided by air distance. Two indices defining the components of sinuosity in percentage were suggested by Mueller (1968):

$$\text{Hydraulic sinuosity} = \frac{(\text{TS} - \text{VS})100}{\text{TS} - 1} \tag{2.6}$$

and

$$\text{Topographic sinuosity} = \frac{(\text{VS} - 1)100}{\text{TS} - 1} \tag{2.7}$$

In a survey of 10 point-bar complexes on the Beatton River in British Columbia, Canada, Nanson and Hickin (1983) found (1) that the rate of meander bend migration reached a maximum value when the value of $r_m/B$ approximated 3 and (2) that it rapidly declines for bends with $r_m/B$ values greater or less than 3, as shown in Fig. 2.3. The migration rate is given in meters per year. The mean radius of curvature $r_m$ is the average of $r'$ and $r''$ shown in Fig. 2.3, where $r'$ is the radius of the circle passing through $x_2y_2$, $x_3y_3$ and $x_4y_4$; $r''$ is for the circle going through $x_1y_1$, $x_3y_3$ and $x_5y_5$; and the digitizing interval, $\Delta x$, for the five sets of coordinates is equal to the mean channel width in the straight reach.

A braided river is usually on a steep slope, characterized by a wide and shallow course that is subdivided into several branches around interlaced islands, as illustrated in Fig. 2.9. The river as a whole has a straight planform, although the anabranches are usually sinuous. Despite the poorly defined and unstable banks,

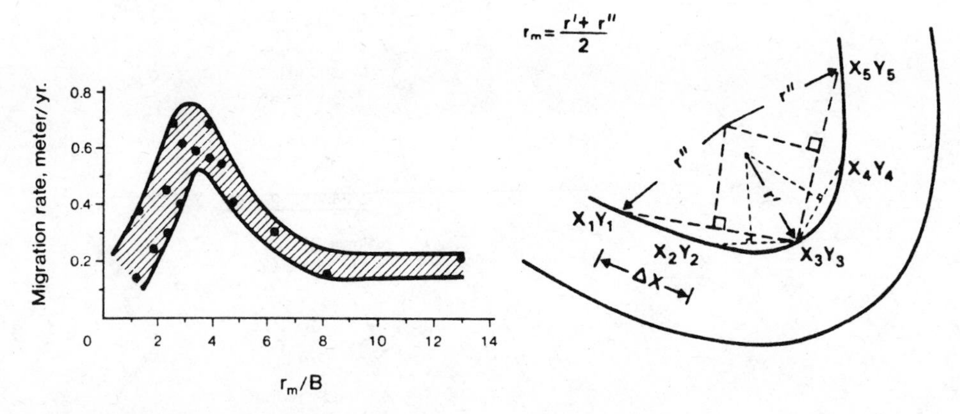


**Figure 2.3** Channel migration rate as function of relative curvature $r_m/B$ and definition of $r_m$ (Nanson and Hickin, 1983).

braided channels can be considered in quasi-equilibrium if there is no aggradation. Lane (1957) concluded that the primary causes that may be responsible for the braided condition are: (1) overloading, that is, the stream may be supplied with more sediment than it can carry, resulting in aggradation, and (2) steep slopes. The braided condition of streams in quasi-equilibrium is associated with steep slopes. Analyses of braiding are given in Chapters 11 and 14.

## River Types

The classification of river types by Brice (1983) is based on four major planform properties that are most readily observed on aerial photographs: sinuosity, point bars, braiding, and anabranching. Four major river types, each of which consists of commonly occurring association of planform properties, are illustrated, in Fig. 2.4, in the direction of increasing slope. Sinuous canaliform rivers, as exemplified in Fig. 2.5, have a flat slope, characterized by narrow crescent-shaped point bars, a notably uniform width, a lack of braiding, and a moderate to high sinuosity. The channel is relatively narrow and deep, with greatest lateral stability and high silt-clay content for the banks.

Sinuous point-bar rivers (see examples in Fig. 2.6) are steeper and have more rapid rates of lateral migration at bends, although straight reaches may remain stable for long period of time. Such rivers tend to have greater width at bend apexes, they also tend to have prominent point bars (see Fig. 2.7) that are typically scrolled and visible at normal stage.

Sinuous braided rivers, exemplified by the one in Fig. 2.8, are steeper and wider than sinuous point-bar rivers with the same discharge, featured by rapid rates of lateral migration and rapid shifts in the position of the thalweg. Such rivers have fairly heavy bed-material load but less silt–clay content. Point bars are more irregular as the braiding increases.

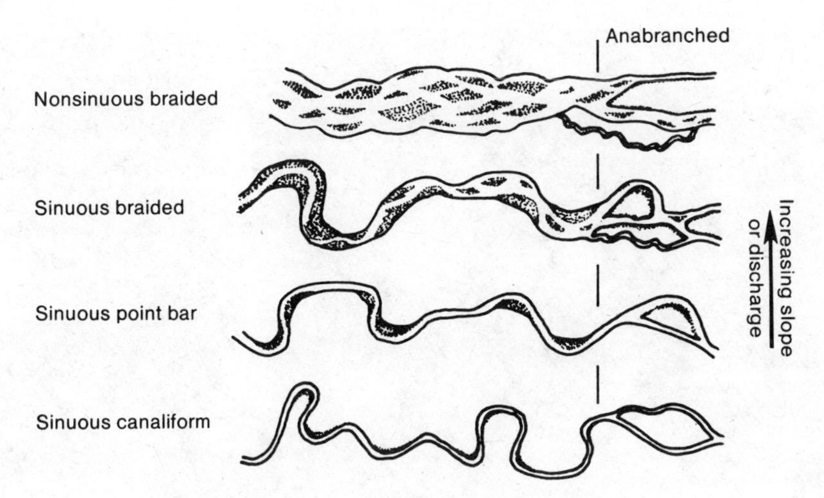


**Figure 2.4** River classification (Brice, 1983).

Nonsinuous braided rivers without point bars exist on steep slopes with heavy bed-material load and low silt–clay content. An example is given in Fig. 2.9. Such rivers are highly braided and have moderate rates of lateral migration at random places where one of the multiple branches impinges against a bank. The branch channels shift at random within the banklines.

The distinctions between sinuous canaliform rivers and nonsinuous braided rivers are illustrated by the River Kwae in Thailand and a desert wash in California, shown in Fig. 2.10. These two rivers are contrasted by the width-depth ratio, slope, channel and bank stability and other features.

Anabranching has diverse origins. It may be caused by: (1) blockage of a main channel by ice jams in northern rivers, (2) geological constraints such as bed rocks, or (3) topographical effects. Field and laboratory findings by Leopold et al. (1964) have shown that anabranches are usually steeper than the undivided channel under the quasi-equilibrium. This may be explained from the viewpoint that the anabranches have a lower transport capacity per unit water discharge and hence need steeper slopes to maintain the uniform sediment rate through divided and undivided reaches. Anabranching therefore represents an adjustment in channel pattern in its adaptation to changes in terrain.

## 2.5 THRESHOLDS IN RIVER MORPHOLOGY

The conditions under which different river patterns and types occur have been pursued by many scientists and engineers. Because of the distinct morphological characteristics among different rivers, many researchers believe that relationships for river morphology are not continuous and that there exist several apparent thresholds or discontinuities between pattern states. Schumm (1974) pointed out

**Figure 2.5** Examples of sinuous canaliform rivers: Red River of the north near Perley, Minnesota (*top*) and Snoqualimie River near Carnation, Washington (*bottom*). Both rivers have uniform width and low rate of lateral erosion. (Courtesy of J. S. Brice.)

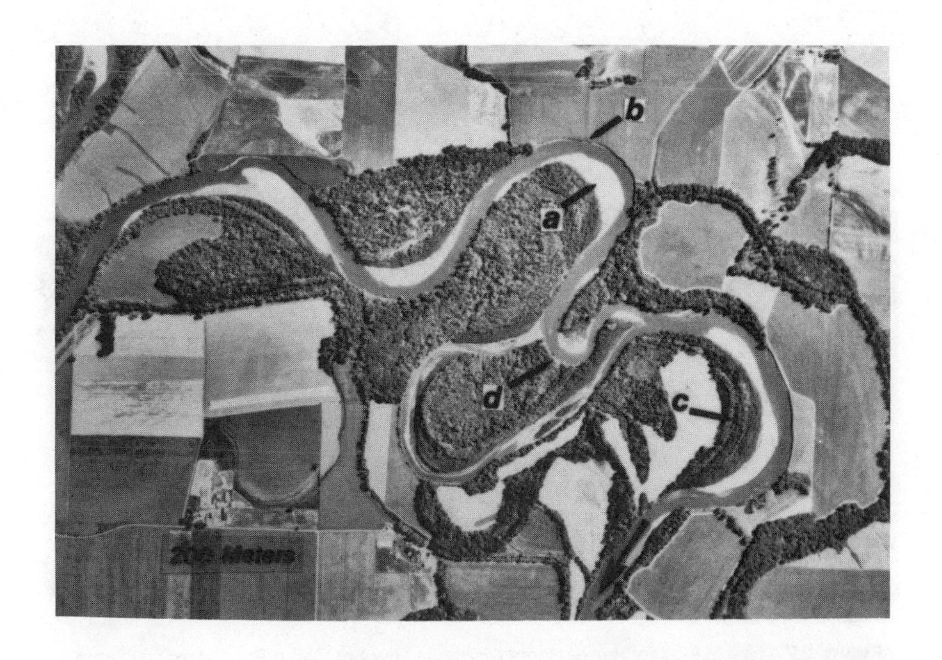


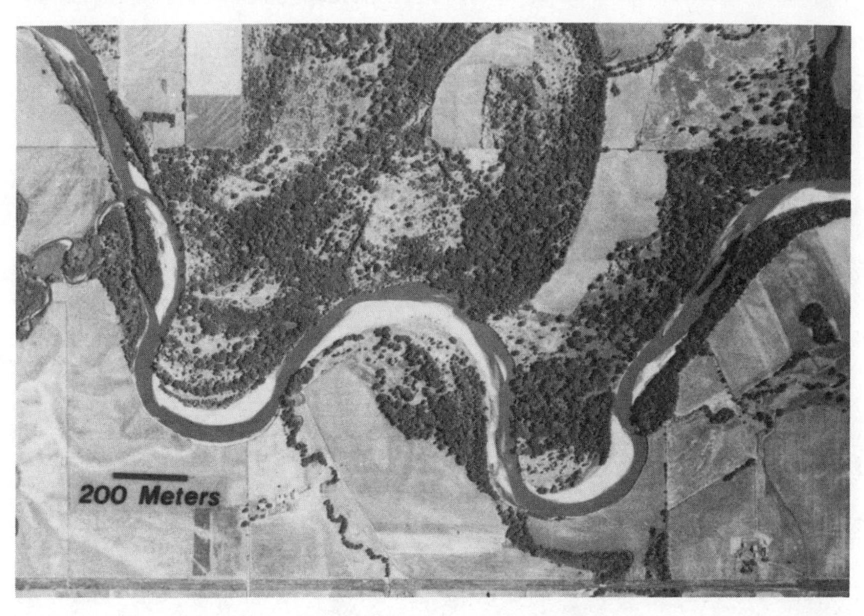


**Figure 2.6** Examples of sinuous point-bar rivers:West Fork White River near Newberry, Indiana (*top*) and Iowa River near Iowa City (*bottom*). Both rivers are at low stage, showing exposed point bars and meander scrolls. (Courtesy of J. S. Brice.)

**Figure 2.7** Exposed point bar of Santa Margarita River at Camp Pendleton, California.

**Figure 2.8** Sinuous braided river: North Canadian River near Guymon, Oklahoma at low stage. Point bars are marked by braided pattern. (Courtesy of J. S. Brice.)

**Figure 2.9** Nonsinuous braided river: Niobrara River near Spencer, Nebraska. (Courtesy of J. S. Brice.)

that regime adjustment involving a small change in slope may lead to a large change in the channel pattern if the slope is close to a critical value.

Certain thresholds were identified by Lane (1957) and by Leopold and Wolman (1957). The discharge–slope relation obtained by Leopold and Wolman (see Fig 2.11) for the threshold separating meandering and steeper braided streams is

$$S = 0.0125Q^{-0.44} \tag{2.8}$$

where $Q$ is the bankfull discharge (in cubic meters per second). This equation was obtained using data of sand-bed and gravel-bed streams. Henderson (1961) attempted to refine this criterion by including the effect of bed-particle size and obtained the following equation for gravel-bed streams:

$$S = 0.0002d^{1.15}Q^{-0.46} \tag{2.9}$$

where $d$ is median particle size (in millimeters).

The meandering–braiding threshold was also observed in experiments by Schumm and Khan (1972). Ackers and Charlton (1970) and Schumm and Khan (1972) also observed in their experiments a lower threshold slope separating straight streams from meandering streams, with slopes of meandering streams above the threshold.

Although the concept of threshold has become firmly embedded in the doctrine of river morphology, certain fundamentals have not been agreed upon among the

**Figure 2.10** A stable sinuous canaliform river (River Kwae in Thailand) is compared with a shifting nonsinuous braided desert wash in California.

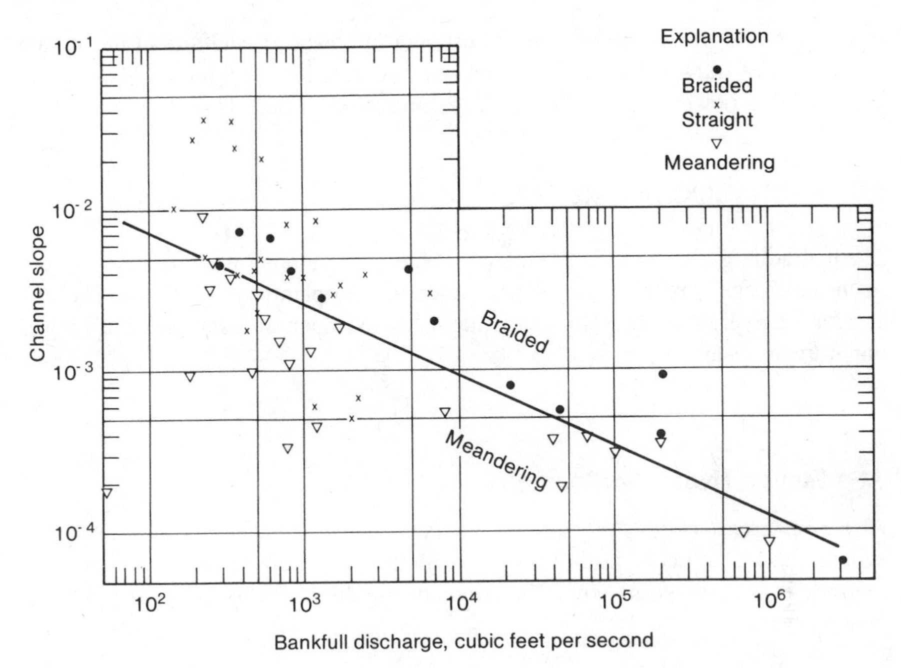


**Figure 2.11** Values of slope and bankfull discharge for natural channels as well as a threshold distinguishing braided from meandering rivers (Leopold and Wolman, 1957).

investigators. The threshold $Q$–$S$ combination separating braided from meandering streams was critically examined by Carson (1984), partly by review of previous data and partly by analysis of new data from New Zealand. He stressed the importance of the inclusion of particle size in pattern discrimination since active gravel streams must plot higher on a $Q$–$S$ diagram than do sand-bed streams because of the greater power requirements for bed-material transport. Within any one size class of bed material, he found no evidence to indicate a clear discrimination between braiding and meandering, instead he found only a weak statistical association between pattern and slope–discharge values. The observation by Carson is more or less confirmed by the analyses of channel geometry presented in Secs. 10.5, 11.3, and 11.4. For fairly steep sand-bed or gravel-bed streams within any bed-material size class, the width–depth ratio is found to increase rapidly with slope. Since braiding is associated with a large width–depth ratio, chances for braiding increase with slope over a transition zone without a clear discrimination. The sinuous braided river, such as that shown in Fig. 2.8, reflects the transition between meandering and braided streams.

However, there exist thresholds in river morphology attributed to the discontinuity in flow resistance between the lower and upper flow regimes. Because this discontinuity must be matched by responses in river morphology, the adjustment in river regime consists of significant changes in channel geometry, channel pattern and silt–clay content, when such a discontinuity is crossed. The change in flow resistance directly reflects the adjustment in power expenditure of the stream as an agent of transportation. Therefore, river morphology and thresholds are analyzed using a power approach that will be presented in Sec. 11.3.

## 2.6 HYDRAULIC GEOMETRY

The hydraulic geometry refers to the interrelationship among water discharge, sediment discharge, stream width, depth, velocity, and planform for rivers. The discussion of hydraulic geometry is distinguished between at-a-station changes and downstream changes.

### At-a-Station Hydraulic Geometry

At a river cross section, the surface width $B$, mean depth $\overline{D}$, mean velocity $U$, and suspended sediment load $Q_s$ change with the discharge. Formulas for these relationships were given as power functions of the discharge by Leopold and Maddock (1953):

$$B = C_a Q^a \tag{2.10}$$

$$\overline{D} = C_b Q^b \tag{2.11}$$

$$U = C_c Q^c \tag{2.12}$$

and

$$Q_s = C_d Q^d \tag{2.13}$$

where $C_a$, $C_b$, $C_c$, $C_d$, $a$, $b$, $c$, and $d$ are numerical constants. Since $Q = B\overline{D}U$, it follows that

$$C_a C_b C_c = 1 \tag{2.14}$$

and

$$a + b + c = 1 \tag{2.15}$$

The variations of $B$, $\overline{D}$, $U$, and $Q_s$ with $Q$ depend on the cross-sectional geometry and are affected by the pattern of scour and fill. For this reason, the numerical constants in the foregoing equations may vary from one stream to the other and also from a riffle section to a pool section. Average values of the exponents $a$, $b$, and $c$ have been obtained by Leopold and Maddock for 20 river cross sections representing a large variety of rivers in the Great Plains and the Southwest. These averages are

$$a = 0.26, \qquad b = 0.40, \qquad c = 0.34 \tag{2.16}$$

The sediment load at a section increases faster than discharge, that is, the exponent $d$ is greater than unity. Values of $d$ typically lie in the range 2–3, although the variation of $Q_s$ with $Q$ can vary considerably from one occasion to the other, as shown in Fig. 2.12. It should be cautioned that the goodness-of-fit implied by the $Q_s$ versus $Q$ relationship is spurious. It is more appropriate to determine the regression of sediment concentration versus discharge.

## Downstream Changes in Hydraulic Geometry

The development of quantitative relationships for downstream changes in hydraulic geometry has been attempted by numerous investigators, for example, Leopold and Maddock (1953), Leopold and Wolman (1957), Blench (1969), Langbein (1964) and Engelund and Hansen (1967), in addition to the regime canal researchers. A stable canal may be considered as a simple type of river.

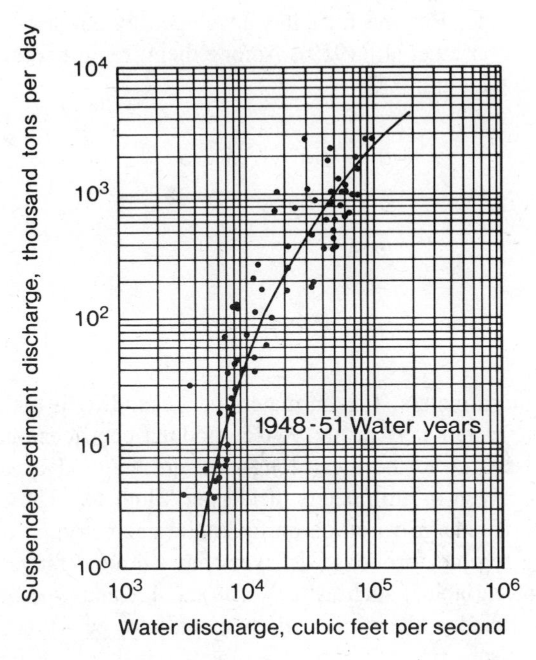


**Figure 2.12** Suspended-sediment discharge in relation to water discharge for Colorado River near Grand Canyon.

Regime formulas for hydraulic geometry in the most general form are given as power functions of the discharge, (mean annual discharge or bankfull discharge), in the same forms as Eqs. 2.10–2.13. Figure 2.13 shows the downstream variations of $B$, $\overline{D}$, and $U$ with mean annual discharge for several river reaches studied by Leopold and Maddock (1953). The lines for each relationship have similar slopes, which indicate similar values of the exponents $a$, $b$, and $c$ for each case. At the same discharge, the values of $B$, $\overline{D}$, and $U$ are usually different because the channel geometry is also a function of other variables such as the sediment size, channel slope, and bank cohesion (silt–clay content).

Other studies of the hydraulic geometry, both empirical and analytical, have also related the width, depth, and velocity as power functions of the discharge. Ming (1983) summarized the respective exponents $a$, $b$, and $c$ and the characteristic discharge (constant, mean annual, or bankfull) used in over 20 investigations in different parts of the world. Values of these exponents seem to be unaffected by the characteristic discharge used; their ranges are:

$$a\text{: } 0.39\text{–}0.60, \qquad b\text{: } 0.29\text{–}0.40, \qquad c\text{: } 0.09\text{–}0.28 \tag{2.17}$$

The coefficients $C_a$, $C_b$, $C_c$, and $C_d$ of certain regime relations for the channel geometry are expressed as functions of appropriate variables, exemplified by the regime relations for canals (as described in Chapter 10) and for rivers (as described in Chapter 11). Regime formulas developed by Russian researchers were introduced by Kondrat'ev et al. (1959). Among them, regime formulas for the Upper Volga and the Oka are:

$$B = 4.67Q^{0.57}K^{0.13}S^{-0.07} \tag{2.18}$$

$$\overline{D} = 0.069Q^{0.22}K^{0.50}S^{-0.24} \tag{2.19}$$

and

$$U = 3.10Q^{0.21}K^{0.37}S^{0.31} \tag{2.20}$$

where $B$ and $\overline{D}$ are in meters, $U$ is in meters per second, $Q$ is the mean long-term discharge in cubic meters per second, $K$ is a modular coefficient equal to the ratio of the passing discharge to the mean long-term discharge given as a percentage, and the slope $S$ is given in millimeters of fall per kilometer. These equations contain the effects of discharge variation on hydraulic geometry.

The foregoing regime formulas for rivers have limited ranges of application because important variables, such as bed-material size and type of bank material, are not included. Besides, the stream width should also be a function of the slope in addition to the discharge. At steep slopes the width should increase rapidly with the slope, and the depth should decrease with the slope, as demonstrated by the braided channel patterns. Because of such limitations, many regime formulas are only applicable to the rivers from which the formulas were developed.

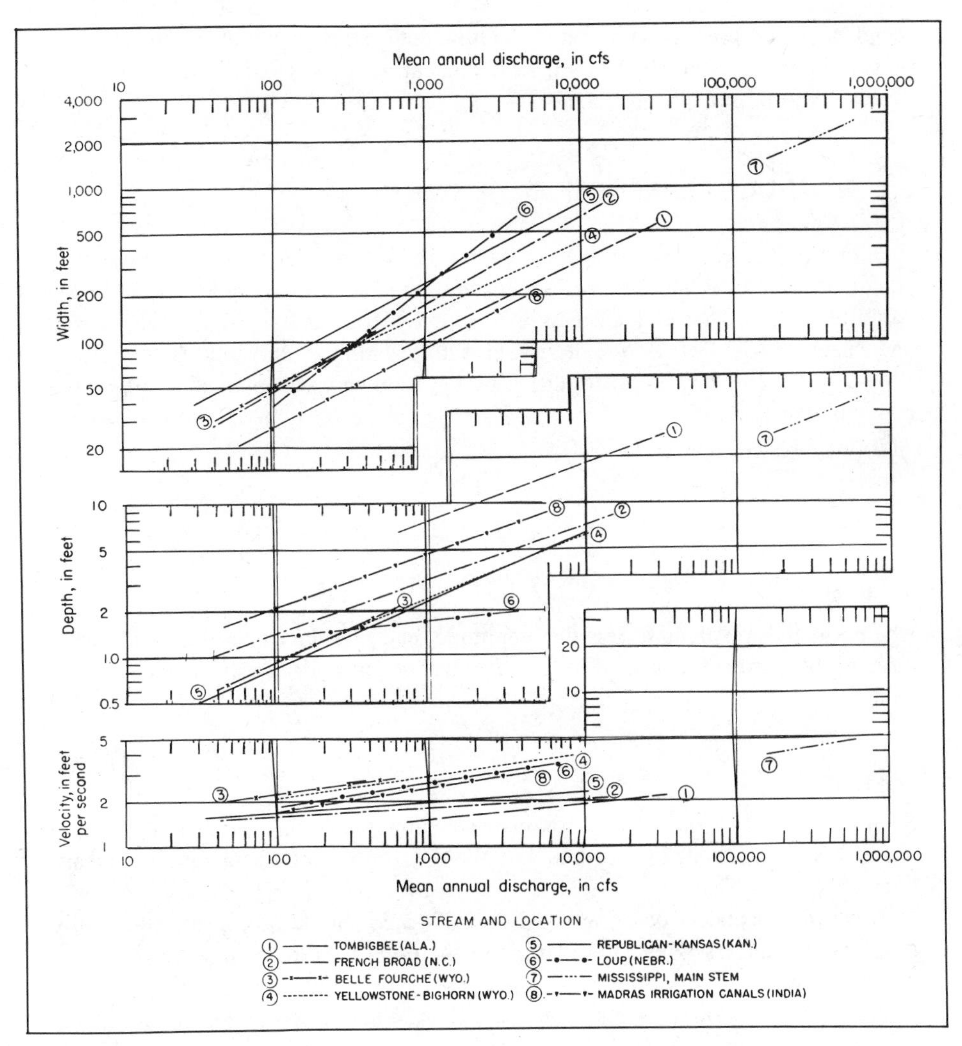


**Figure 2.13** Width, depth, and velocity in relation to mean annual discharge as discharge increases downstream in various river systems (Leopold and Maddock, 1953).

River morphology is correlated with the type of sediment forming the channel boundary, which generally reflects the type of sediment transported. Schumm (1968) related the shape of the cross section to the percentage of silt–clay content measured from the channel perimeter. Silt–clay refers to the sediment smaller than 0.074 mm (200 mesh size). Using the data from the sand-bed rivers in the Great Plains of the United States and New South Wales, Australia, Schumm found that the channel width–depth ratio $F$ is inversely related to the percentage of silt-clay in the channel perimeter $M$ as follows:

$$F = 225M^{-1.08} \tag{2.21}$$

It indicates that streams with smaller width–depth ratios have higher silt–clay contents. Analytical correlation of silt–clay content with thresholds in river morphology is given in Sec. 11.3.

## 2.7 MEANDER PLANFORM

A definition sketch of a regular meander path is shown in Fig. 2.14. Empirical equations usually relate the wavelength and amplitude of meander bends to the bankfull width of the channel (Inglis, 1949; Leopold and Wolman, 1957, 1960; and Zeller, 1967); they also relate wavelength to radius of curvature (Leopold and Wolman, 1960). They are exemplified by the following equations due to Leopold and Wolman (1960):

$$\lambda = 10.9B^{1.01} \tag{2.22}$$

$$a = 2.7B^{1.1} \tag{2.23}$$

$$\lambda = 4.7r_c^{0.98} \tag{2.24}$$

where $\lambda$ is the wavelength, $a$ is the amplitude, and $r_c$ is the center radius of curvature. All of them are measured in feet. If the exponents are approximated as unity, then

$$r_c = 2.4B \tag{2.25}$$

Such a relation indicates the approximate maximum curvature for meander bends, which has been found, by Leopold and Wolman, to be consistent for river meanders.

However, meanders described by Eqs 2.22–2.25 should be considered as fully developed meanders (Hey, 1976) because there exist so many sinuous rivers for which the relative curvature $r_c/B$ is considerably greater than 2.4 or 3, corresponding to a sinuosity of less than 1.5. Using data obtained from rivers in the

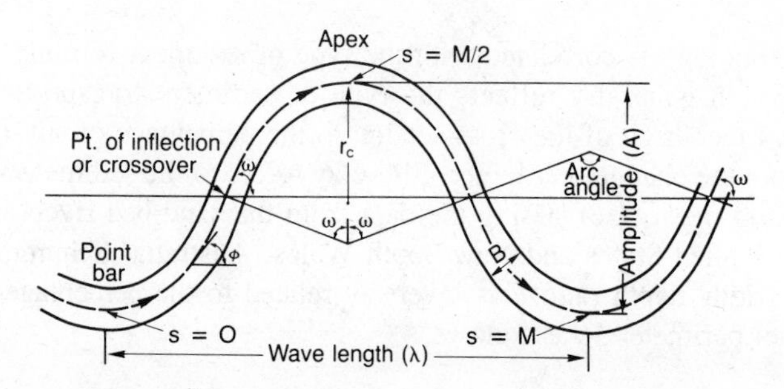


**Figure 2.14** Definition sketch of meander geometry.

United Kingdom, Hey presented the relationship between radius of curvature, bankfull width, and arc angle, as shown in Fig. 2.15. At the same bankfull width, it shows that the radius is not unique and that it increases with the arc angle. However, the meander arc length or riffle–riffle spacing, which is measured along the meander path, appears to be well correlated with the channel width.

Yalin (1971) proposed that meanders could be attributed to the law governing the spatial correlation among perturbations in turbulent flow. He considered that macroturbulent eddies, presumably secondary flow cells, would produce a sequence of periodical reverses along the direction of flow, and he suggested that the wavelength could be defined by

$$\lambda = 2\pi B \tag{2.26}$$

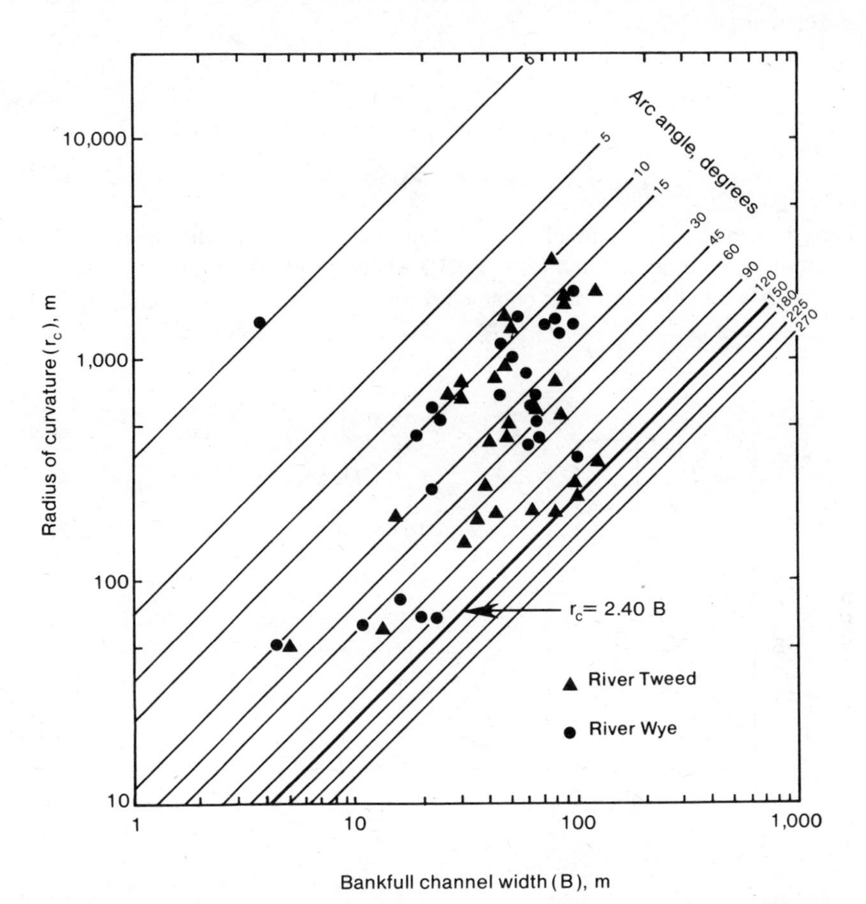


**Figure 2.15** Relationship between radius of curvature, bankfull width, and arc angle (Hey, 1976). Reprinted by permission from *Nature*, Vol. 262, p. 483. Copyright (c) 1976 Macmillan Journals Ltd.

Extensive field findings by Keller and Melhorn (1978) and by Hey (1976) have shown that the arc length measured along the channel may be given by the relation

$$M = 2\pi B \tag{2.27}$$

Figure 2.16 shows frequency distributions of pool spacing made by Roy and Abrahams (1980) for seven alluvial streams and four bedrock streams reported by Keller. Normal curves were fitted for data from both types of stream. A dramatic meander pattern of gooseneck loops is shown in Fig. 2.17.

Anderson's (1967) approach on meander wavelength was based on the oscillation of the transverse velocity component which contributes to the rise in water surface at the bank toward which it is directed. The period of oscillation, compared to a pendulum swing, depends on the mass of fluid involved and a damping constant. Using laboratory data, Anderson verified his relationship and obtained the following equations

$$\frac{\lambda}{A^{1/2}} = 72\mathbf{F}^{0.52} \tag{2.28}$$

$$\lambda = 39Q^{0.39} \tag{2.29}$$

where $A$ is the cross-sectional area of flow and $\mathbf{F}$ is the Froude number. In an evaluation made by Edgar and Rao (1983), Anderson's relationships were substantiated statistically by the results obtained from their analysis of data collected from

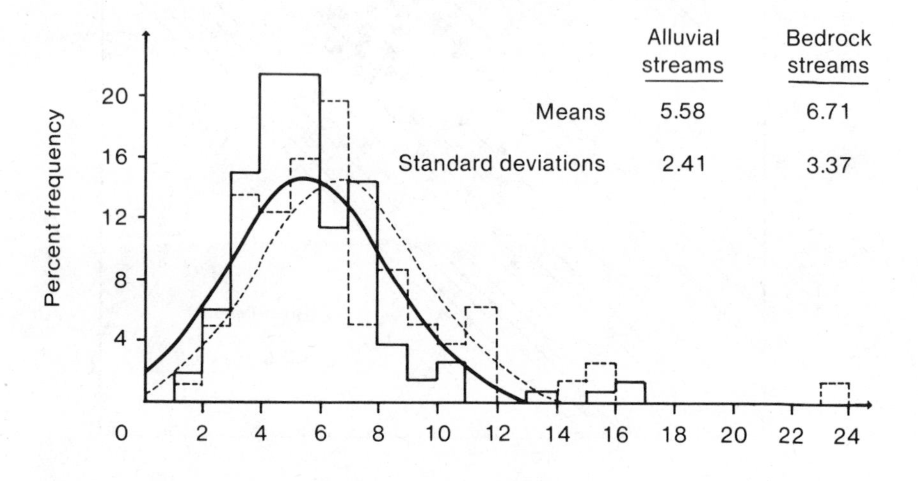


**Figure 2.16** Frequency distributions of pool spacing for alluvial streams (solid line) and bedrock streams (dashed line). Normal curves were fitted using pooled standard deviation of data from both types of streams (Roy and Abrahams, 1980).

**Figure 2.17** Meander loops of San Juan River near Mexican Hat, Utah.

61 laboratory channels with constant discharge in noncohesive, coarse-grained materials. They concluded that relative meander length $\lambda/A^{1/2}$ varies approximately as the 0.5 power of the Froude number and that meander wavelength varies approximately as the 0.39 power of discharge. However, they found Anderson's hypothesis inadequate to explain the characteristics of meanders observed in the 100 field channels analyzed. This study indicates that experimental results obtained under laboratory conditions are not always transferable to seemingly analogous field situations.

## 2.8 GEOMORPHIC ANALYSIS OF RIVER CHANNEL RESPONSES

For a river reach in equilibrium, its transport capacities for water and sediment are in balance with the rates supplied. Mobile rivers with migration can be considered in some form of equilibrium. Adjustments of equilibrium can be induced by climatic, hydrologic, and tectonic events; they may also be results of such human interferences as damming, diversion, mining, cutoffs, and so on. As equilibrium is disturbed by any of such factors, changes will occur in order to restore equilibrium. In a stream undergoing changes in morphology, channel storage (or depletion) of sediment has great impact on sediment delivery or yield (Lane et al., 1982). Case histories of river channel adjustments can be found in abundance. For example, channel changes of the Gila River in Arizona reported by Burkham (1972) were dramatic in character. This stream channel was fairly stable, the average width was less than 90 m, and the sinuosity was 1.2 in 1903. Major destruction of the flood plain took place during 1905–1917 and was caused mainly by large floods, and the average stream width increased to about 610 m while the sinuosity decreased to 1.0. The narrow channel was gradually restored by flood-plain reconstruction during low flows from 1918–1970. River channel adjustments are also exemplified by the dramatic widening of the Cimarron River in Kansas, from an average of 15 m before 1874 to 366 m between 1914 and 1931, by the major flood of 1914 reported by Schumm and Lichty (1963). On the other hand, the conversion of the broad North Platte and South Platte Rivers and the Arkansas River to relatively insignificant streams was due to flood control works and diversions for irrigation (Schumm, 1974, 1977).

From the engineering viewpoint, it is important to predict river channel adjustments to different control or regulation schemes in order to avoid certain potential problems. For example, significant widening of the lower Mississippi River in response to cutoffs was unexpected and has resulted in extensive levee construction in order to maintain the unnatural alignment (see Sec. 11.4). Analytical methods to evaluate change have been limited mainly because of the complexity of the phenomenon. Thornes (1977) pointed out that the difficulties in the analysis were *indeterminacy* and *equifinality* which mean respectively, that a channel has many different ways of responding to changes and that the same result could be caused in different ways. Another difficulty is due to the existence of thresholds or discontinuities between equilibrium states; therefore, river adjustment may involve a complete transformation of river morphology—that is, river metamorphosis—if such a threshold is crossed. It is clear that identification of thresholds is of fundamental importance in river control and engineering.

Quantitative analyses of river channel changes are given in Sec. 11.5 and in Chapters 13 and 14. Geomorphic analyses of the adjustments in regime are described herein. As long as the analysis is on the adjustments of equilibrium but not on the transient behavior, appropriate regime relationships for rivers are applicable. The discharge is regarded as the dominant variable to which the dependent variables, such as width, depth, velocity, and meander wavelength, adjust. Most empirical relationships in this chapter may be used in the analysis of regime adjustments.

Geomorphic principles are useful for qualitative analysis of river response without describing the transient behavior. Geomorphic approach is often employed at the conception and planning stage of a river project. The well-known geomorphic relationship by Lane (1955), depicting the concept of equilibrium, is

$$Q_s d \propto QS \tag{2.30}$$

where $Q_s$ is the sediment discharge and $d$ is median sediment size. This principle is illustrated in Fig. 2.18 as a relationship of balance. If one or more variables are altered, adjustments in one or more of the others are necessary to restore equilibrium. This relationship, which is useful in predicting certain qualitative trends of adjustment, does not include the channel geometry nor does it provide thresholds between equilibrium states. Quantification of Lane's relationship and its applications to determine river channel changes are presented in Sec. 11.5.

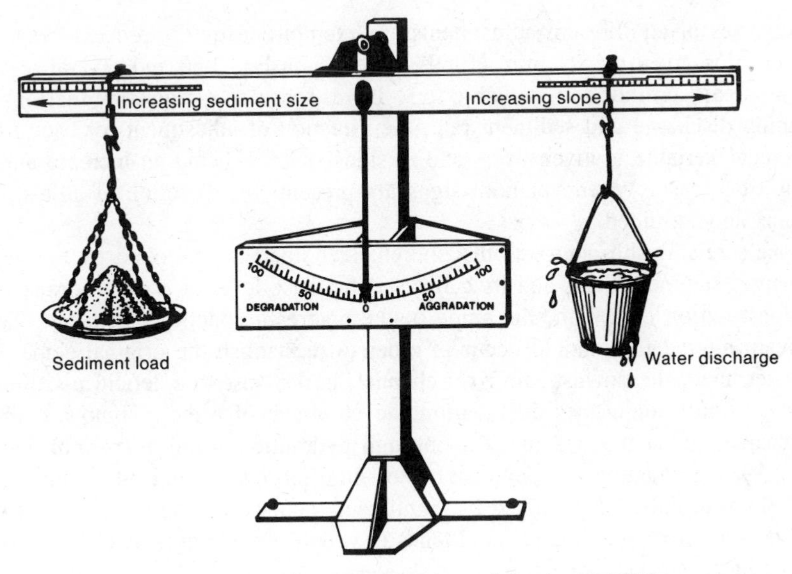


**Figure 2.18** Stable channel balance (after Lane, 1955).

**TABLE 2.1 Qualitative Model of Channel Metamorphosis Developed by Schumm 1969)[a]**

| | |
|---|---|
| Increase in discharge alone: $Q^+ \sim B^+D^+F^+\lambda^+S^-$ | Decrease in discharge alone: $Q^- \sim B^-D^-F^-\lambda^-S^+$ |
| Increase in bed-material discharge: $Q_S^+ \sim B^+D^-F^+\lambda^+S^+P^-$ | Decrease in bed-material discharge: $Q_S^- \sim B^-D^+F^-\lambda^-S^-P^+$ |

Discharge and bed material discharge increase together; for example, during urban construction, or during early stages of afforestation:

$$Q^+Q_S^+ \sim B^+D^\pm F^+\lambda^+S^\pm P^-$$

Discharge and bed-material discharge decrease together; for example, downstream from a reservoir:

$$Q^-Q_S^- \sim B^-D^\pm F^-\lambda^-S^\pm P^+$$

Discharge increases as bed-material discharge decreases; for example, increasing humidity in an initially subhumid zone:

$$Q^+Q_S^- \sim B^\pm D^+F^-\lambda^\pm S^-P^+$$

Discharge decreases as bed-material discharge increases; for example, increased water use combined with land-use pressure:

$$Q^-Q_S^+ \sim B^\pm D^-F^+\lambda^\pm S^+P^-$$

[a] *F*, Width–depth ratio; *P*, sinuosity.

Analyses of equilibrium adjustments are exemplified by the process–response relationships given by Schumm (1969), based upon Eq. 2.30 and several regime relations, as summarized in Table 2.1. In response to given combinations of changing discharge and sediment rate, the direction of adjustment for each morphological variable is given by + and − signs, with + being an increase and − being a decrease. Whenever both signs are present, the direction of adjustment remains undetermined.

Lane's relationship suggests that the channel slope is controlled by the water discharge, sediment load, and its caliber. If the base level of a river is raised by the construction of a dam, the slope of the upstream channel is reduced. Then upstream aggradation should occur in order to reestablish the original slope. On the other hand, the downstream river channel, in response to a deficit in sediment supply, usually undergoes degradation, development of a more sinuous course, and coarsening of the bed material through hydraulic sorting as exemplified in Fig. 2.19. Aggradation and degradation, or short-term scour and fill, of the channel bed are usually accompanied by significant changes in channel width. Generally speaking, an aggrading channel tends to widen and a degrading channel tends to slide back into its banks.

**Figure 2.19** San Luis Rey River downstream of check dam at Shearer Crossing near Escondido, California. Channel responses to sediment deficit include degradation, development of more sinuous course, and coarsening of bed material.

## REFERENCES FOR PART I

Ackers, P. and Charlton, F. G., "The Geometry of Small Meandering Channels," *Proc. Inst. Civ. Eng.*, Paper 73285, pp. 289–317, 1970.

Anderson, A. G., "On the Development of Stream Meanders," Proceedings of the Twelfth Congress, IAHR, Colorado State University, Fort Collins, Colorado, 1967.

ASCE Task Committee on Preparation of Sedimentation Manual, "Sediment Measurement Techniques: A. Fluvial Sediment," *J. Hydraul. Div. ASCE*, 95(HY5), pp. 1477–1514, September 1969.

Blench, T., *Mobil-Bed Fluviology*, University of Alberta Press, Edmonton, 1969.

Brice, J. C., "Planform Properties of Meandering Rivers," *River Meandering*, Proceedings of the October 24–26, 1983 Rivers '83 Conference, ASCE, New Orleans, Louisiana, pp. 1–15, 1983.

Burkham, D. E., "Channel Changes of the Gila River in Safford Valley, Arizona 1846–1970," *USGS Professional Paper 655–G*, 1972, 24 pp.

Carlston, C. W., "The Relation of Free Meander Geometry to Stream Discharge and Its Geomorphic Implications," *Am. J. Sci.*, **263**, pp. 864–885, 1965.

Carson, M. A., "The Meandering-Braided River Threshold: A Reappraisal," *J. of Hydrol.*, **73**, pp. 315–334, 1984..

Chang, H. H., "Minimum Stream Power and River Channel Patterns," *J. of Hydrol.*, **41**, pp. 303–327, 1979.

Edgar, D. E. and Rao, A. R., "An Empirical Analysis of Meandering and Flow Characteristics in Laboratory and Natural Channels," *River Meandering*, Proceedings of the October 24–26, 1983 Rivers '83 Conference, ASCE, New Orleans, Louisiana, 1983.

Engelund, F. and Hansen, E., *A Monograph on Sediment Transport in Alluvial Streams*, Teknisk Vorlag, Copenhagen, 1967.

Hack, J. T., "Studies of Longitudinal Stream Profiles in Virginia and Maryland," *USGS Professional Paper 294–B*, pp. 53–63, 1957.

Henderson, F. M., "Stability of Alluvial Channels," *J. Hydraul. Div. ASCE*, **87**(HY6), pp. 109–138, November 1961.

Hey, R. D., "Geometry of River Meanders," *Nature*, **262**, pp. 482–484, August 5, 1976.

Inglis, C. C., "The Behaviour and Control of Rivers and Canals," Central Water and Power Research Station, Poona, India, Research Publication 13, 1949.

Keller, E. A. and Melhorn, W. N., "Rhythmic Spacing and Origin of Pools and Riffles," *Geol. Soc. Am. Bull.*, **89**, pp. 723–730, 1978.

Kennedy, J. F. and Brooks, N. H., "Laboratory Study of Alluvial Streams at Constant Discharge," *Proceedings*, *Federal Interagnecy Sedimentation Conference*, Miscellaneous Publication No. 970, Agricultural Research Service, pp. 320–330, 1963.

Kondrat'ev, N. E., ed., *River Flow and River Channel Formation*, 1959, translated from Russian, published for the National Science Foundation and the Department of the Interior by the Israel Program for Scientific Translations, Jerusalem, 1962.

Lane, E. W., "The Importance of Fluvial Geomorphology in Hydraulic Engineering," *Proc. ASCE*, **81**, Paper 745, pp. 1–17, 1955.

Lane, E. W., "A Study of the Shape of Channels formed by Natural Streams Flowing in Erodible Material," *M.R.D. Sediment Series No. 9*, U.S. Army Engineering Division, Missouri River, Corps of Engineers, 1957.

Lane, L. J., Chang, H. H., Graf, W. L., Grissinger, E. H., Guy, H. P., Osterkamp, W. R., Parker, G., and Trimble, S. W., "Relationships between Morphology of Small Streams and Sediment Yield," *J. Hydraul. Eng. ASCE*, **108**(11), pp. 1328–1365, November 1982.

Langbein, W. B., "Geometry of River Channels," *J. Hydraul. Div. ASCE*, **90**(HY2), Proc. Paper 3846, pp. 301–312, March 1964.

Leopold, L. B. and Langbein, W. B., "River Meanders," *Sci. Am.*, **214**, pp. 60-70, June 1966.

Leopold, L. B. and Maddock, T. Jr., "The Hydraulic Geometry of Stream Channels and Some Physiographic Implications," *USGS Professional Paper 252*, 1953, 57 pp.

Leopold, L. B. and Wolman, M. G., "River Channel Patterns: Braided, Meandering and Straight," *USGS Professional Paper 282–B*, 1957, pp. 45–62.

Leopold, L. B. and Wolman, M. G., "River Meanders," *Geol. Soc. Am. Bull.*, **71**, pp. 769–794, 1960.

Leopold, L. B., Wolman, M. G., and Miller, J. P., *Fluvial Processes in Geomorphology*, W. H. Freeman, San Francisco, California, 1964, 522 pp.

Mackin, J. H., "Concept of the Graded River," *Geol. Soc. Am. Bull.*, **59**, pp. 463-512, May 1948.

Ming, Z. F., "Hydraulic Geometry of Alluvial Streams," *J. Sediment Res.*, **4**, pp. 75–84, 1983 (in Chinese).

Mueller, J. E., "Introduction to Hydraulic and Topographic Sinuosity Indexes," *Ann. Assoc. Am. Geogr.*, **58**, pp. 371–385, 1968.

Nanson, G. C. and Hickin, E. J., "Channel Migration and Incision on the Beatton River," *J. Hydraul. Eng. ASCE*, **109**(3), pp. 327–337, March 1983.

Roy, A. G. and Abrahams, A. D., Discussion of "Rhythmic Spacing and Origin of Pools and Riffles," *Geol. Soc. Am. Bull.*, **91**, pp. 248–250, April 1980.

Shulits, S., "Rational Equation for River-Bed Profile," *Am. Geophys. Union Trans.*, **22**, pp. 622–630, 1941.

Schumm, S. A., "River Adjustment to Altered Hydrologic Regimen--Murrumbidgee River and Paleochannels, Australia," *USGS Professional Paper 598*, 1968, 65 pp.

Schumm, S. A., "River Metamorphosis," *J. Hydraul. Div. ASCE*, **95**(HY1), Paper 6352, pp. 255–273, January 1969.

Schumm, S. A., "Fluvial Geomorphology: Historical Perspoctive," *River Mechanics*, H.W. Shen, ed., P. O. Box 606, Ft. Collins, Colorado, 1971.

Schumm, S. A., "Geomorphic Thresholds and Complex Response of Drainage System," in *Fluvial Geomorphology*, M. Morisawa, State University of New York, Binghamton, New York, 1974, pp. 299–310.

Schumm, S. A., *The Fluvial System*, John Wiley & Sons, New York, 338 pp., 1977.

Schumm, S. A. and Beathard, R. M., "Geomorphic Thresholds: An Approach to River Management," *Rivers 76*, **1**, Third Symposium of the Waterways, Harbors and Coastal Engineering Division, ASCE, pp. 707–724, 1976.

Schumm, S. A. and Khan, H. R., "Experimental Study of Channel Patterns," *Geol. Soc. Am. Bull.*, **83**, pp. 1755–1770, 1972.

Schumm, S. A. and Lichty, R. W., "Channel Widening and Floodplain Construction along Cimarron River in Southwestern Kansas," *USGS Professional Paper 352–D*, 1963, pp. 71–88.

Thornes, J. B., "Hydraulic Geometry and Channel Changes," in *River Channel Changes*, K. J. Gregory, ed., John Wiley & Sons, New York, 1977, pp. 91–100.

Williams, G. P., "Bank-full Discharge of Rivers," *Water Resour. Res.*, **14**(6), pp. 1141–1154, 1978.

Yalin, M. S., "On the Formation of Dunes and Meanders," *Proceedings of the 14th International Congress of the Hydraulic Research Association*, Paris, 1971, pp. 1–8.

Zeller, J., "Meandering Channels in Switzerland," *Int. Assoc. Sci. Hydrol.*, **75**, pp. 174–186, 1967.

# PART II

# FOUNDATIONS OF FLUVIAL PROCESSES

# 3

# HYDRAULICS OF FLOW IN RIVER CHANNELS

An important aspect of fluvial processes in rivers is the hydraulics of river flow with which sediment transport, river channel formation, and transient changes are interrelated. Since river flow is an open-channel flow, principles and equations in nonprismatic open-channel hydraulics are generally employed in river hydraulics. Because open-channel flow is a free-surface flow, an important task in river hydraulics is to determine the free surface so that other hydraulic parameters, such as hydraulic radius, velocity, and energy gradient may also be obtained. River channels are seldom straight over a length of a few channel widths. Secondary currents or transverse flow, inherent in curved channels, play an important role in all aspects of river hydraulics. Basic equations and analytical methods in open-channel flow are presented in this chapter. The mechanics of transverse flow in curved channels is given in Chapter 8.

Open-channel flow may be classified as uniform flow, gradually varied flow, and rapidly varied flow on the basis of its longitudinal variation. It is designated as steady flow or unst ady flow, depending on the time variation. Uniform flow is defined as one in which the physical properties, such as velocity, depth, roughness, and so on, stay constant along the flow direction. Human-made channels often have uniform flow, but this is not so for natural rivers. Uniform flow equations are still used for river channels under the assumption that the energy loss at a section for gradually varied flow is the same as that for a uniform flow having the velocity and hydraulic radius of the section (Chow, 1959). According to this assumption, uniform-flow relationships may be used to evaluate the energy gradient of a gradually varied flow at a given cross section.

This chapter covers a variety of topics in open-channel flow. Among them, shear-stress distribution, uniform-flow formulas, boundary-layer regions, turbulent shear flows, fixed-bed friction factors, composite roughness, and side-wall corrections are for steady uniform flow. The energy equation and its use to determine the water-surface profile of gradually varied flow is also described. The continuity and momentum equations for unsteady flow are derived, and a method of solution is presented.

## 3.1 SHEAR-STRESS DISTRIBUTION

The shear stress is the force per unit area in the flow direction. Its distribution in a steady and uniform two-dimensional flow in a channel is analyzed herein. With reference to Fig. 3.1, the internal shear $\tau$ at any level $z$ above the channel bed can be obtained by considering a control volume, with its surface defined by $ABCD$, and a unit width perpendicular to the surface. The $x$-coordinate is in the flow direction along the channel slope $S$, and the $z$-coordinate is perpendicular to flow. Forces acting on the element with a length $dx$ in the $x$-direction include the following:

1. The hydrostatic pressure forces on $AB$ and $CD$, respectively.
2. The shear stress $\tau$ acting on $BC$.
3. The $x$-component of the fluid weight $W_x$ for the volume.

Under the uniform-flow condition, the hydrostatic forces are in balance, and hence $W_x$ ($= WS$) must be counterbalanced by the shear force, that is,

$$WS - \tau\,dx = 0 \tag{3.1}$$

Replacing the weight $W$ by $\gamma(D - z)dx$ ($\gamma$ is the specific weight of fluid and $D$ is the flow depth) in this equation yields

$$\tau = \gamma(D - z)S \tag{3.2}$$

Therefore, the shear stress increases linearly with the depth.

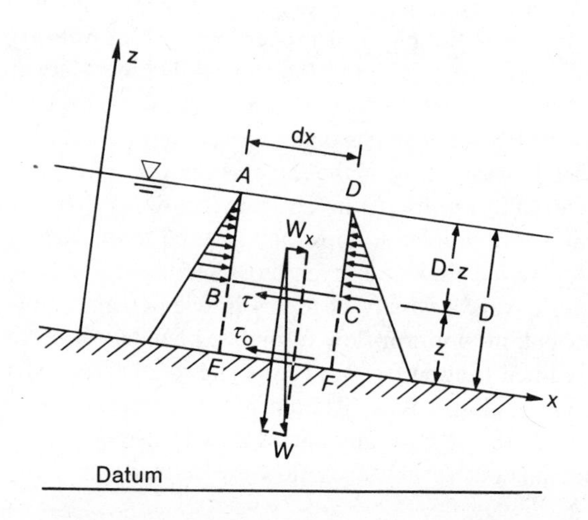


**Figure 3.1** Schematic of forces on control volume.

If the channel flow is three-dimensional, the average shear on the channel boundary $\tau_0$ can be obtained following a similar approach. Consider the control volume *AEFD*, with a differential reach length $dx$ and a cross-sectional area of channel $A$. The weight of fluid in the volume is $\gamma A\, dx$. The $x$-component of $W$ is counterbalanced by the boundary shear force, that is,

$$\gamma AS\, dx - \tau_0 P\, dx = 0 \tag{3.3}$$

where $P$ is the wetted perimeter. Therefore,

$$\tau_0 = \gamma RS \tag{3.4}$$

where $R = A/P$ is the hydraulic radius.

## 3.2 UNIFORM-FLOW FORMULAS

The uniform flow formula, also known as the flow-resistance formula, is used to determine the normal depth (flow depth of uniform flow) in a channel at a given discharge. It is also used in conjunction with other basic equations to determine the water-surface profile for a gradually varied flow. Therefore, the flow-resistance formula is a stage-discharge predictor. The development of flow-resistance relationships traces back to ancient canal builders, who were interested in knowing how the hydraulic variables of flow rate, flow depth, channel geometry, roughness, slope, and so on, would be related. The continued effort by engineers and scientists has produced hundreds of flow-resistance formulas. Selected relationships for rigid channels are included in this chapter, and those for alluvial channels are presented in Chapter 6. The commonly used uniform-flow formulas, including the Manning formula, the Chezy formula, and the Darcy–Weisbach formula, are described in the following subsections.

### The Manning Formula

The Manning formula is perhaps the most widely used uniform-flow formula; it has the following form in metric (SI) units

$$U = \frac{1}{n} R^{2/3} S^{1/2} \tag{3.5}$$

where $U = Q/A$ is the cross-sectional average velocity (in meters per second) and $n$ is the roughness coefficient. In English units, this formula is

$$U = \frac{1.486}{n} R^{2/3} S^{1/2} \tag{3.6}$$

The roughness coefficient has the necessary dimension to maintain the dimensional homogeneity of the equation. The Manning formula denotes that the velocity in a channel is directly related to the hydraulic radius and channel slope but is inversely proportional to the roughness coefficient.

Selected $n$ values for commonly encountered channel conditions are listed in Table 3.1. Those for alluvial rivers should be considered as averages, since such flow-induced roughness varies with the flow condition. The roughness coefficient in channels with vegetation is also flow-dependent because vegetation generates greater resistance to flow at low velocity while standing upright, and it causes lower resistance at high flow while bent. The application of Manning's formula should be limited to fully rough flow, for which the value of $n$ is generally independent of the depth.

## The Chezy Formula

The Chezy formula has the form

$$U = C(RS)^{1/2} \tag{3.7}$$

where $C$ is the Chezy coefficient. Empirical relations for determining the $C$ value are given in standard textbooks on open-channel hydraulics, for example, Chow (1959), Henderson (1969), and French (1985). This value has been found to vary with the flow depth even for rigid channels.

**TABLE 3.1 Values of Manning's Roughness Coefficient *n* (after Henderson, 1969)**

| Material | Value |
|---|---|
| Glass, plastic, machined metal | 0.010 |
| Dressed timber, joints flush | 0.011 |
| Sawn timber, joints uneven | 0.014 |
| Cement plaster | 0.011 |
| Concrete, steel troweled | 0.012 |
| Concrete, timber forms, unfinished | 0.014 |
| Untreated gunite | 0.015–0.017 |
| Brickwork or dressed masonry | 0.014 |
| Rubble set in cement | 0.017 |
| Earth, smooth, no weeds | 0.020 |
| Earth, some stones and weeds | 0.025 |
| Natural river channels | |
| Clean and straight | 0.025–0.030 |
| Winding, with pools and shoals | 0.033–0.040 |
| Very weedy, winding, and overgrown | 0.075–0.150 |
| Clean straight alluvial channels | $0.031d^{1/6}$ ($d = d_{75}$, in feet) |

### The Darcy–Weisbach Formula

The Darcy–Weisbach formula, originally developed for pipes, defines the friction factor $f$ as a dimensionless parameter

$$f = \frac{4\tau_0}{\frac{1}{2}\rho U^2} \tag{3.8}$$

where $\tau_0$ is the average shear stress at the boundary and $\rho$ is the mass density of fluid. If the shear stress is replaced by $\gamma RS$, this equation becomes

$$U = \left(\frac{8gRS}{f}\right)^{1/2}, \quad \text{or} \quad \frac{U}{U_*} = \left(\frac{8}{f}\right)^{1/2} \tag{3.9}$$

where $U_* = (gRS)^{1/2}$ is the friction, or shear, velocity. Because the friction factor $f$, unlike Chezy's $C$ and Manning's $n$, is dimensionless, the Darcy–Weisbach formula is desirable and is used extensively in different fields for fluid friction. For open-channel flow, however, the friction factor is not a function of the channel surface roughness alone but it is usually depth-dependent. Methods for determining the friction factor of fixed-bed channels are given in Secs. 3.5 and 3.6.

## 3.3 BOUNDARY-LAYER REGIONS

The flow near a smooth solid boundary has three regions: laminar sublayer, transition, and fully turbulent zone (see Fig. 3.2). The laminar sublayer is a thin layer from the boundary, in which the Reynolds number $uz/\nu$ ($u$ is velocity at distance $z$ from boundary and $\nu$ is kinematic viscosity) is small enough and the flow is laminar.

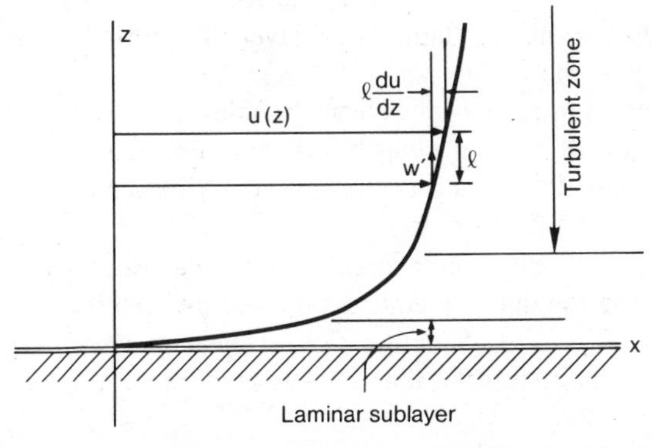


**Figure 3.2** Regions of boundary layer and explanation of mixing length.

Within the laminar sublayer, the velocity may be assumed to vary linearly with $z$, and thus the local boundary shear $\tau_0$ is given by Newton's viscosity law

$$\tau_0 = \mu \frac{du}{dz}, \quad \text{or} \quad u = \frac{\tau_0 z}{\mu} \tag{3.10}$$

where $\mu$ is the molecular viscosity. At the edge of the laminar sublayer, $z = \delta$ ($\delta$ is the laminar sublayer thickness), Eq. 3.10 gives the velocity as $\tau_o \delta / \mu$. The Reynolds number at which laminar flow starts its transition into turbulent flow has been experimentally determined to have the value of about 11.6; that is, $uz/\nu = 11.6$. Thus, the laminar sublayer thickness is

$$\delta = 11.6 \frac{\nu}{U_*} \tag{3.11}$$

Although the laminar sublayer exists over smooth surfaces, it may be disturbed on rough surfaces if the roughness elements protrude into the turbulent zone. On the basis of the roughness size, boundary surfaces are classified as one of the following three regimes: the hydraulically smooth regime, the transition regime, and the hydraulically rough regime.

For the hydraulically smooth regime, the shear Reynolds number has been experimentally determined to be less than about 5, that is,

$$\frac{k_s U_*}{\nu} < 5 \tag{3.12}$$

where $k_s$ is the mean diameter of sand grains used by Nikuradse and is know as *Nikuradse's sand roughness*. A list of equivalent $k_s$ values for different surfaces is given in Table 3.2. Under this regime, the roughness elements are so small that they are contained within the laminar sublayer. The turbulent shear flow is therefore independent of the roughness.

For the transition regime, the shear Reynolds number is in the range of about 5–70. The roughness elements partly protrude through the laminar sublayer to affect the turbulent flow. Such protrusions contribute additional flow resistance which is mainly a result of the form drag.

For the hydraulically rough regime, the shear Reynolds number is greater than about 70. The roughness elements protrude through the laminar sublayer into the turbulent core, thus contributing flow resistance associated mainly with the form drag. The velocity profile, that is, flow resistance, is a strong function of the boundary roughness.

**TABLE 3.2 Values of $k_s$ (in feet) for Concrete and Masonry Surfaces (after Henderson, 1969)**

| | |
|---|---|
| 0.0005 | Concrete class 4 (monolithic construction, cast against oiled steel forms with no surface irregularities). |
| 0.001 | Very smooth cement-plastered surfaces, all joints and seams hand-finished flush with surface. |
| 0.0016 | Concrete cast in lubricated steel molds, with carefully smoothed or pointed seams and joints. |
| 0.002 | Wood-stove pipes, planed-wood flumes, and concrete class 3 (cast against steel forms, or spun-precast pipe). Smooth troweled surfaces. Glazed sewer pipe. |
| 0.005 | Concrete class 2 (monolithic construction against rough forms or smooth-finished cement-gun surface, the latter often termed *gunite* or *shot concrete*). Glazed brickwork. |
| 0.008 | Short lengths of concrete pipe of small diameter without special facing of butt joints. |
| 0.01 | Concrete class 1 (precast pipes with mortar squeeze at the joints). Straight, uniform earth channels. |
| 0.014 | Roughly made concrete conduits. |
| 0.02 | Rubble masonry. |
| 0.01–0.03 | Untreated gunite. |

## 3.4 TURBULENT SHEAR FLOW IN CHANNELS

In a turbulent shear flow, the shear stress at any level $z$ from the solid boundary may be considered as consisting of two components, that is,

$$\tau = \tau_l + \tau_t \tag{3.13}$$

where $\tau_l$ is the shear stress caused by molecular viscosity $\mu$ and $\tau_t$ is the shear stress caused by turbulent eddies. The quantity $\tau_l$ is given by Newton's viscosity law (Eq. 3.10), from which the shear stress is directly proportional to the velocity gradient $du/dz$. The turbulent shear is given by

$$\tau_t = -\rho\overline{u'w'} \tag{3.14}$$

where $\overline{u'w'}$ is the time average of the product of the fluctuating velocities $u'$ and $w'$ in $x$ and $z$ directions, respectively. Definitions of the time-average velocity $u$ and its fluctuating velocity $u'$ are given in Fig. 3.3. The instantaneous velocity is the sum of these two quantities. Prandtl suggested a relation for the turbulent shear in terms of the time-average velocity $u$ as

$$\tau_t = \rho l^2\left(\frac{du}{dz}\right)^2 \tag{3.15}$$

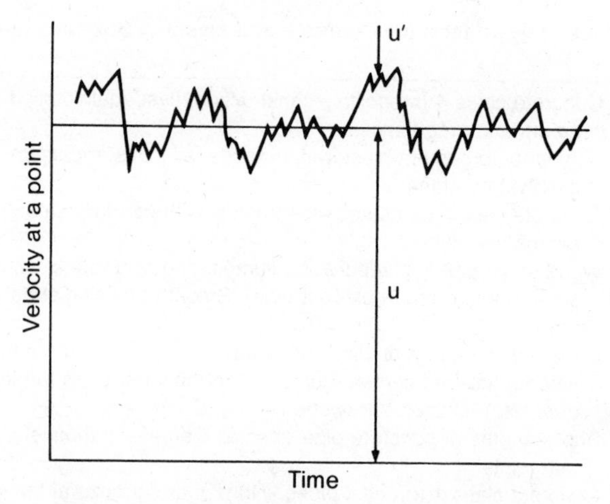


**Figure 3.3** Schematic of velocity in turbulent flow.

where $l$ is the *mixing length* proposed by Prandtl and is briefly described below. As illustrated in Fig. 3.2, a particle is brought through a certain vertical distance $l$ as a result of a positive $w'$. Concomitant to this movement, the velocity of the particle changes by an amount of the order of $-l\, du/dz$. The negative sign indicates a magnitude less than the time average. It is assumed that $u'$ is due to this velocity difference and that $w'$ has the same order of magnitude as $u'$, then

$$u' \simeq -l\frac{du}{dz}, \qquad w' \simeq l\left|\frac{du}{dz}\right| \tag{3.16}$$

From Eqs. 3.14–3.16, the turbulent shear is expressed in terms of the mixing length and velocity gradient as

$$\tau_t = \rho l^2 \left|\frac{du}{dz}\right| \frac{du}{dz} \tag{3.17}$$

The exact form of variation of $l$ with depth is not known, but, near a solid boundary, $l$ may be assumed to be proportional to the distance from the boundary, that is,

$$l = \kappa z \tag{3.18}$$

where the constant $\kappa$ is called the *von Karman constant*. In homogeneous fluid, this constant has the approximate value of 0.4 for any degree of roughness provided the roughness element is sufficiently small in comparison to the boundary-layer thickness or depth. A discussion on the value of the von Karman constant for sediment-laden fluid is given in Sec. 7.6.

Now, substituting $l = \kappa z$ into Eq. 3.15 yields

$$\tau_t = \rho \kappa^2 z^2 \left( \frac{du}{dz} \right)^2 \tag{3.19}$$

This equation for turbulent shear flow is valid only in the turbulent core outside the laminar sublayer layer. It can also be written in the form

$$\kappa z \frac{du}{dz} = U_*$$

where $U_* = (\tau_0/\rho)^{1/2}$ is the friction, or shear, velocity. Integrating this equation yields

$$\frac{u}{U_*} = \frac{1}{\kappa} \ln z + C \tag{3.20}$$

where the constant $C$ is to be evaluated from the condition at the boundary.

Equation 3.20 gives the velocity profile of turbulent shear flow as a function of the boundary condition reflected in the value of $C$. The velocity profiles pertaining to three different boundary conditions–hydraulically smooth, hydraulically rough, and transition–are described as follows.

1. *Hydraulically Smooth Surface*. There exists a laminar sublayer on a smooth surface. At $z = \delta$, the velocity $u$ is $\tau_0 \delta/\mu$ from Eq. 3.10. Substituting these values into Eq. 3.20 yields the C value, with which the velocity profile for the turbulent region is obtained as follows

$$\frac{u}{U_*} = 5.5 + 2.5 \ln \frac{U_* z}{\nu} \tag{3.21}$$

Note that this velocity distribution is independent of the size and nature of the roughness. For the value of $U_* z/\nu$ between 70 and 700, the logarithmic velocity profile of Eq. 3.21 can be approximated by the following power-law equation

$$\frac{u}{U_*} = 8.74 \left( \frac{U_* z}{\nu} \right)^{1/7} \tag{3.22}$$

2. *Hydraulically Rough Surface*. When the surface roughness protrudes into the fully turbulent zone, the laminar sublayer is destroyed and the roughness directly affects the velocity profile or flow resistance. In the absence of the laminar sublayer, the velocity distribution is not affected by the viscosity but is influenced by $\tau_0$, $\rho$, and $k_s$. For this case, the boundary condition is taken at $z = k_s$,

where the velocity is assumed to be $u_1$. After evaluating the constant $C$ using the boundary condition, Eq. 3.20 becomes

$$\frac{u}{U_*} = \frac{1}{\kappa} \ln \frac{z}{k_s} + \frac{u_1}{U_*} \tag{3.23}$$

This relationship should be independent of the Reynolds number. The variables that can affect $u_1$ are $\tau_0$, $\rho$, and $k_s$. Since these quantities are constants, the term $u_1/U_*$ is also a constant. The value of this constant is experimentally determined to be about 8.5; therefore

$$\frac{u}{U_*} = 8.5 + 2.5 \ln \frac{z}{k_s} \tag{3.24}$$

3. *Transition Surface*. In the transition regime, the velocity depends on the viscosity as well as on $\tau_0$, $\rho$, and $k_s$. These variables can form two dimensionless parameters, say, $u_1/U_*$ and $U_* k_s/\nu$ ; therefore,

$$\frac{u_1}{U_*} = F\left(\frac{U_* k_s}{\nu}\right) \tag{3.25}$$

Substituting this relationship into Eq. 3.20 gives

$$\frac{u}{U_*} = F\left(\frac{U_* k_s}{\nu}\right) + \frac{1}{\kappa} \ln \frac{z}{k_s} \tag{3.26}$$

This equation may be reduced to Eq. 3.21 for the smooth surface or to Eq. 3.24 for the rough surface.

## 3.5 FIXED-BED FLOW RESISTANCE

Flow resistance may be assumed to consist of two components: grain roughness and form roughness. The former is due to the shear force and the latter is attributed to the pressure difference in the presence of larger elements such as bed forms. Any such roughness may be in terms of Manning's $n$, Chezy's $C$, friction factor $f$, or other variables. Relationships that are applicable to gravel and sand grain roughness are presented in this section; those pertaining to alluvial bed forms are given in Chapter 6.

In channels paved with sand or gravel, the resistance to flow in the absence of bed forms can be considered to be mainly caused by grain roughness. Formulas that relate grain roughness to Manning's $n$ are given in the following text. First, the well-known Strickler's (1923) formula, which defines Manning's $n$ as a function of the particle size, is

$$n = \frac{d^{1/6}}{21.1} \tag{3.27}$$

where $d$ represents the diameter of uniform sand (in meter). Since the value of the exponent is $\frac{1}{6}$, the roughness coefficient is not very sensitive to the particle size. If $d$ is in feet, Eq. 3.27 becomes

$$n = \frac{d^{1/6}}{25.7} \tag{3.28}$$

Strickler's formula is based on data from gravel-bed rivers and fixed-bed channels, with grains pasted on the bottom and walls. Substituting Strickler's formula for $n$ into Manning's formula yields the so-called *Manning–Strickler formula*

$$\frac{U}{U_*} = 6.74\left(\frac{R}{d}\right)^{1/6} \tag{3.29}$$

where $U_* = (gRS)^{1/2}$ is the shear velocity. The quantities $R$ and $d$ have the same unit in this dimensionless equation.

Many other formulas fall in the same form as the Strickler formula. For example, Meyer-Peter and Muller (1948) developed the following relationship for sand mixtures

$$n = \frac{(d_{90})^{1/6}}{26} \tag{3.30}$$

where $d_{90}$ is the size (in meters) for which 90% of the material is finer. In the case of canals paved with gravel, Lane and Carlson (1953) suggested the following formula based on their study of the San Luis Canals in Colorado:

$$n = \frac{(d_{75})^{1/6}}{39} \tag{3.31}$$

where $d_{75}$ is the size (in inches) for which 75% of the material is finer. Additional formulas of the Strickler type are given in Sec. 3.6, which covers the flow-resistance relation for gravel-bed streams.

The friction factor for turbulent flow in fixed-bed channels stems from the resistance in sand-roughened pipe experiments by Nikuradse. In his classic work, roughness was created by gluing uniform sands to the pipe wall. Five sieve-sorted sands, with mean diameters ranging from 0.1 to 1.6 mm, were used in pipes with diameters of 2.474, 4.94, and 9.94 cm, to give six values of relative roughness (grain diameter over pipe diameter, $k_s/D$).

Brownlie (1981) reexamined the Nikuradse data and provided a Moody-type diagram for a range of data believed to be valid. He first used an alternate form by transforming the variables of the friction factor, Reynolds number $\mathbf{R}_*$ ($= U_*D/\nu$), and relative roughness $k_s/D$ in the Moody diagram into two parameters $1/f^{1/2} - 2\log(\frac{1}{2}D/k_s)$ and $\mathbf{R}_*k_s/D$ ($=U_*k_s/\nu$) so that a single curve may be plotted as shown in Fig. 3.4. In plotting this figure, data from 90 runs were selected randomly from the 362 that were published. From the curve fitted to the data points in Fig. 3.4, the Moody-type flow-resistance chart was derived as shown in Fig. 3.5. This chart provides a useful tool for computing flow resistance due to grain roughness in turbulent flow. When used for open channels, the pipe diameter $D$ should be replaced by $4R$, with $R$ being the hydraulic radius.

Figure 3.5 is based on three equations for the domains shown in Fig. 3.4. For hydraulically smooth surfaces, the roughness as a function of $\mathbf{R}_*$, independent of $k_s/D$, is given by

$$\frac{1}{f^{1/2}} = 0.103 + 2\log \mathbf{R}_* \quad \text{for} \quad \log\frac{\mathbf{R}_*k_s}{D} < 0.5 \tag{3.32}$$

For the transition from smooth to rough regime, the relation is

$$\frac{1}{f^{1/2}} - 2\log\frac{1}{2}\frac{D}{k_s} = \sum_{i=0}^{6} A_i\left(\log\frac{\mathbf{R}_*k_s}{D}\right)^i \quad \text{for} \quad 0.5 \leqslant \log\frac{\mathbf{R}_*k_s}{D} \leqslant 2.0 \tag{3.33}$$

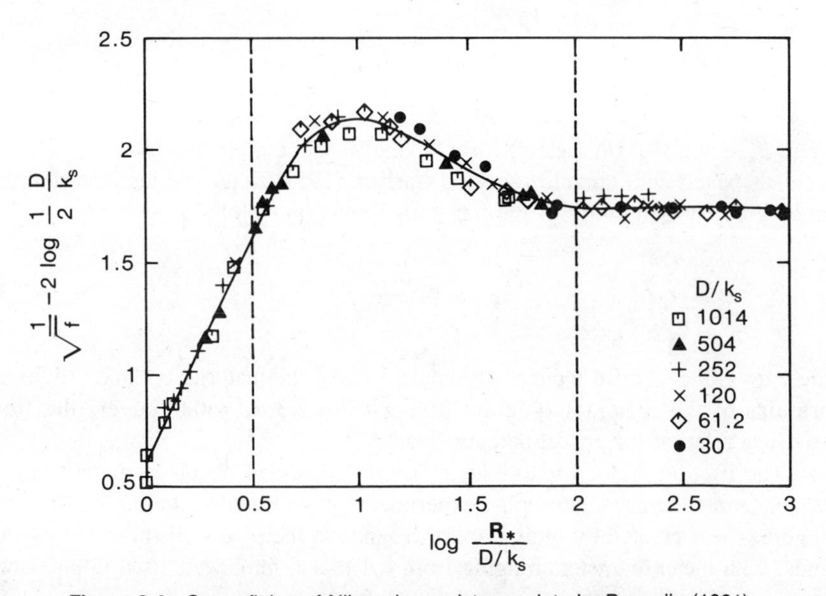


**Figure 3.4** Curve fitting of Nikuradse resistance data by Brownlie (1981).

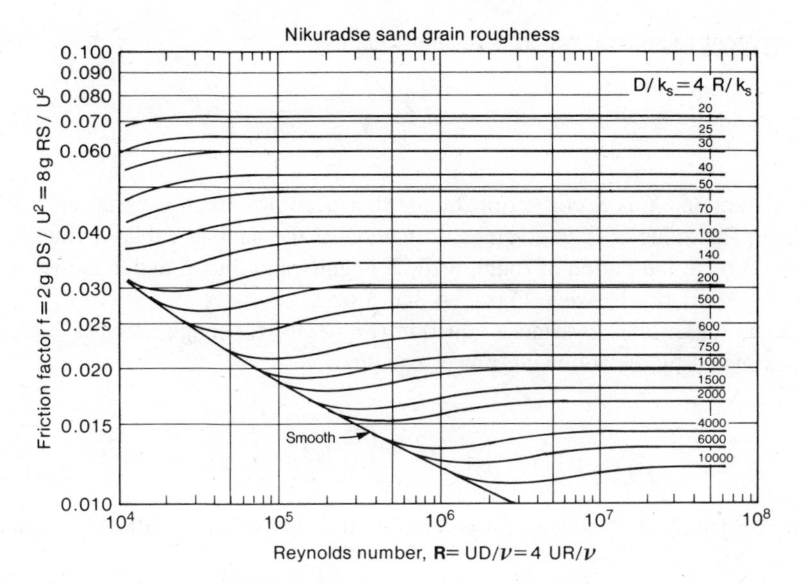


**Figure 3.5** Friction factor diagram for pipes of diameter *D* or channels of hydraulic radius *R* (Brownlie, 1981).

The coefficients $A_0$ through $A_6$ are given as 1.3376, −4.3218, 19.454, −26.480, 16.509, −4.9407 and 0.57864, respectively. For the fully rough regime, the friction factor, as a function of relative roughness only, is given by

$$\frac{1}{f^{1/2}} - 2 \log \frac{1}{2} \frac{D}{k_s} = 1.74 \quad \text{for} \quad \log \frac{\mathbf{R}_* k_s}{D} > 2.0 \tag{3.34}$$

In the case of a sand bed, the equivalent roughness $k_s$ of a plane bed is usually only related to the largest particles of the bed material, without considering the gradation and shape of particles and flow condition. Einstein (1950) first defined the equivalent roughness $\Delta$ and related it to $d_{65}$. Using the equivalent roughness, Einstein (1950) presented the following logarithmic equation for velocity profile in a plane bed:

$$\frac{u}{U_*} = 5.75 \log\left(30.2 \frac{z}{\Delta}\right) \tag{3.35}$$

which is equivalent to Eq. 3.24. He presented another equation for mean velocity:

$$\frac{U}{U_*} = 5.75 \log\left(12.27 \frac{D}{\Delta}\right) \tag{3.36}$$

where $U_* = (gRS)^{1/2}$ is the shear velocity resulting from grain roughness and $\Delta$ is the apparent roughness which is related to $k_s$ as

$$\Delta = \frac{k_s}{X} = \frac{d_{65}}{X} \tag{3.37}$$

The parameter $X$ is a correction factor that accounts for the variation in flow regime. The value of $X$ is given as a function of the laminar sublayer thickness $\delta$ in Fig. 3.6. In the region of rough wall, $X$ is unity and thus $\Delta$ and $k_s$ are identical. Note the similarity between Figs. 3.4 and 3.6.

The logarithmic resistance equation (Eq. 3.36) for rough surfaces is closely approximated by the following power equation:

$$\frac{U}{U_*} = 7.66\left(\frac{R}{k_s}\right)^{1/6} \tag{3.38}$$

Note the similarity between this equation and the Manning–Strickler equation (Eq. 3.29).

Expressions of equivalent roughness used by other investigators, as summarized by van Rijn (1982), are given as follows:

Ackers–White (1973): $k_s = 1.25d_{35}$
Hey (1979): $k_s = 3.5d_{85}$
Engelund–Hansen (1967): $k_s = 2d_{65}$
Kamphuis (1974): $k_s = 2.5d_{90}$
Mahmood (1971): $k_s = 5.1d_{84}$
van Rijn (1982): $k_s = 3d_{90}$

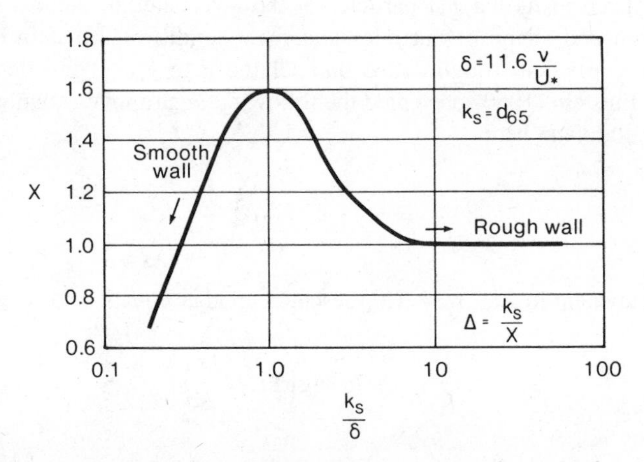


**Figure 3.6** Correction factor in logarithmic velocity distribution (Einstein, 1950).

The Engelund–Hansen resistance equation for rough surfaces is

$$\frac{U}{U_*} = 6 + 2.5 \ln \frac{R}{k_s} \tag{3.39}$$

## 3.6 FLOW RESISTANCE IN GRAVEL-BED RIVERS

Flow resistance in gravel-bed rivers is primarily a result of grain roughness since dunes tend to be poorly developed. For this reason, resistance formulas developed for gravel-bed river are similar to those for fixed-bed grain roughness. Of course the total roughness in a river consists of contributions from other sources such as bars, curvature, and other forms of irregularity. Numerous resistance equations have been developed in terms of Manning's coefficient or the friction factor, see, for example, Lane and Carlson (see Eq. 3.31), Limerinos (1970), Federal Highway Administration (1975), Hey (1979), and Bray (1979).

The Federal Highway Administration formula was developed for rock riprap as a curve fit of many sets of data, as shown in Fig. 3.7. This Manning–Strickler-type equation is

$$\frac{ng^{1/2}}{D^{1/6}} = 0.225\left(\frac{d_{50}}{D}\right)^{1/6}, \quad \text{or} \quad n = 0.0395(d_{50})^{1/6} \tag{3.40}$$

where $D$ is the depth and $d_{50}$ is expressed in feet.

From the Colebrook–White equation, Hey (1979) developed the following relationship for flow resistance in gravel-bed rivers.

$$\frac{1}{f^{1/2}} = 2.03 \log\left(\frac{aR}{3.5d_{84}}\right) \tag{3.41}$$

where $d_{84}$ is used as the representative roughness height of nonuniform gravel material, probably because of the development of wake interference losses behind large roughness elements. The coefficient $a$ is used to define the effect of cross-sectional geometry on flow resistance. Its value ranges from 11.1 to 13.46, in reverse relation to the width–depth ratio. This suggests slightly different $a$ values for riffle and pool sections. The Hey equation is substantiated with data from certain British rivers with discharges from $< 1$ up to 444 cubic meters per second. It therefore should not be used for very large rivers.

Using the data from 67 gravel-bed rivers in Alberta, Canada, Bray (1979) obtained the best-fit coefficients for a logarithmic resistance equation:

$$\frac{1}{f^{1/2}} = 0.248 + 2.36 \log\left(\frac{D}{d_{50}}\right) \tag{3.42}$$

where $D$ is the depth of flow. He found that there is no significant difference

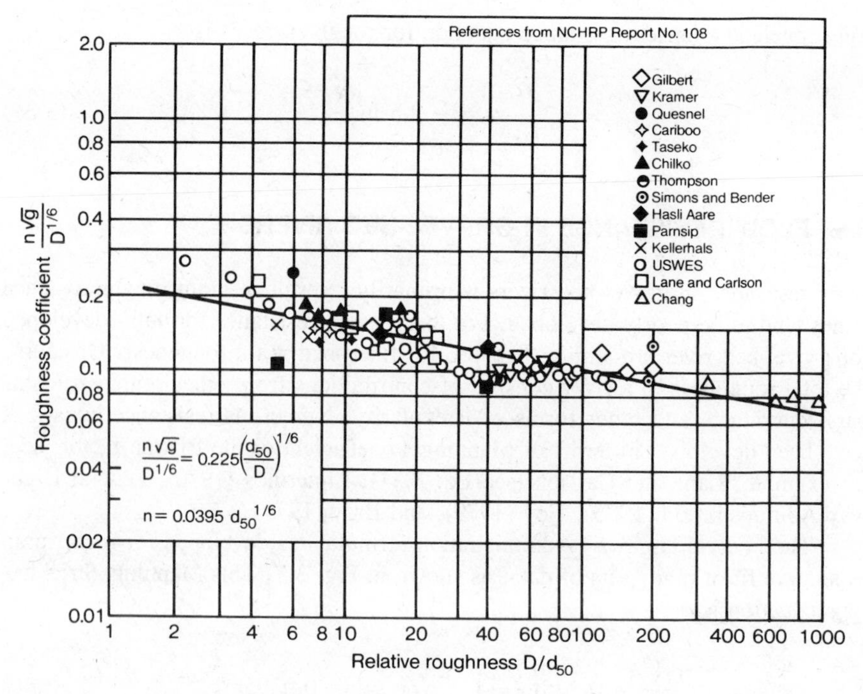


**Figure 3.7** Variation of Manning's $n$ with relative roughness for channel bed of rock riprap (Federal Highway Administration, 1975).

between the relationships obtained by using $d_{50}$, $d_{65}$, or $d_{90}$ as the characteristic size for the bed material. The best-fit power form of resistance equation based on $d_{50}$ obtained by Bray is

$$\frac{1}{f^{1/2}} = 1.36\left(\frac{D}{d_{50}}\right)^{0.281} \tag{3.43}$$

For the same $D/d_{50}$, Eqs. 3.42 and 3.43 produce very similar results. Note that the exponent in Eq. 3.43 has the value of 0.281—in contrast to 0.167 in the Strickler formula, which only accounts for grain roughness.

The application of a flow-resistance formula for normal depth determination is illustrated by the following example. Since the depth is usually contained in this type of formula as an implicit variable, it is usually determined following a trial-and-error procedure. Except for channels with regular geometry, it is necessary to relate, in advance, the geometric parameters of cross-sectional area $A$, and wetted perimeter $P$ to the depth in an application.

► ***Example 3.1.*** Determine the normal depth of a channel which is on the slope of 0.0012 and at the discharge of 54 $m^3/sec$ using the logarithmic Bray formula

(Eq. 3.42). The channel is trapezoidal in shape and has gravel material with a median size of 45 mm on the bed and banks. The bottom width $b$ of the trapezoid is 25 m, and the side slopes are three horizontal units to one vertical unit ($z = 3$).

*Solution*: In the first trial, the depth of 1.5 m is assumed. With this depth, one has

$$A = bD + zD^2 = 44.25 \text{ m}^2$$
$$P = b + 2(1 + z^2)^{1/2}D = 34.49 \text{ m}$$
$$R = A/P = 1.28 \text{ m}$$
$$U = Q/A = 1.33 \text{ m/sec}$$

Now the friction factor is given by

$$f = \frac{8gRS}{U^2} = 0.0677$$

Substituting this value into Eq. 3.42 yields $D = 1.09$ m. Since this value is different from the assumed value of 1.5 m, a new value for $D$ must be selected and then the computation is repeated. This procedure continues until the assumed value and the computed value match. The final results so obtained are as follows:

$$D = 1.43 \text{ m}$$
$$A = 41.88 \text{ m}^2$$
$$R = 1.23 \text{ m}$$
$$U = 1.29 \text{ m/sec}$$
$$f = 0.0695$$

## 3.7 COMPOSITE ROUGHNESS AND SIDE-WALL CORRECTIONS

Surface roughness can be different for the bed and side walls of a channel. For example, the sand bed in a laboratory flume can be quite rougher, because of bed forms, than the side walls. In natural rivers, the banks may be either rougher or smoother than the bed.

### Composite Roughness

Different methods for obtaining the composite roughness of the channel section have been proposed, as summarized by Chow (1959) and ASCE (1975). The principal assumption is that the channel cross-sectional area can be divided into subareas $A_1, A_2, \ldots, A_N$ with wetted perimeters $P_1, P_2, \ldots, P_N$ (see Fig. 3.8).

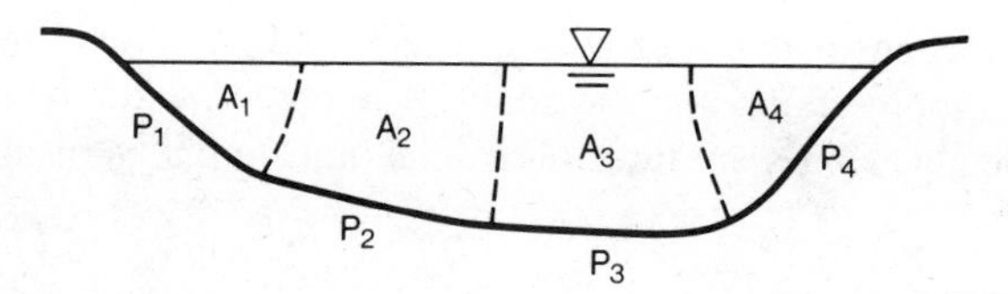


**Figure 3.8** Subareas of cross section with composite roughness.

In the approach by Horton (1933) and Einstein and Banks (1950), each subarea is assumed to have the same mean velocity and energy gradient. Assume that the Manning formula can be applied to the subareas as well as to the section as a whole, then

$$A_i = \left(\frac{U}{S^{1/2}}\right)^{3/2} (n_i)^{3/2} P_i \tag{3.44}$$

where the subscript $i$ is the subarea index. From the geometrical requirement

$$A = \sum_{i=1}^{N} A_i \tag{3.45}$$

the composite roughness $n$ for the section is obtained:

$$n = \frac{\left[\sum_{i=1}^{N} P_i (n_i)^{3/2}\right]^{2/3}}{P} \tag{3.46}$$

If the Darcy–Weisbach formula is used under the same assumptions, then

$$\frac{U^2}{S} = \frac{8gA}{fP} = \frac{8gA_i}{f_i P_i} \tag{3.47}$$

Substituting this equation into the geometrical requirement (Eq. 3.45) yields

$$Pf = \sum_{i=1}^{N} P_i f_i \tag{3.48}$$

For the simple case of different roughness for the bed and side walls, Eq. 3.48 becomes

$$Pf = P_b f_b + P_w f_w \tag{3.49}$$

where the subscripts $b$ and $w$ refer to the bed and side walls, respectively. This equation provides the friction factor $f$, $f_b$, or $f_w$ when other quantities are known.

## Side-Wall Corrections

For flow in a flume with smooth side walls at a given Reynolds number **R** and friction factor $f$, a correction procedure was developed by Vanoni and Brooks (1957) whereby the friction factor of the bed can be determined from Eq. 3.49. This procedure, later modified by Brownlie (1981), is capable of considering variable roughness height $k_s$ for the side walls. In this method, the cross-sectional area is subdivided into two parts $A_b$ and $A_w$. From the Reynolds numbers of these sections,

$$\mathbf{R} = \frac{4RU}{\nu}, \qquad \mathbf{R}_w = \frac{4R_w U}{\nu}, \qquad \mathbf{R}_b = \frac{4R_b U}{\nu} \tag{3.50}$$

and the definition of the friction factor, the following can be obtained:

$$\frac{\mathbf{R}_w}{f_w} = \frac{\mathbf{R}}{f}, \qquad \frac{R_w}{f_w} = \frac{R_b}{f_b} = \frac{R}{f} \tag{3.51}$$

The friction factors $f$, $f_w$, and $f_b$ can be obtained from Fig. 3.5. The procedure to calculate $R_w$ and $R_b$ is as follows:

1. Calculate **R** and $f$ for the total section from the experimental data, and compute $\mathbf{R}_w/f_w$ which, according to Eq. 3.51, is equal to $\mathbf{R}/f$.
2. Plot $\mathbf{R}_w/f_w = \mathbf{R}/f$ = constant on Fig. 3.5 as a straight line with a slope of 1 in log units. The intercept of this line at $f = 0.01$ is at $0.01\ \mathbf{R}/f$. The desired values of $f_w$ and $\mathbf{R}_w$ lie on this line.
3. Select a trial value for $R_w$, compute $4R_w/k_{sw}$, and determine $f_w$ from Fig. 3.5.
4. Compute $R_w = (R/f)f_w$ and compare it with the selected value. Return to Step 3. The solution should converge after two or three iterations.
5. Obtain $f_b$ and $R_b$ directly from Eqs. 3.49 and 3.51. The bed shear stress is computed from $\tau_b = \tau R_b S$.

## Composite Roughness for Natural River Cross Sections

A river cross section is a compound section if it consists of distinct subsections, such as the main channel, the left side channel, and the right side channel, as illustrated in Fig. 3.9. Because of the difference in flow duration, these subsections are often different in roughness, with the side channels usually rougher. It is convenient to define a natural cross section, as well as a geometrically regular section, by a series of points across the boundary. Such a section, as exemplified in Fig. 3.9, is called a digitized cross section. A river reach may be defined by multiple digitized cross sections. Such a scheme is employed in many computer programs, such as HEC-2 (1982) and FLUVIAL (described in Chapter 13).

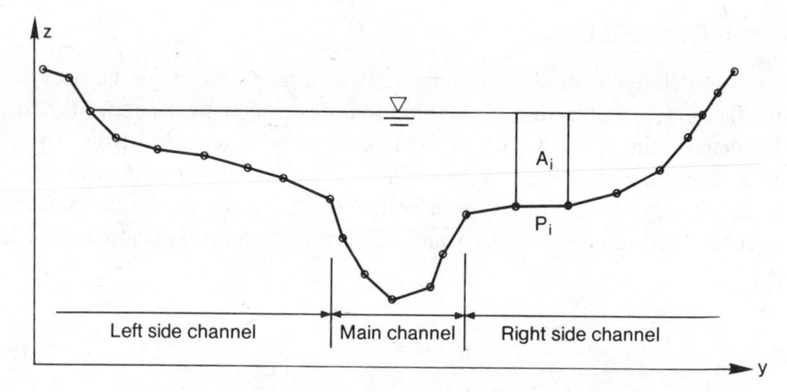


**Figure 3.9** Digitized cross section.

For a river channel, the Manning formula may be applied separately to each subsection in determining the stage, mean velocity, and composite roughness of the total section. For a digitized cross section, the subsections are conveniently defined by two or more adjacent points on the boundary. Let $u_i$ be the mean velocity of a subsection with an area $A_i$, then

$$Q = \sum_{i=1}^{N} u_i A_i \tag{3.52}$$

Assume that the Manning formula is applicable to each subsection; then

$$u_i = \frac{1}{n_i}\left(\frac{A_i}{P_i}\right)^{2/3} S^{1/2} \tag{3.53}$$

The velocities so obtained are approximations because this approach is not truly two-dimensional. Substituting Eq. 3.53 into Eq. 3.52 yields

$$Q = \sum_{i=1}^{N} \frac{1}{n_i} A_i (R_i)^{2/3} S^{1/2} \tag{3.54}$$

## 3.8 ENERGY EQUATION AND WATER-SURFACE PROFILES

The flow-resistance equation is used to determine the normal depth of uniform flow. For gradually varied flow, the energy equation is also employed in order to obtain the water-surface profile in a channel of given configuration, discharge, and roughness. From the water-surface profile, other hydraulic parameters such as velocity, hydraulic radius, and energy gradient may be obtained. Under the assumptions of hydrostatic pressure distribution and small channel slope, the total energy head for a cross section consists of the potential energy head and the kinetic energy head, that is,

$$H = Z + \alpha \frac{U^2}{2g} \tag{3.55}$$

where $H$ is the total energy head, $Z$ is the water-surface elevation and $\alpha$ is the *energy coefficient*. The kinetic energy head based on the mean velocity, $U^2/2g$, is multiplied by $\alpha$ in order to account for the flow nonuniformity in the cross section. The value of $\alpha$ is given by

$$\alpha = \frac{\int u^3\, dA}{U^3 A} \simeq \frac{\sum u_i^3 A_i}{U^3 A} \tag{3.56}$$

where $u_i$ is the mean velocity of a subarea $A_i$ that is usually between two velocity contours. The numerator of Eq. 3.56 is the actual energy flux for the section, and the denominator is the energy flux based on the mean velocity. The value of $\alpha$ increases with flow nonuniformity. For natural streams without ice cover, the $\alpha$ value ranges from 1.15 to 1.50, with the average being 1.30 (Kolupaila, 1956).

Differentiating Eq. 3.55 with respect to the distance along the channel yields

$$\frac{dH}{dx} = \frac{d}{dx}\left(Z + \alpha \frac{U^2}{2g}\right) = -S \tag{3.57}$$

where $S$ is the energy gradient, which is equal to $-dH/dx$ by definition. Written in finite-difference form for discrete cross sections along the channel, Eq. 3.57 becomes

$$Z_j + \alpha_j \frac{U_j^2}{2g} - \left(Z_{j+1} + \alpha_{j+1} \frac{U_{j+1}^2}{2g}\right) = S\,\Delta x \tag{3.58}$$

where $j$ is the cross-section index that is counted from upstream to downstream, and $\Delta x$ is the distance between sections $j$ and $j + 1$. The energy gradient $S$ is for the reach $\Delta x$, and it is usually taken as the geometric mean of those of the adjacent sections, that is ,

$$S = (S_j S_{j+1})^{1/2} \tag{3.59}$$

The energy gradient $S$ is the head loss per unit channel length attributed to boundary resistance, flow contraction and expansion, secondary currents, and so on, that is,

$$S = \frac{H_l' + H_e + H_l''}{\Delta x} = S' + S_e + S'' \tag{3.60}$$

where $H_l'$ is the head loss due to boundary resistance, $H_e$ is due to flow contraction and expansion, and $H_l''$ is due to secondary currents. The energy loss due to secondary currents is described in Sec. 8.6. The boundary loss at a section may be evaluated using a flow-resistance equation. If the Manning formula is employed, the energy gradient at a section is given by

$$S' = \frac{n^2|U|U}{R^{4/3}} \tag{3.61}$$

The contraction and expansion head loss may be expressed as

$$H_e = K_e\left|\alpha_j\frac{U_j^2}{2g} - \alpha_{j+1}\frac{U_{j+1}^2}{2g}\right| \tag{3.62}$$

where $K_e$ is the energy loss coefficient. According to Chow (1959), $K_e$ = 0 to 0.1 and 0.2 for gradually converging and diverging reaches, respectively, and it is about 0.5 for abrupt expansions and contractions.

## Standard-Step Method

Equation 3.58 provides the basis of computation for gradually varied flows. Among various available methods summarized by Chow (1959), the standard-step method is most commonly used for natural channels. Depending on the flow condition (supercritical or subcritical), the computation must be made in different directions. For subcritical flow that is under downstream control, the computation starts from downstream and proceeds upstream. For supercritical flow that is under upstream control, the computation is from upstream toward downstream. Only one flow regime, subcritical or supercritical, may be considered in the computation in using the energy equation alone. In order to handle both subcritical and supercritical flows, it is necessary to apply the relationships for alternate depths and conjugate depths (for hydraulic jumps).

The standard-step method employs the energy equation for water-surface computation for both geometrically regular and natural channel configurations. The channel geometry is defined at discrete cross sections. Pertinent hydraulic parameters such as flow area, wetted perimeter and roughness must be expressed in terms of the stage for each cross section. The water-surface elevation is computed from section to section following the appropriate direction. In going from section $j + 1$ to section $j$, for example, the stage at section $j$, namely $Z_j$, is computed from the known stage $Z_{j+1}$. Since $Z_j$ is implicit in Eq. 3.58, it is determined following a trial-and-error approach outlined as follows:

*Step 1:* Compute the flow area, velocity, and energy gradient at section $j + 1$ based on the given stage.

*Step 2:* Assume a value for the stage at section $j$, namely $Z_j$.

*Step 3:* Compute flow area, velocity, and energy gradient at section $j$.

*Step 4:* Compute $S$ according to Eq. 3.59.

*Step 5:* Substitute the values of necessary variables into Eq. 3.58. If this equation is balanced, then the assumed value for $Z_j$ is good. Otherwise, repeat Steps 2 to 5 until Eq. 3.58 is in balance.

Hand calculation of water-surface profile using the standard-step method is less tedious for regular channels. In the case of natural channels, it is much more convenient to use the computer solution. In fact, most water-surface profiles, for human-made and natural channels, are computed using the computer nowadays. Among the programs for water-surface profiles, the HEC-2 (1982) program developed by the U.S. Army Corps of Engineers, the WSP2 (1975) developed by the U. S. Soil Conservation Service, and the E431 (Shearman, 1976) program developed by the U. S. Geological Survey are perhaps the most widely used. During the continuous development, the capacity of each program has been expanded to cover broad cases that may be encountered in channels, particularly for different physical conditions at bridges and culverts. Each program allows different combinations of low flow, pressure flow, and weir flow at bridges and culverts under a variety of configurations.

Although these computer programs are powerful tools, the user is reminded of the fact that in using a one-dimensional program, necessary adjustments must be made in preparing the input data to order to approximate the one-dimensionality. For example, ineffective flow areas in the channel should be excluded.

## 3.9 UNSTEADY OPEN-CHANNEL FLOW

River flows are generally unsteady, characterized by time variations of the discharge, velocity, and other flow and channel properties. Unsteady flow in an open channel is associated with channel storage (and depletion) of water that occurs with the rise and fall in stage. Because of channel storage, the discharge varies from section to section, even in the absence of lateral inflow or outflow. Flow unsteadiness may also be attributed to transient changes in river channel geometry and alluvial bed roughness, but these effects are significant only under rare circumstances such as during a dam breach.

Mathematical formulation of unsteady open-channel flow, that is, water routing, is based on the continuity and momentum equations of flow. Their derivations and methods of solution are given in this section.

### Continuity Equation

The continuity equation of flow is derived from the condition that the net influx of mass through the control surface equals the rate of increase of mass inside the control volume, that is,

$$\oiint_{\text{c.s.}} \rho \mathbf{U} \cdot d\mathbf{A} = \frac{\partial}{\partial t} \iiint_{\text{c.v.}} \rho \, dV \tag{3.63}$$

where **U** is the velocity vector, **A** is the surface area, $\rho$ is the mass density, and $V$ is the volume. For the one-dimensional channel flow shown schematically in Fig. 3.10, the mass influxes include the upstream inflow $\rho AU$ and the lateral inflow $\rho q dx$, where $q$ is the lateral inflow rate per unit channel length. The mass efflux is given by $(A + \partial A/\partial x \, dx)(\rho U + \partial \rho U/\partial x \, dx)$. Therefore, Eq. 3.63 can be written as

$$\rho AU + \rho q \, dx - \left(A + \frac{\partial A}{\partial x} dx\right)\left(\rho U + \frac{\partial \rho U}{\partial x} dx\right) = \frac{\partial}{\partial t} \rho A \, dx$$

After neglecting second-order terms and simplifying, it becomes

$$\frac{\partial A}{\partial t} + \frac{\partial Q}{\partial x} - q = 0 \tag{3.64}$$

This is the general one-dimensional continuity equation for unsteady flow, in which the rate of change in channel storage of water $\partial A/\partial t$ is related to the spatial variation in discharge $\partial Q/\partial x$ and lateral inflow rate. In this equation, the variable $A$ is a function of stage and

$$\frac{\partial A}{\partial t} = \frac{\partial A}{\partial Z}\frac{\partial Z}{\partial t} = B\frac{\partial Z}{\partial t} \tag{3.65}$$

where $B$ is the surface width. Substituting Eq. 3.65 into Eq. 3.64 yields

$$\frac{\partial Z}{\partial t} + \frac{1}{B}\frac{\partial Q}{\partial x} - \frac{q}{B} = 0 \tag{3.66}$$

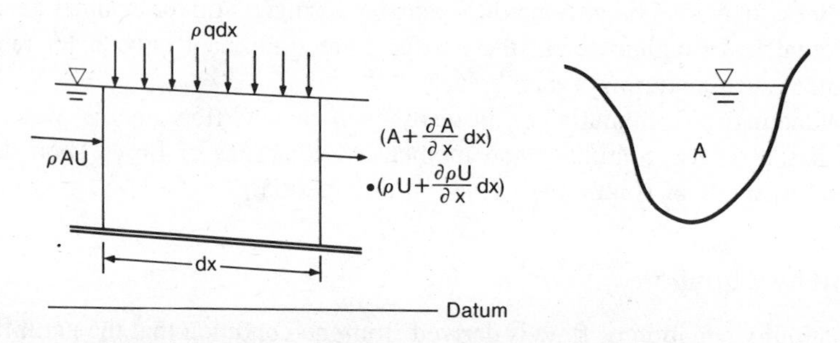


**Figure 3.10** Control volume for continuity equation.

For steady flow, the first term in the above equation vanishes, and, in the absence of lateral inflow, this equation is reduced to the familiar continuity equation

$$Q = AU = \text{constant}$$

## Momentum Equation

The one-dimensional momentum equation for unsteady flow is employed for water routing in river channels. Its derivation, methods of solution, and sample applications are given by Chen (1973), Fread (1974), Liggett (1975), and Cunge et al. (1980), among others. In the following derivation, the pressure is assumed to be hydrostatic and the slope is assumed to be small. With reference to a control volume, the momentum equation in the $x$ (flow) direction has the form

$$F_x + \iiint_{\text{c.v.}} \rho g_x \, dV = \oiint_{\text{c.s.}} \rho u(\mathbf{U} \cdot d\mathbf{A}) + \frac{\partial}{\partial t}\left(\iiint_{\text{c.v.}} \rho u \, dV\right) \tag{3.67}$$

where $F_x$ is the sum of the surface forces in the $x$ direction, $g_x$ is the $x$ component of gravitational acceleration, and $u$ is the $x$ component of the velocity vector $\mathbf{U}$. In this equation, the sum of the surface forces (first term) and body force (second term) is equated to the momentum flux across the control surface (third term) plus the rate of change of momentum inside the control volume (fourth term).

Surface forces acting on the control volume in the $x$ direction are a result of hydrostatic pressure and boundary friction. For the control volume of channel element shown in Fig. 3.11, surface forces in the $x$ direction consist of the hydrostatic force on the upstream surface $\gamma A\overline{D}$, hydrostatic force on the downstream

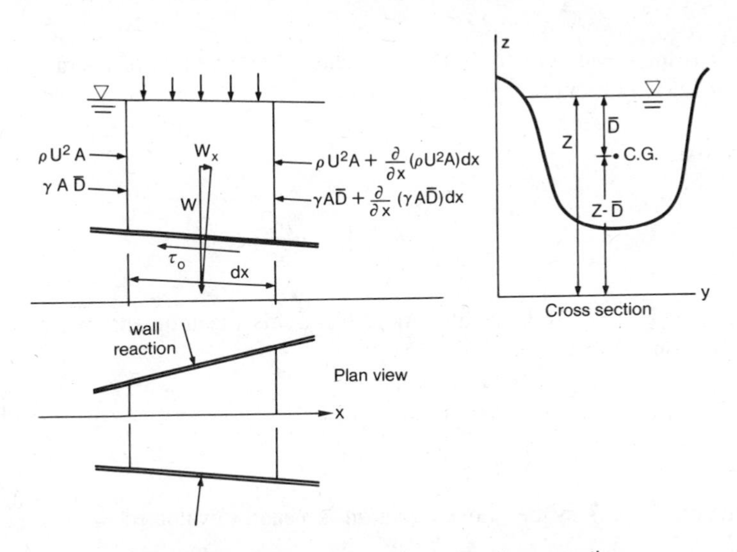


**Figure 3.11** Control volume for momentum equation.

surface $\gamma A\overline{D} + \partial/\partial x\,(\gamma A\overline{D})\,dx$, and the channel wall reaction $\gamma\overline{D}\,\partial A/\partial x\,dx$, that is,

$$F_x = \gamma A\overline{D} - \left[\gamma A\overline{D} + \frac{\partial}{\partial x}(\gamma A\overline{D})\,dx\right] + \gamma\overline{D}\frac{\partial A}{\partial x}\,dx = -\gamma A\frac{\partial\overline{D}}{\partial x}\,dx \qquad (3.68)$$

where $\overline{D}$ is the depth at the centroid of the section. The $x$ component of the body force is given by

$$-\gamma A\frac{\partial(Z - \overline{D})}{\partial x}\,dx \qquad (3.69)$$

From the average boundary shear stress $\tau_0 = \gamma RS$, the boundary friction is obtained as $-\gamma AS\,dx$. Adding the surface and body forces gives the following expression for the total force.

$$-\gamma A\left(\frac{\partial Z}{\partial x} + S\right)dx \qquad (3.70)$$

The third term in Eq. 3.67 is the net momentum flux through the control surface. For the control volume of Fig. 3.11, the following relation can be written :

$$\oiint_{\text{c.s.}} \rho u(\mathbf{U}\cdot d\mathbf{A}) = \rho U^2A + \frac{\partial}{\partial x}(\rho U^2A)\,dx - \rho U^2A - \rho qU_l\,dX$$

$$= \frac{\partial}{\partial x}(\rho QU)\,dx - \rho qU_l\,dx \qquad (3.71)$$

where $U$ is the mean velocity in the $x$ direction, $q$ is the rate of lateral inflow per unit channel length, and $U_l$ is the $x$ component of the lateral inflow velocity.

For the one-dimensional channel, the fourth term in Eq. 3.67 can be expressed as

$$\frac{\partial}{\partial t}\left(\iiint_{\text{c.v.}} \rho u\,dV\right) = \frac{\partial\rho Q}{\partial t}\,dx \qquad (3.72)$$

Substituting Eqs. 3.70-3.72 into Eq. 3.67 yields the momentum equation for unsteady channel flow:

$$\frac{\partial Q}{\partial t} + \frac{\partial}{\partial x}\left(\frac{Q^2}{A}\right) + gA\frac{\partial Z}{\partial x} + gAS - qU_l = 0 \qquad (3.73)$$

The friction slope $S$ in the above equation is usually evaluated using a uniform-flow equation. If the Manning equation is used, then

$$S = \frac{n^2 Q^2}{A^2} R^{-4/3}$$

## Methods of Solution

Various schemes for solving the unsteady-flow equations of continuity and momentum can be classified into the following major categories: (1) explicit finite-difference scheme, (2) implicit finite-difference scheme, and (3) method of characteristics. Each scheme is based on a space–time domain illustrated in Fig. 3.12, in which the time domain is represented by discrete time steps, and the space domain is symbolized by discrete cross sections along the channel. The differential equations of continuity and momentum are solved at these discrete points following successive time steps. The solution produces the discharge, stage, velocity, and so on, of each cross section at different time steps.

In the explicit finite-difference scheme, the unknowns are expressed directly and the solution is advanced point by point from one time step to the next. Unfortunately, there exist serious computational stability problems. The method of characteristics provides a high degree of accuracy, except the solution does not fall on the point of interest and is therefore not suitable when fixed grid is preferred.

The implicit finite-difference scheme is applicable to a fixed grid, and it has possible unconditional stability. For these important reasons, it has been adopted in virtually all major modeling systems. In this scheme, Eqs. 3.64 and 3.73, which constitute a system of nonlinear partial differential equations, are written as

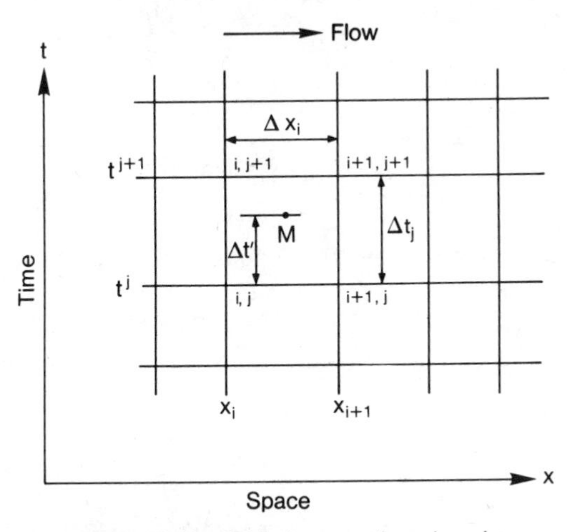


**Figure 3.12** Grid of space–time domain.

difference equations at a number of discrete points (cross sections) in the space–time domain. With appropriate initial and boundary conditions, the discharge $Q$ and the stage $Z$ at each cross section at a time step will be obtained as solutions of the system of equations. Other hydraulic parameters at a cross section, such as mean velocity and energy gradient, are obtained directly from the computed $Q$ and $Z$ values. The successive variations of $Q$ and $Z$ with time are obtained at discrete time steps.

In the space–time domain shown in Fig. 3.12, the subscript $i$ is used to designate the discrete points (cross sections) along the $x$ axis, and the superscript $j$ is for the time step along the $t$ axis. Note that the space and time increments $\Delta t$ and $\Delta x$ need not be constants. A centered difference scheme in space will be used. However, if a centered difference scheme in time is used, oscillations in computation frequently occur. To avoid oscillations and maintain stability in computation, Preissmann (1960) introduced a weighting factor defined as $\Delta t'/\Delta t$ for point $M$ in Fig. 3.12. With a value between 0.5 and 1 for the weighting factor, oscillations can be minimized without sacrificing the accuracy. The most appropriate value for the weighting factor has been determined to be between 0.55 and 0.65 based on numerical tests.

The partial derivative at a point $M$ of a variable $\phi$, which can be $Q$, $Z$, $A$, $U$, and so on, with respect to $x$ can be expressed as

$$\frac{\partial \phi}{\partial x} = [\theta(\phi_{i+1}^{j+1} - \phi_i^{j+1}) + (1 - \theta)(\phi_{i+1}^j - \phi_i^j)]\frac{1}{\Delta x_i} \tag{3.74}$$

where $\theta = \Delta t'/\Delta t$ is the weighting factor. The derivative of $\phi$ with respect to $t$ is

$$\frac{\partial \phi}{\partial t} = (\theta_{i+1/2}^{j+1} - \phi_{i+1/2}^j)\frac{1}{\Delta t_j} \tag{3.75}$$

where

$$\phi_{i+1/2} = \frac{1}{\Delta x_i}\int_0^{x_i} \phi(x)\,dx \tag{3.76}$$

If the function $\phi$ represents $A$ or $Q$, a good approximation to $\phi_{i+1/2}$ is $(\phi_i + \phi_{i+1})/2$. However, if $\phi$ represents $S$, either Eq. 3.76 or Eq. 3.59 should be used, especially in less uniform reaches. Replacing all the terms in Eq. 3.64 by these finite-difference representations yields the following difference equation:

$$F_i = [(1 - \theta)(Q_{i+1}^j - Q_i^j) + (Q_{i+1}^{j+1} - Q_i^{j+1})]\frac{\Delta t_j}{\Delta x_i}$$
$$+ \tfrac{1}{2}(A_{i+1}^{j+1} + A_i^{j+1} - A_{i+1}^j - A_i^j) - q_{i+1/2}^{j+1/2}\Delta t_j = 0 \tag{3.77}$$

In progressing from one time step to the next, the quantities at time step $j$ are known, but those at $j + 1$ are to be determined. Since $A$ is a unique function of $Z$ for the prescribed channel geometry, Eq. 3.77 can be written as a function of $Q$ and $Z$ at the time step $j + 1$, that is,

$$F_i(Q_i^{j+1}, Z_i^{j+1}, Q_{i+1}^{j+1}, Z_{i+1}^{j+1}) = 0 \tag{3.78}$$

Similarly, replacing all the terms in Eq. 3.73 by the finite-difference representations yields the following equation given in functional form.

$$G_i(Q_i^{j+1}, Z_i^{j+1}, Q_{i+1}^{j+1}, Z_{i+1}^{j+1}) = 0 \tag{3.79}$$

There are $N - 1$ points like $M$ in the space–time domain; therefore, Eqs. 3.78 and 3.79 provide $2(N - 1)$ equations containing $2N$ unknowns in $Q$ and $Z$. Other variables such as $A$ and $S$ are dependent on $Q$ and Z.

Two additional equations are required to complete the system of equations; they are from the upstream and downstream boundary conditions. At the upstream boundary, the discharge may be given as a function of time if the hydrograph is specified, that is,

$$G_0 = Q_1^{j+1} - Q'(t^{j+1}) = 0 \tag{3.80}$$

where $Q'(t^{j+1})$ is the discharge at the upstream entrance at time $t^{j+1}$. Alternatively, if the stage–time relationship is known, then this boundary condition can be written as

$$G_0 = Z_1^{j+1} - Z'(t^{j+1}) = 0 \tag{3.81}$$

where $Z'(t^{j+1})$ is the stage at the upstream boundary at time $t^{j+1}$.

At the downstream end, if the stage (base level) is known as a function of time, the boundary condition is

$$F_N = Z_N^{j+1} - Z''(t^{j+1}) = 0 \tag{3.82}$$

where $Z''$ is the given stage. If the stage–discharge relationship is known, then

$$F_N = Z_N^{j+1} - Z''(Q_N^{j+1}) = 0 \tag{3.83}$$

Equations 3.78–3.83 constitute a system of $2N$ nonlinear equations that may be solved using the generalized iteration method of Newton described in standard textbooks on numerical analysis. In this method, the solution of the system of nonlinear equations is reduced to successive solutions of linear systems, thus eliminating the difficulty of solving the nonlinear equations.

In the first approximation, trial values for the unknowns $Q$ and $Z$ at time step $j + 1$ are assigned and substituted into the system of equations. Such trial values may be taken as those at time step $j$. With the trial values, the right-hand side of each equation will generally not be zero but will have some residual value. If each residual value is designated by $R$, this system of equations at the $k$th step of successive solutions may be written as

$$\begin{cases} G_0(Z_1^k, Q_1^k) = R_1^k \\ F_1(Z_1^k, Q_1^k, Z_2^k, Q_2^k) = R_2^k \\ G_1(Z_1^k, Q_1^k, Z_2^k, Q_2^k) = R_3^k \\ \cdots\cdots \\ F_i(Z_i^k, Q_i^k, Z_{i+1}^k, Q_{i+1}^k) = R_{2i}^k \\ G_i(Z_i^k, Q_i^k, Z_{i+1}^k, Q_{i+1}^k) = R_{2i+1}^k \\ \cdots\cdots \\ F_N(Z_N^k, Q_N^k) = R_{2N}^k \end{cases} \tag{3.84}$$

In the next approximation for iteration $k + 1$, the values of $Z$ and $Q$ are then given by

$$Z_i^{k+1} = Z_i^k + dZ_i \tag{3.85}$$

and

$$Q_i^{k+1} = Q_i^k + dQ_i \tag{3.86}$$

The correction values $dZ$ and $dQ$ are obtained in solving the following set of $2N$ linear equations:

$$\begin{cases} \dfrac{\partial G_0}{\partial Z_1} dZ_1 + \dfrac{\partial G_0}{\partial Q_1} dQ_1 = -R_1^k \\ \dfrac{\partial F_1}{\partial Z_1} dZ_1 + \dfrac{\partial F_1}{\partial Q_1} dQ_1 + \dfrac{\partial F_1}{\partial Z_2} dZ_2 + \dfrac{\partial F_1}{\partial Q_2} dQ_2 = -R_2^k \\ \cdots\cdots \\ \dfrac{\partial F_i}{\partial Z_i} dZ_i + \dfrac{\partial F_i}{\partial Q_i} dQ_i + \dfrac{\partial F_i}{\partial Z_{i+1}} dZ_{i+1} + \dfrac{\partial F_i}{\partial Q_{i+1}} dQ_{i+1} = -R_{2i}^k \\ \dfrac{\partial G_i}{\partial Z_i} dZ_i + \dfrac{\partial G_i}{\partial Q_i} dQ_i + \dfrac{\partial G_i}{\partial Z_{i+1}} dZ_{i+1} + \dfrac{\partial G_i}{\partial Q_{i+1}} dQ_{i+1} = -R_{2i+1}^k \\ \cdots\cdots \\ \dfrac{\partial F_N}{\partial Z_N} dZ_N + \dfrac{\partial F_N}{\partial Q_N} dQ_N = -R_{2N}^k \end{cases} \tag{3.87}$$

The above system of $2N$ linear algebraic equations have $2N$ unknowns, $dZ_i$ and $dQ_i$. The coefficients of the equations are the partial derivatives that are evaluated at the $k$th iteration, by differentiating the appropriate equation and substituting the values of $Z$, $Q$, and so on, at the $k$th iteration. Fread (1971) developed a direct solution technique for solving the system of linear equations. The solution yields the differentials $dZ_i$ and $dQ_i$. Then the assumed values of $Z$ and $Q$ are corrected according to Eqs. 3.85 and 3.86. The iteration continues until the desired accuracy is obtained.

To use this technique for flood routing, the boundary conditions and the initial conditions of flow in the channel must be specified. The initial values of $Z_i$ and $Q_i$ are used as the first trial values in the iterative solution at each time step.

Inaccuracy may exist in a finite-difference solution because continuous functions are approximated by discretized relations. *Convergence* refers to the property of a finite-difference scheme to produce solutions for the governing equations that approach the exact solutions of those same equations as the computational grid is reduced, that is, as $\Delta x$ and $\Delta t$ approach zero. Since many differential equations cannot be solved exactly, the exact solutions for such equations are not available. But the convergence of solutions has been used as a criterion to which each scheme should be submitted. The behavior of a scheme as a function of $\Delta t$ and $\Delta x$ can be assessed by numerical experiments using different combinations of $\Delta t$ and $\Delta x$. The criteria for convergence can thus be developed based on the numerical results.

## REFERENCES


Ackers, P. and White W., "Sediment Transport: New Approaches and Analysis," *J. Hydraul. Div. ASCE*, **99**(HY11), pp. 2041-2060, November 1973.

ASCE, *Sedimentation Engineering*, Manuals and Reports on Engineering Practice, No. 54, Vito A. Vanoni, ed., 1975.

Bray, D. I., "Estimating Average Velocity in Gravel-Bed Rivers," *J. Hydraul. Div. ASCE*, **105**(HY9), Proc. Paper 14810, pp. 1103-1122, September 1979.

Brownlie, W. R., "Re-Examination of Nikuradse Roughness Data," *J. Hydraul. Div. ASCE*, **107**(HY1), pp. 115-119, January 1981.

Chen, Y.-H., "Mathematical Modeling of Water and Sediment Routing in Natural Channels," Ph.D. Thesis, Department of Civil Engineering, Colorado State University, Fort Collins, Colorado, March 1973.

Chow, V. T., *Open Channel Hydraulics*, McGraw-Hill, New York, 1959.

Cunge, J. A., Holly, F. M. Jr., and Verwey, A., *Practical Aspects of Computational River Hydraulics*, Pitman, London, 1980.

Einstein, H. A., "The Bed Load Function for Sediment Transportation in Open Channels," Technical Bulletin 1026, U.S. Department of Agriculture, 1950.

Einstein, H. A. and Banks, R. B., "Fluid Resistance of Composite Roughness," *Trans. Am. Geophys. Union*, **31**(4), pp. 603-610, August 1950.

Engelund, F. and Hansen, E., *A Monograph on Sediment Transport in Alluvial Streams*, Teknisk Vorlag, Copenhagen, Denmark, 1967.

Federal Highway Administration, "Design of Stable Channels with Flexible Linings," Hydraulic Engineering Circular No. 15, U.S. Department of Transportation, October 1975.

Fread, D. L., "Discussion on Implicit Flood Routing in Natural Channels by M. Amein and C. S. Fang (Dec. 1970)," *J. Hydraul. Div. ASCE*, **97**(HY7), pp. 1156-1159, July 1971.

Fread, D. L., "Numerical Properties of Implicit Four-Point Finite Difference Equations of Unsteady Flow," National Weather Service, NOAA, *Technical Memorandum NWS Hydro-18*, March 1974.

French, R. H., *Open Channel Hydraulics*, McGraw-Hill, New York, 1985.

"HEC-2 Water Surface Profiles: Users Manual," Generalized Computer Program, The Hydrological Engineering Center, U.S. Army Corps of Engineers, September 1982.

Henderson, F. M., *Open Channel Flow*, 3rd edition, Macmillan, New York, 1969.

Hey, R. D., "Flow Resistance in Gravel-Bed Rivers," *J. Hydraul. Div. ASCE*, **105**(HY4), Proc. Paper 14500, pp. 365-379, April 1979.

Horton, R. E., "Separate Roughness Coefficients for Channel Bottom and Sides," *Eng. News Rec.*, **111**(22), pp. 652-653, November 30, 1933.

Kamphuis, J.W., *J. Hydraul. Res.*, **12**(2), 1974.

Kolupaila, S., "Methods of Determination of the Kinetic Energy Factor," *Port Eng.*, (*Calcutta*), **5**(1), pp. 12-18, January 1956.

Lane, E. W. and Carlson, E. J., "Some Factors Affecting the Stability of Canals Constructed in Coarse Granular Materials," International Association for Hydraulic Research, Minneapolis, Minnesota, September 1953.

Liggett, J. A., "Basic Equations of Unsteady Flow," in *Unsteady Flow in Open Channels*, Water Resources Publications, Fort Collins, CO, 1975, Chapter 2.

Limerinos, J. T., "Determination of the Manning Coefficient for Measured Bed Roughness in Natural Channels," *USGS Water Supply Paper 1898-B*, 1970.

Mahmood, K., "Water Management Technical Report, No. 11," Colorado State University, Fort Collins, Colorado, February 1971.

Meyer-Peter, E. and Muller, R., "Formulas for Bed-Load Transport," *Proc. Second Meeting Int. Assoc. Hydraul. Res.*, Paper No. 2, pp. 39-64, 1948.

Preissman, A., "Propagation des Intumescneces dans les Canaux et Rivieres," *Proceedings, 1st Congres de l'Association Francaise de Calcul*, Grenoble, France, 1960, pp. 433-442.

Shearman, J. O., "Computer Applications for Step-Backwater and Floodway Analysis; Computer Program E431, User's Manual," Open File Report 76-499, U.S. Geological Survey, 1976.

Strickler, A., "Beitrage zur Frage der Geschwindigkeitsformel und der Rauhigkeitszahlen fur Strome, Kanale und geschlossene Leitungen" ("Some Contributions to the Problem of the Velocity Formula and Roughness Factors for Rivers, Canals, and Closed Conduits"), Mitteilungen des eidgenossischen Amtes fur Wasserwirtschaft, Bern, Switzerland, No. 16, 1923.

Vanoni, V. A. and Brooks, N. H., "Laboratory Studies of the Roughness and Suspended Load of Alluvial Streams," Sedimentation Laboratory Report No. E68, California Institute of Technology, Pasadena, CA, 121 pp., 1957.

van Rijn, L. C., "Equivalent Roughness of Alluvial Bed," *J. Hydraul. Div. ASCE*, **108**(HY10), pp. 1215-1218, October 1982.

"WSP2 Computer Program: A Water-Surface Profile Computer Program for Determining Flood Elevations and Flood Areas for Certain Flow Rates," U.S. Soil Conservation Service, Technical Release 61, 1976.

# 4

# PHYSICAL PROPERTIES OF SEDIMENT

Properties of sediment include those of the particles and those of the sediment mixture. The former consists of the particle size, shape, fall velocity, and so on, and the latter includes the size distribution, specific weight, angle of repose, and so forth. Such physical properties are important in the sedimentation processes of entrainment, transportation, and deposition. Certain correlations have also been established between river morphology and sediment properties. Only fundamental properties of sediment are presented in this chapter.

## 4.1 SIZE OF SEDIMENT PARTICLES

The size of sediment particles is the physical property upon which different classes or grades are classified. Many classification systems are available. The grade scale proposed by the Subcommittee on Sediment Terminology of the American Geophysical Union (Lane, 1947) is given in Table 4.1. It contains six consecutive size classes: boulders, cobbles, gravel, sand, silt, and clay. In this scale, the sizes are arranged in a geometric series with a ratio of 2; they also correspond closely to the mesh opening in sieves. The size dividing cobbles from gravel is 64 mm; it is 2 mm between gravel and sand and is 0.062 mm between sand and silt.

Natural sediment particles are of irregular shape for which the size is often expressed by the diameter of an equivalent sphere. The following diameters are generally used.

*Nominal Diameter*. This is the diameter of a sphere having the same volume as the particle.

*Sieve Diameter*. This is the diameter of a sphere equal to the length of the side of a square sieve opening through which the given particle will just pass.

*Sedimentation Diameter*. This is the diameter of a sphere having the same specific weight and the same terminal velocity as the given particle in the same fluid under the same conditions.

**TABLE 4.1 Sediment Grade Scale**

| Size | | | | Approximate Sieve Mesh Openings per Inch | | |
|---|---|---|---|---|---|---|
| Millimeters | | Microns | Inches | Tyler | U.S. Standard | Class |
| 4000–2000 | | | 160–80 | | | Very large boulders |
| 2000–1000 | | | 80–40 | | | Large boulders |
| 1000–500 | | | 40–20 | | | Medium boulders |
| 500–250 | | | 20–10 | | | Small boulders |
| 250–130 | | | 10–5 | | | Large cobbles |
| 130–64 | | | 5–2.5 | | | Small cobbles |
| 64–32 | | | 2.5–1.3 | | | Very coarse gravel |
| 32–16 | | | 1.3–0.6 | | | Coarse gravel |
| 16–8 | | | 0.6–0.3 | $2\frac{1}{2}$ | | Medium gravel |
| 8–4 | | | 0.3–0.16 | 5 | 5 | Fine gravel |
| 4–2 | | | 0.16–0.08 | 9 | 10 | Very fine gravel |
| 2–1 | 2.00–1.00 | 2000–1000 | | 16 | 18 | Very coarse sand |
| $1-\frac{1}{2}$ | 1.00–0.50 | 1000–500 | | 32 | 35 | Coarse sand |
| $\frac{1}{2}-\frac{1}{4}$ | 0.50–0.25 | 500–250 | | 60 | 60 | Medium sand |
| $\frac{1}{4}-\frac{1}{8}$ | 0.25–0.125 | 250–125 | | 115 | 120 | Fine sand |
| $\frac{1}{8}-\frac{1}{16}$ | 0.125–0.062 | 125–62 | | 250 | 230 | Very fine sand |
| $\frac{1}{16}-\frac{1}{32}$ | 0.062–0.031 | 62–31 | | | | Coarse silt |
| $\frac{1}{32}-\frac{1}{64}$ | 0.031–0.016 | 31–16 | | | | Medium silt |
| $\frac{1}{64}-\frac{1}{128}$ | 0.016–0.008 | 16–8 | | | | Fine silt |
| $\frac{1}{128}-\frac{1}{256}$ | 0.008–0.004 | 8–4 | | | | Very fine silt |
| $\frac{1}{256}-\frac{1}{512}$ | 0.004–0.0020 | 4–2 | | | | Coarse clay |
| $\frac{1}{512}-\frac{1}{1024}$ | 0.0020–0.0010 | 2–1 | | | | Medium clay |
| $\frac{1}{1024}-\frac{1}{2048}$ | 0.0010–0.0005 | 1–0.5 | | | | Fine clay |
| $\frac{1}{2048}-\frac{1}{4096}$ | 0.0005–0.00024 | 0.5–0.24 | | | | Very fine clay |

*Fall Diameter*. This is the diameter of a sphere having a specific gravity of 2.65 and having the same terminal velocity as the particle when each is allowed to settle alone in quiescent distilled water of infinite extent at 24°C.

A sediment sample is made up of grains having a range of sizes and other characteristics. The size distribution is obtained by the separation of a sample into a number of size classes, known as the *mechanical analysis*. The results of such analyses are represented statistically. The size distribution of a sediment sample is usually presented as a cumulative size–frequency curve plotted on probability paper (Fig. 4.1). The percentage by weight of sediment finer or smaller than a certain size is plotted against the size.

The grain size corresponding to the 50% finer point is obtained from the curve as the median size. Under the assumption of log-normal size distribution, the geometric mean size may be obtained at the intersection of the 50% line and a

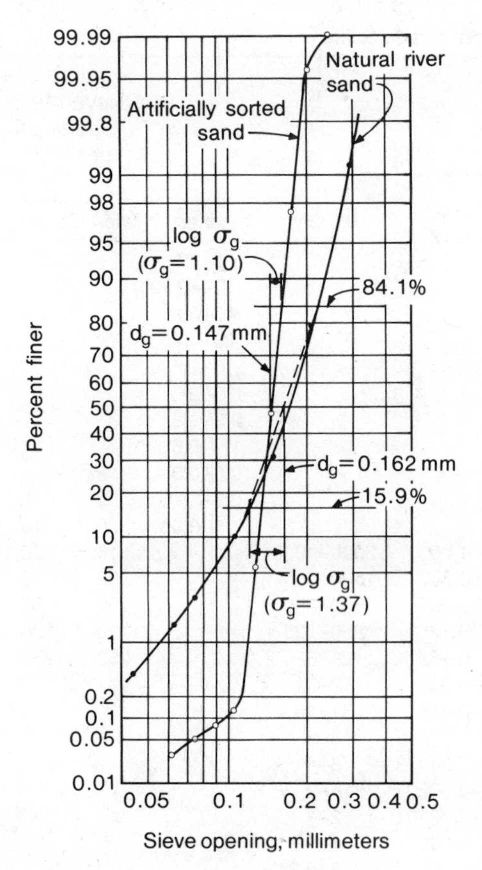


**Figure 4.1** Cumulative logarithmic probability size–frequency graphs for sands (after ASCE, 1975).

line passing through the curve at 15.9% and 84.1%. It may also be obtained from

$$d_g = (d_{15.9} d_{84.1})^{1/2} \tag{4.1}$$

where $d_g$ is the geometric mean size, and $d_{15.9}$ and $d_{84.1}$ are the grain sizes for which 15.9% and 84.1% by weight, respectively, of the sediment is finer. For log-normal distribution, the geometric standard deviation of the distribution $\sigma_g$ is given by

$$\sigma_g = \left(\frac{d_{84.1}}{d_{15.9}}\right)^{1/2} \tag{4.2}$$

where $d_{84.1}$ and $d_{15.9}$ may also be obtained from the cumulative size–frequency curve in Fig. 4.1. For sediment distributions that are not log-normal, the values of

$d_g$ and $\sigma_g$ are still determined the same way in practice, except that $d_{16}$ and $d_{84}$ are used in place of $d_{15.9}$ and $d_{84.1}$, respectively. Of course, the representation for such distributions can be refined by considering the skewness, kurtosis, and so on.

## 4.2 SHAPE FACTOR OF SEDIMENT PARTICLES

The shape, or overall geometric form, of a particle may be described by the *sphericity*, which is defined as the ratio of the surface area of a sphere having the same volume as the particle to the surface area of the given particle. Because of the difficulties in measuring the surface area or the sharpness of the edges, McNown and Malaika (1950) expressed the shape of particles by a shape factor, SF, defined as

$$\mathrm{SF} = \frac{c}{(ab)^{1/2}} \tag{4.3}$$

where $a$, $b$, and $c$ are, respectively, the lengths of the longest, intermediate, and shortest mutually perpendicular axes of the particle. The average value of shape factor has been found to be about 0.7 for natural sands (Interagency Committee, 1957).

## 4.3 FALL VELOCITY

Definitions for the standard fall velocity and standard fall diameter given by the Interagency Committee are as follows:

*Standard Fall Velocity of a Particle*. This is the average rate of fall that a particle would attain if falling alone in quiescent distilled water of infinite extent at 24°C.

*Standard Fall Diameter of a Particle*. This is the diameter of a sphere that has the same specific weight and has the same standard fall velocity as the given particle.

The standard fall diameter is a more precise representation of the particle size as compared with the sedimentation diameter, since the latter also varies with the water temperature. However, the sedimentation diameter is not very sensitive to temperature, and it may be considered as a constant within a certain range of temperature variation.

## Fall Velocity of Spheres

At a constant fall velocity, a particle is under the balanced forces of its submerged weight and the upward drag force. In fluid dynamics, the drag coefficient $C_D$ is defined as the ratio of drag per unit area to the dynamic pressure, that is,

$$C_D = \frac{F_D/A}{\frac{1}{2}\rho w_s^2} \tag{4.4}$$

where $F_D$ is the drag force equal to the submerged weight $\frac{1}{6}\pi d^3 g(\rho_s - \rho)$, $A = \pi/4\, d^2$ is the projected area of particle normal to the flow direction, $d$ is the diameter of sphere, $\rho$ is the mass density of fluid, and $w_s$ is the fall velocity. The drag coefficient varies with the Reynolds number $w_s d/\nu$, and the variation shown in Fig. 4.2 has three different regions that correspond to three distinct flow patterns around the particle shown in Fig. 4.3. In the first region, where the Reynolds number is less than about 1, the laminar boundary layer is not separated from the particle, and the drag coefficient reflecting only skin-friction drag is given by Stokes' law:

$$C_D = \frac{24}{\mathbf{R}} \tag{4.5}$$

where $\mathbf{R}$ is the Reynolds number. From Eqs. 4.4 and 4.5, the following relation for fall velocity is obtained.

$$w_s = \frac{gd^2}{18}\left(\frac{\rho_s - \rho}{\mu}\right) \tag{4.6}$$

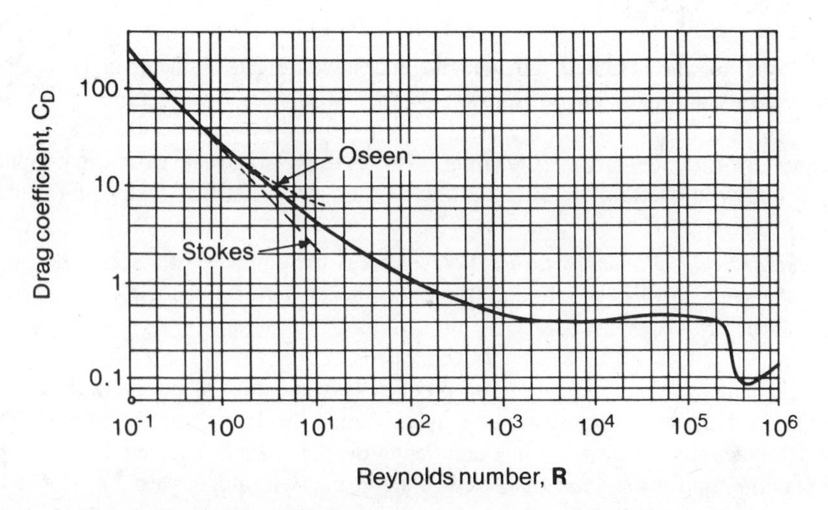


**Figure 4.2** Drag coefficient of spheres as function of Reynolds number.

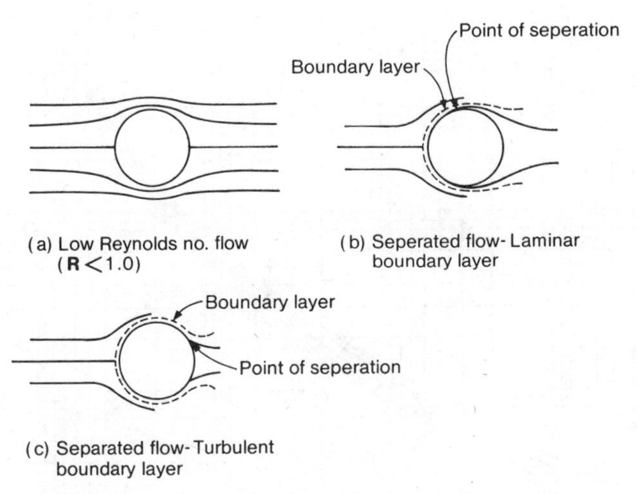


**Figure 4.3** Flow patterns surrounding a falling particle.

Generally speaking, Stokes' law is applicable to quartz particles, in the silt- and clay-size range, falling in water. Because of the very small fall velocity, silt and clay are usually not found in appreciable quantities in streambeds. They are often referred to as *wash load* because these fine materials tend to wash on through the system. Outside the Stokes range, the drag coefficient cannot be expressed analytically, but it has been determined experimentally as a function of **R**, as shown in Fig. 4.2.

The second region in Fig. 4.2 is roughly between the Reynolds numbers of 1 and $2 \times 10^5$. A falling particle in this region is characterized by a laminar boundary layer with flow separation, as shown in Fig. 4.3. The significant drop in drag at the **R** value of about $2 \times 10^5$ represents the change in boundary-layer separation from laminar to turbulent flow. The separated region, and therefore the drag, decreases as the laminar flow changes into turbulent.

The fall velocity $w_s$ can be obtained from Fig. 4.2 for the given temperature, diameter, and specific gravity of particle. Since $w_s$ is included in both $C_D$ and **R**, it can be obtained following a trial-and-error approach. The procedure is to Assume a value for $w_s$, calculate **R**, read $C_D$ from Fig. 4.2, and calculate $w_s$ from Eq. 4.4. These steps are repeated until the assumed $w_s$ and the calculated $w_s$ values are the same. The results so obtained are summarized in Fig. 4.4.

### Fall Velocity of Sand Grains

The fall velocity in relation to sieve diameter for naturally worn quartz particles falling in distilled water for a range of water temperature and particle shape factor, SF, is given in Fig. 4.5. The sieve diameter is used in this case because sizes of sand are usually determined in a sieve analysis and therefore given in terms of the

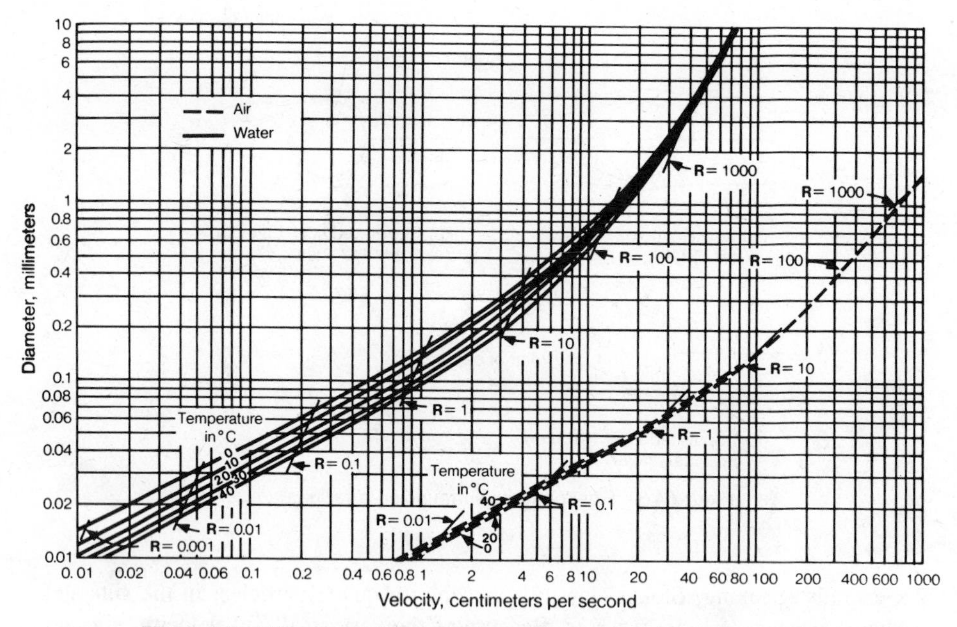


**Figure 4.4** Fall velocity of quartz spheres in air and water (Rouse, 1937).

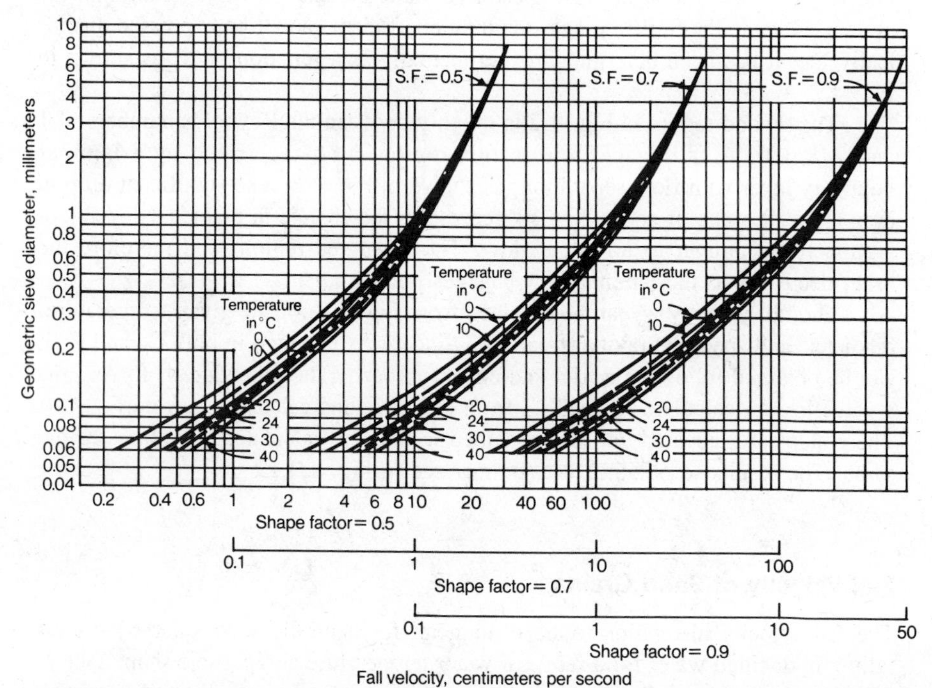


**Figure 4.5** Fall velocity of naturally worn quartz sand particles in distilled water as function of diameter and shape factor (Interagency Committee, 1957)

sieve diameter. For practical purposes, the sieve diameter for a size fraction of particles retained in one of a set of sieves is taken as the arithmetic or geometric mean of the openings of the adjacent sieve sizes. A relationship between sieve diameter and standard fall diameter is shown in Fig. 4.6 for naturally worn quartz particles falling in distilled water and for a range of water temperature and particle shape factor SF.

The fall velocity is also affected by several other factors, including adjacent particles, sediment concentration, flow turbulence, size of the settling tank, and so on. If more than a single particle settle in an unbound fluid system, a mutual interaction exists. It has been observed that the settling velocity increases if only a few closely spaced particles move. However, if many particles are dispersed throughout the fluid, the drag is increased and the settling velocity is reduced. In the case of clay and silt in suspension, electrochemical forces tend to hold particles together to form flocs, which settle faster as a group than do the individual particles. More detailed information on the fall velocity of particles can be found in other books devoted to sedimentation.

## 4.4 ANGLE OF REPOSE FOR SEDIMENTS

The angle of repose, $\theta$, is the angle of slope formed by particulate material under the critical equilibrium condition of incipient sliding (Interagency Committee,

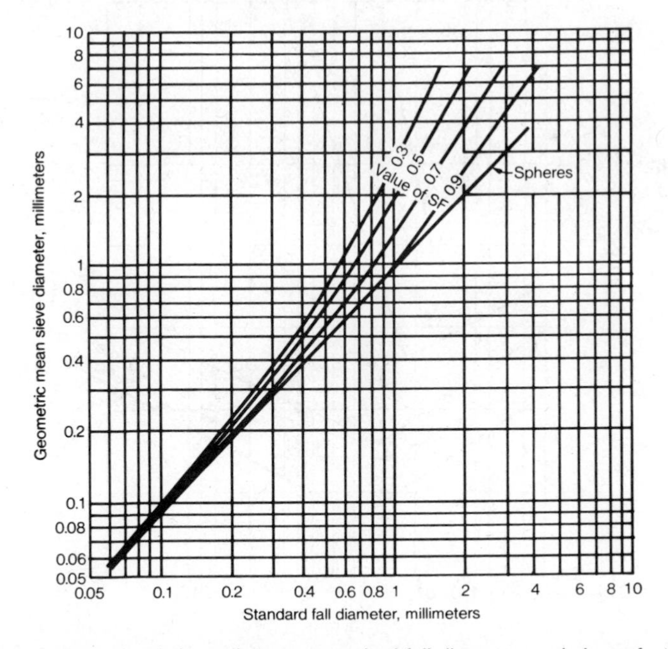


**Figure 4.6** Relationship of sieve diameter, standard fall diameter, and shape factor for naturally worn sand particles (Interagency Committee, 1957).

1957). This angle for submerged material is introduced in formulas for the initiation of sediment motion in Chapter 5. It can be measured by the critical toe angle of the submerged cone after introducing sediment into still water. The variation of $\theta$ with the particle size was prepared by Simons and Senturk (1976) for the uniform materials of lignite, bakelite, pumice, and sand and gravel, as shown in Fig. 4.7. For each type of material, the angle of repose decreases with an increase in size until a minimum is reached, at which time the angle of repose increases again. For sand and gravel, the minimum value of $\theta$ occurs at the $d$ value of about 2.4 mm. If the beginning of motion for sand mixtures is studied, it should be divided into different size fractions by sieving. After measuring the angle of repose for each size fraction, a relationship similar to that in Fig. 4.7 may be established. If the angle of repose of the mixture is plotted in the figure, two diameters ($d_1$ and $d_2$) are obtained. In this case, the smaller size, instead of the median size, should be used as the representative size in the study of the initiation of motion for the mixture.

The data for the angle of repose of riprap are contained in two published sets of curves by Lane (1955) and Simons (1957), the latter is shown in Fig. 4.8. The angle of repose is given as a direct function of the median stone size and angularity. The Lane and Simons curves differ more for smaller stone sizes than for the larger ones.

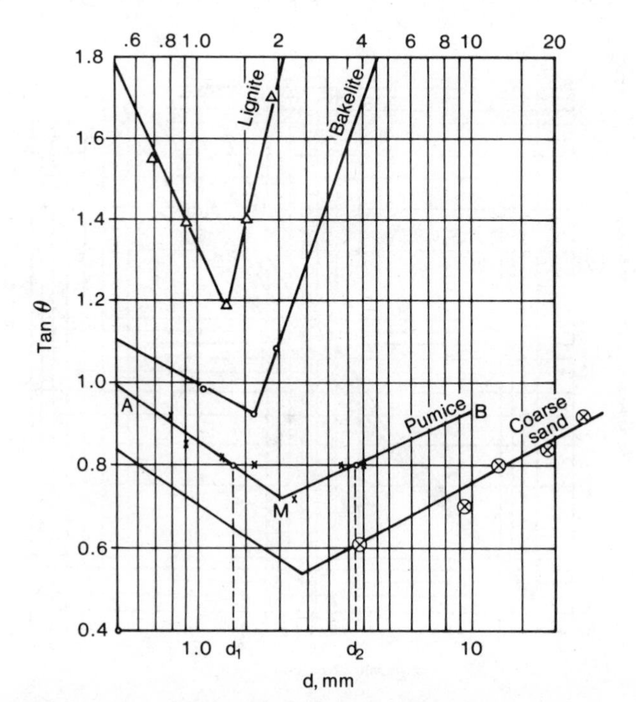


**Figure 4.7** Variation of angle of repose, $\theta$, with size for uniform materials (Simons and Senturk, 1976).

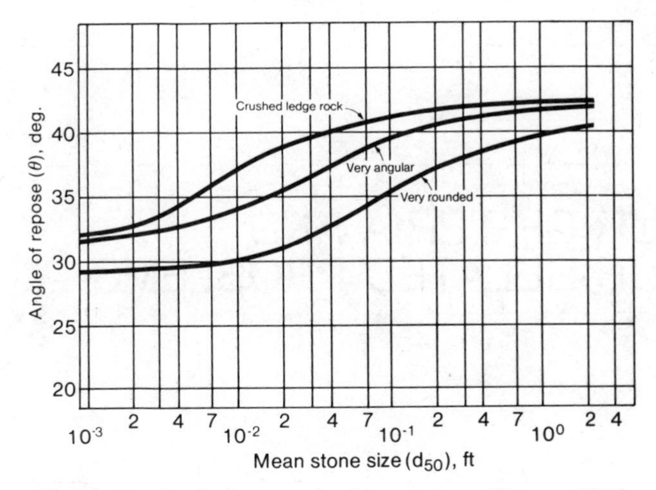


**Figure 4.8** Angle of repose for dumped riprap (Simons, 1957).

## REFERENCES


ASCE, *Sedimentation Engineering*, Manuals and Reports on Engineering Practice, No. 54, Vito A. Vanoni, ed., 1975.

Interagency Committee, "Some Fundamentals of Particle Size Analysis, A Study of Methods Used in Measurement and Analysis of Sediment Loads in Streams," Report No. 12, Subcommittee on Sedimentation, Interagency Committee on Water Resources, St. Anthony Falls Hydraulic Laboratory, Minneapolis, Minnesota, 1957.

Lane, E. W., et al., "Report of the Subcommittee on Sediment Terminology," *Trans. Am. Geophys. Union*, **28**(6), pp.936-938, 1947.

Lane, E. W., "Design of Stable Channels," *Trans. ASCE*, **120**, pp. 1234-1279, 1955.

McNown, J. S. and Malaika, J., "Effect of Particle Shape on Settling Velocity at Low Reynolds Number," *Trans. Am. Geophys. Union*, **31**, pp. 74-82, 1950.

Rouse, H., "Nomograph for the Settling Velocity of Spheres," Division of Geology and Geography Exhibit D of the Report of the Commission on Sedimentation, 1936-1937, National Research Council, October 1937, pp. 57-64.

Simons, D. B., "Theory and Design of Stable Channels in Alluvial Material," Ph.D. Thesis, Department of Civil Engineering, Colorado State University, 1957.

Simons, D. B. and Senturk, F., *Sediment Transport Technology*, Water Resources Publications, P.O. Box 2841, Littleton, Colorado, 1976.

# 5

# SCOUR CRITERIA AND SCOUR-RELATED PROBLEMS

*Scour* refers to the removal of material by running water. Scour criteria are involved with physical conditions pertaining to the threshold of motion for the material. Therefore, determination of such criteria is the prerequisite for scour control and sediment transport.

From the engineering viewpoint, one is always interested in determining the potential scour so that provisions can be made in design and construction. *Local scour* refers to scour caused by a local obstruction. It occurs at bridge piers, abutments, and other objects that obstruct the flow in different ways. Methods for estimating such scour are presented. *General scour* is provoked by the imbalance in sediment transport and may include constriction scour at a bridge opening or other encroachments; general aggradation and degradation of the stream bed, and scour induced by the curvature effect. Methods for analyzing the general scour are described in Chapter 13.

The engineer is often called to design bank protection and conveyance channels that must maintain stability while subject to scour. Methods for designing such structures to safeguard against scour are also presented.

## 5.1 CRITICAL SHEAR

The motion of a particle is under the interaction of two opposing forces: the applied force and the resisting force. The former is caused by the hydrodynamics of flow; the latter is associated with the submerged weight. The particle will be moved or entrained if the applied forces overcome the resistance. At the critical condition for entrainment, that is, threshold of movement, the applied forces are just balanced by the resisting force.

In a flowing stream, the forces acting on a grain of noncohesive sediment lying on the stream bed consists of the hydrodynamic drag, the hydrodynamic lift, and the submerged weight, as shown schematically in Fig. 5.1. The drag $F_D$ is in the direction of flow, and the lift $F_L$ is normal to the flow. The lift is not considered in most analytical treatments because of the difficulty in determining its mag-

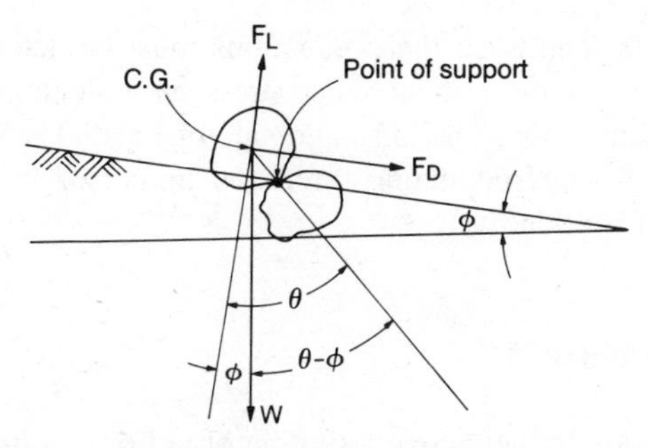


**Figure 5.1** Schematic of forces on sediment grain on sloping bed.

nitude. However, the lift is directly related to the drag. Since all relations contain coefficients that must be determined experimentally, the effects of lift are thus considered.

The hydrodynamic drag $F_D$ may be expressed as $c_1 \tau_0 d^2$, where $c_1$ is a constant, $\tau_0$ is bed shear stress or unit tractive force, $d$ is the grain diameter, and $c_1 d^2$ is the effective surface area upon which the shear stress is exerted. Under turbulent flow, the inertial effects predominate and the resultant of the drag acts at the center of gravity. The resisting force is from the submerged weight $W = c_2(\gamma_s - \gamma)\, d^3$, where $c_2$ is a constant, and $\gamma$ and $\gamma_s$ are the specific weights for water and sediment, respectively. When the grain is just about to move, it is supported at a downstream point and it loses firm contact below the particle.

The slope angle of the bed surface in Fig. 5.1 is designated as $\phi$. Angle $\theta$ is the angle of repose, or friction angle, between the immersed grains; it is with reference to the sloping bed along which movement may occur. At the threshold of movement, the resultant of these two forces is along the direction of the friction angle; that is, the ratio of forces on the grain acting parallel to bed to those acting normal to the bed is equal to tan $\theta$.

$$\tan \theta = \frac{F_D + W \sin \phi}{W \cos \phi} \tag{5.1}$$

Replacing $F_D$ and $W$ by their respective expressions given above yields the critical shear stress $\tau_c$

$$\tau_c = \frac{c_2}{c_1}(\gamma_s - \gamma) d \cos \phi (\tan \theta - \tan \phi) \tag{5.2}$$

For a horizontal bed, $\phi$ is zero and this equation becomes

$$\frac{\tau_c}{(\gamma_s - \gamma) d} = c \tan \theta \tag{5.3}$$

The coefficients $c$, $c_1$ and $c_2$ in these equations must be determined experimentally. It can be seen that the critical shear stress on a sloping stream bed is less than that on a horizontal bed. The left-hand side of Eq. 5.3 represents the ratio of two opposing forces—hydrodynamic force and immersed weight—which governs the initiation of motion.

## 5.2 SHIELDS DIAGRAM

Major variables that affect the incipient motion of uniform sediment on a level bed include $\tau_c$, $d$, $\gamma_s - \gamma$, $\rho$ and $\nu$. From dimensional analysis, they may be grouped into the following dimensionless parameters

$$F\left[\frac{\tau_c}{(\gamma_s - \gamma)d}, \frac{(\tau_c/\rho)^{1/2}d}{\nu}\right] = 0 \tag{5.4}$$

that is,

$$\frac{\tau_c}{(\gamma_s - \gamma)d} = F\left(\frac{U_{*c}d}{\nu}\right) \tag{5.5}$$

where $U_{*c} = (\tau_c/\rho)^{1/2}$ is the critical friction velocity. The left-hand side of this equation is the dimensionless critical shear stress and is often referred to as the *critical Shields stress*, $\tau_{*c}$. The right-hand side is called the *critical boundary Reynolds number* and is denoted by $\mathbf{R}_{*c}$. When any bed shear stress $\tau_0$, other than $\tau_c$, is used in the two quantities in Eq. 5.5, they become the Shields stress and boundary Reynolds number and are designated as $\tau_*$ and $\mathbf{R}_*$, respectively.

Figure 5.2 shows the functional relationship of Eq. 5.5 established based on experimental data, obtained by Shields (1936) and other investigators, on flumes with a flat bed. It is generally referred to as the *Shields diagram*. Each data point corresponds to the condition of incipient sediment motion or vanishing bed load.

The general shape of the Shields curve is similar to the drag coefficient ($C_D$)–Reynolds number ($\mathbf{R}$) curve plotted in Fig. 4.2; it is also similar to the friction coefficient ($f$)–boundary Reynolds number ($\mathbf{R}_*$) curve for boundary friction plotted in Fig. 3.5. Common to these relations are the laminar flow region, the transition region, and the turbulent region. In the laminar region of Fig. 5.2, where $\mathbf{R}_*$ is less than about 2, the particle size is less than the thickness of the laminar sublayer and, hence, is enclosed in a thin laminar film. Since the boundary is hydraulically smooth, the movement is mainly caused by viscous action; the critical Shields stress is inversely related to $\mathbf{R}_{*c}$ or $\tau_{*c} = C/\mathbf{R}_{*c}$, where $C$ is a constant. In the transition region of intermediate boundary Reynolds numbers, the grain size has a magnitude similar to that of the laminar sublayer; therefore, the movement is partially influenced by viscosity. The critical Shields stress has a minimum value of 0.03 at the $\mathbf{R}_{*c}$ value of about 10.

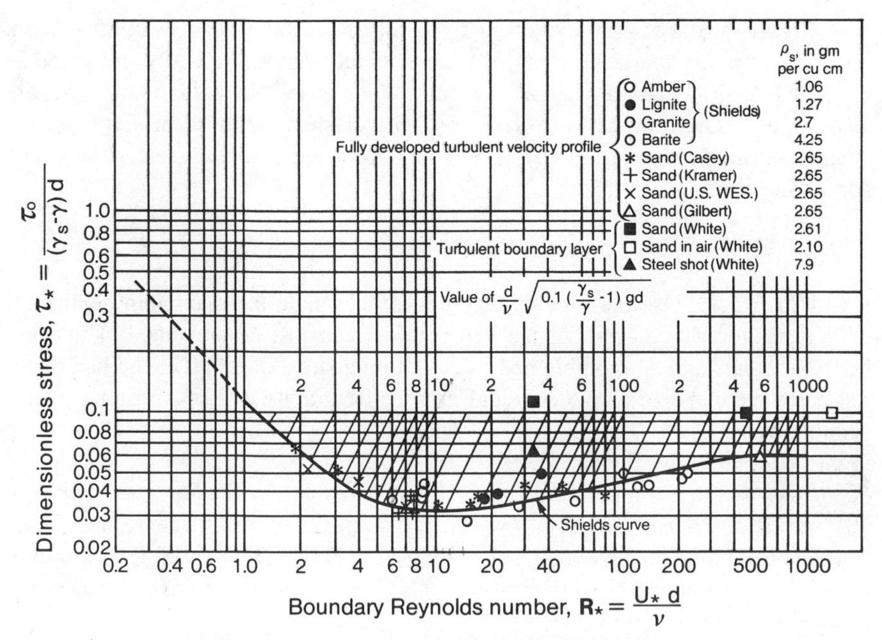


**Figure 5.2** Shields diagram for incipient motion.

In the turbulent region of large Reynolds numbers ($\mathbf{R}_* > 400$), the laminar sublayer is interrupted by the grain size. For this hydraulically rough boundary, the critical Shields stress has a constant value of 0.06, independent of the Reynolds number. Lower values for $\tau_{*_c}$ have been suggested by other investigators; for example, Zeller (1963) suggested a lower value of 0.047 in this region.

The Shields diagram contains the critical shear stress $\tau_c$ as an implicit variable that cannot be obtained directly. To overcome this difficulty, the ASCE Sedimentation Manual (1975) utilizes a third dimensionless parameter

$$\frac{d}{\nu}\left[0.1\left(\frac{\gamma_s}{\gamma}-1\right)gd\right]^{1/2}$$

which appears as a family of parallel lines in the diagram. From the value of the third parameter, the value of the critical Shields stress is obtained at an intersection with the Shields curve from which $\tau_c$ can be calculated.

The Shields diagram has gained wide acceptance. However, it is not without criticism. One important limitation on any predictive method for incipient motion is related to the fact that in the turbulent flow there exists momentary shear stress or velocity which may be considerably different from the average value. Besides, the lift is not explicitly represented in the method for the Shields diagram. The fact is that the lift force on a particle fluctuates with the orientation of the particle.

Figure 5.3 is a graph relating $\tau_c$ to mean sediment size calculated from Shields curve for quartz sand and different temperatures, originally presented in the ASCE Sedimentation Manual. Several values of the Reynolds number $\mathbf{R}_*$ are also shown. The temperature effect on critical shear is more pronounced for smaller sediment sizes from 0.1 to 0.5 mm for which the $\mathbf{R}_*$ value is less than approximately 5.

▶ ***EXAMPLE 5.1.*** Use the Shields diagram to determine the maximum depth of a wide canal for which scour of the bed material can just be prevented. The canal has rigid banks and an erodible bed; it is laid on a slope of 0.0005. The bed material has a median size of 2.5 mm and its specific gravity is 2.65. Assume a temperature of 10°C.

*Solution*. For the temperature of 10°C, the kinematic viscosity $\nu = 1.306 \times 10^{-6}$ m²/sec. Then, $d/\nu[0.1(\gamma_s/\gamma - 1)gd]^{1/2} = 121.8$. With the value of this parameter, the dimensionless stress is obtained as 0.043 from the Shields curve in Fig. 5.2. Therefore, $\tau_c = 0.043\ (\gamma_s - \gamma)d = 1.74$ N/m². From $\tau_c = \gamma DS$, the depth $D$ is 0.36 m.

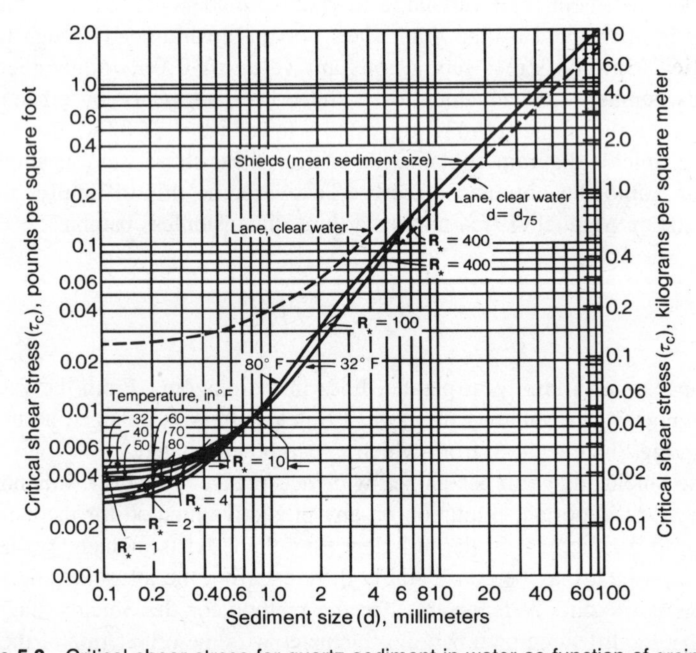


**Figure 5.3** Critical shear stress for quartz sediment in water as function of grain size.

## 5.3 OTHER SCOUR CRITERIA BASED ON SHEAR STRESS

Lane (1955) recommended values of $\tau_c$ for design of irrigation canals to convey clear water with noncohesive beds, as shown in Fig. 5.3. For coarse material, the recommended $\tau_c$ is in terms of $d_{75}$, which is regarded as being more characteristic of the riprap mixture. The line for coarse material as shown is independent of the Reynolds number and is given only as a function of the stone size by

$$\tau_c = 0.0164 d_{75} \tag{5.6}$$

where $d_{75}$ is in millimeters and $\tau_c$ is in pounds per square foot. For coarse material, the critical shear stress $\tau_c$ obtained based on the Shields diagram and Lane's recommended value are quite similar. But these values are quite different for smaller sediment sizes. The reason for this discrepancy is not totally clear, probably because the canals studied by Lane were actually operated with a mobile bed.

For the incipient motion of coarse material, the dimensionless shear stress becomes independent of the Reynolds number, so that

$$\frac{\tau_c}{(\gamma_s - \gamma)d} = \text{Constant} \tag{5.7}$$

Therefore, the critical shear for rock having a given specific weight is directly proportional to the effective rock size. For example, the Highway Research Board (1970) has determined the empirical equation for the critical shear as

$$\tau_c = 4 d_{50} \tag{5.8}$$

where $\tau_c$ is in pounds per square foot and $d_{50}$ is in feet. This relationship is shown in Fig. 5.4, in which the value of $d_{50}$ in the data used ranges from about $5 \times 10^{-4}$ to about 0.5 ft. The Highway Research Board has also compared several published formulas for $\tau_c$ as a function of $d_{50}$, as shown in Fig. 5.5. Certain relationships have included the hydraulic radius $R$ as an additional variable. It should be noted that many of these lines have slopes steeper than 45°, which means that $\tau_c$ tends to increase faster than Eq. 5.8 indicates. This trend of variation may be attributed to the higher values of Shields stress at very large values of $\mathbf{R}_*$, suggested by Shen and Wang (1984).

## 5.4 CRITICAL SHEAR ON SIDE SLOPES

A noncohesive particle on the stream bed is subject to the tractive force, or boundary shear, acting in the direction of flow. If this particle is on a side slope, then there is an additional gravitational force component parallel to the slope which tends to roll the particle down. Figure 5.6 shows a trapezoidal channel with a side slope angle $\phi$. Point $A$ on the side slope is acted upon by the tractive force

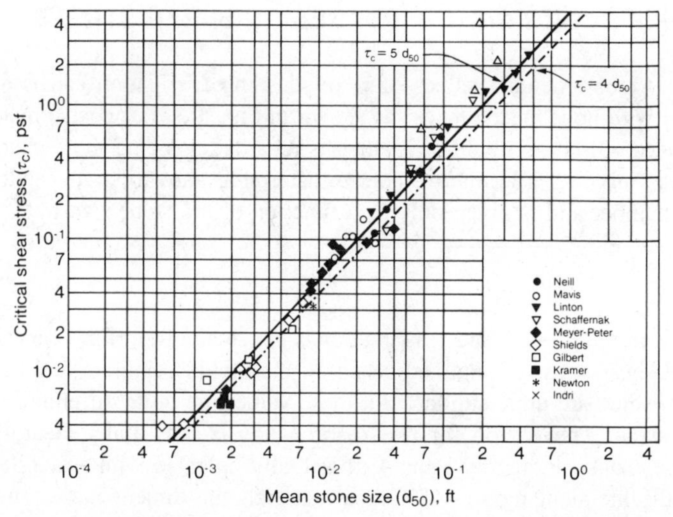


**Figure 5.4** Critical boundary shear for riprap in terms of stone size (after Highway Research Board, 1970).

$F_D$ and its submerged weight $W$. The latter is resolved into a component normal to the side slope $W \cos \phi$ and another component parallel to the slope $W \sin \phi$. The resultant of the applied forces is given by $[F_D^2 + (W \sin \phi)^2]^{1/2}$, and the resistance force is the normal force $W \cos \phi$ multiplied by the coefficient of friction, $\tan \theta$, where $\theta$ is the angle of repose or the internal friction angle. At the threshold of motion, the applied forces tending to cause the movement is equal to the resistance to motion, that is,

$$W \cos \phi \tan \theta = [F_D^2 + (W \sin \phi)^2]^{1/2} \tag{5.9}$$

Solving for the tractive force that causes impending motion on a sloping surface yields

$$(F_D)_s = W \cos \phi \tan \theta \left(1 - \frac{\tan^2\phi}{\tan^2\theta}\right)^{1/2} \tag{5.10}$$

where $(F_D)_s$ is the critical tractive force on the sloping surface. When motion of a particle is impending on the level surface where $\phi = 0$, the following is obtained:

$$(F_D)_b = W \tan \theta \tag{5.11}$$

Now let $K$ be the ratio of these two critical tractive forces $(F_D)_s$ and $(F_D)_b$, then we obtain the following from Eqs. 5.10 and 5.11:

$$K = \frac{(F_D)_s}{(F_D)_b} = \cos \phi \left(1 - \frac{\tan^2\phi}{\tan^2\theta}\right)^{1/2} = \left(1 - \frac{\sin^2\phi}{\sin^2\theta}\right)^{1/2} \tag{5.12}$$

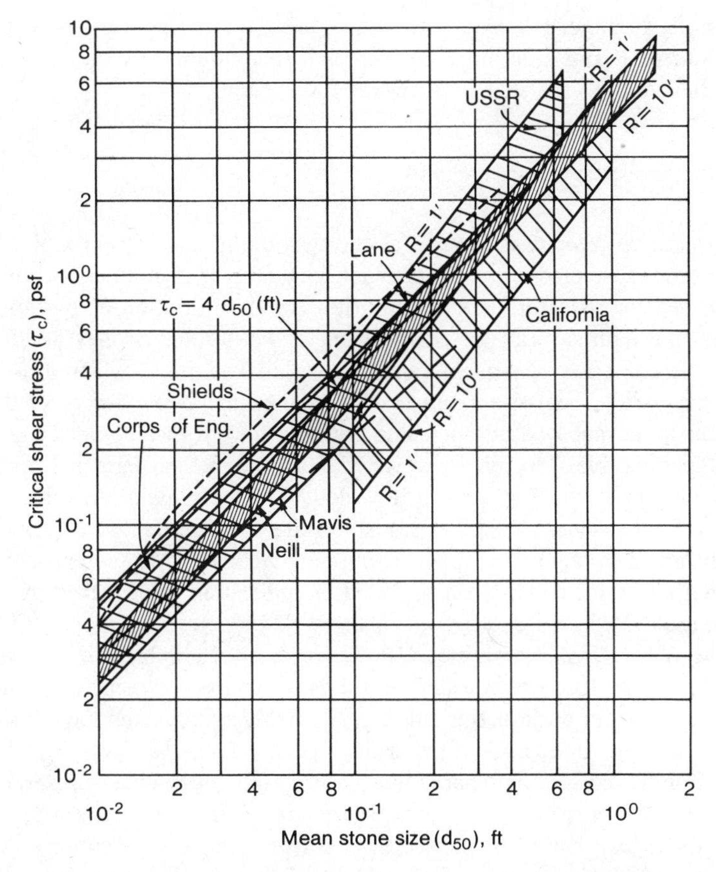


**Figure 5.5** Comparison of various published formulations of critical boundary shear (after Highway Research Board, 1970).

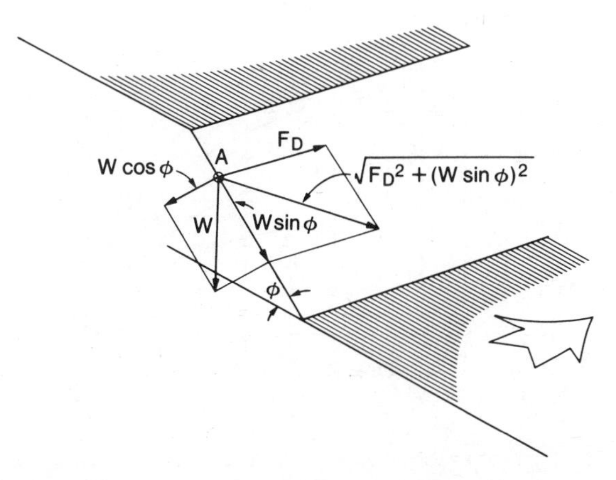


**Figure 5.6** Forces on sediment particle on side slope.

This ratio is an important parameter used in design; it is also shown graphically as a function of $\theta$ and side slope in Fig. 5.7. As would be expected, the ratio approaches unity as the side slope moves toward zero.

## 5.5 PERMISSIBLE VELOCITY

The *permissible velocity* is defined as the maximum mean velocity of a channel that will not cause erosion of the channel boundary. It is often called the *critical velocity* because it refers to the condition for the initiation of motion. Canal designers are familiar with the classic work of Fortier and Scobey (1926) on the permissible velocities of straight canals shown in Table 5.1, which also includes the corresponding shear-stress values converted by the U. S. Bureau of Reclamation. Canals designed based upon this criterion have a mean velocity no greater than the permissible velocity. However, this simple method for canal design does not consider the channel shape and flow depth. At the same mean velocity, channels of different shapes or depths may have quite different tractive forces acting on the boundaries. In other words, the permissible velocity is depth-dependent, and a correction factor for depth is usually used in application as described in a following paragraph.

The ASCE Task Committee (1967) presented a graphical relationship (see Fig 5.8) showing the critical water velocities for quartz sediment as a function of mean grain size. In constructing this figure, the data points and the curves of the upper, mean, and lower limit of the critical mean velocity are taken from the work of Hjulstrom (1935), who prepared the curves based on the data of several investigators. The curves are for flows with depths greater than 1 m. The data for mean sediment size less than 0.01 mm were taken from Fortier and Scobey (1926). For

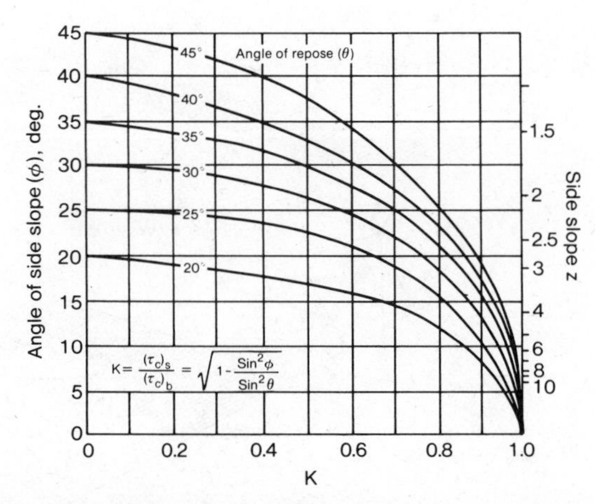


**Figure 5.7** Ratio of critical shear on side slope to critical shear on bottom for noncohesive sediment.

**TABLE 5.1 Maximum Permissible Velocities by Fortier and Scobey (1926) for Straight Channels of Small Slope After Aging**

| Material | $n$ | Clear Water $U$ (ft/sec) | Clear Water $\tau_0$ (lb/ft$^2$) | Water Transporting Colloidal Silts $U$ (ft/sec) | Water Transporting Colloidal Silts $\tau_0$ (lb/ft$^2$) |
|---|---|---|---|---|---|
| Fine sand, colloidal | 0.020 | 1.50 | 0.027 | 2.50 | 0.075 |
| Sandy loam, noncolloidal | 0.020 | 1.75 | 0.037 | 2.50 | 0.075 |
| Silt loam, noncolloidal | 0.020 | 2.00 | 0.048 | 3.00 | 0.11 |
| Alluvial silts, noncolloidal | 0.020 | 2.00 | 0.048 | 3.50 | 0.15 |
| Ordinary firm loam | 0.020 | 2.50 | 0.075 | 3.50 | 0.15 |
| Volcanic ash | 0.020 | 2.50 | 0.075 | 3.50 | 0.15 |
| Stiff clay, very colloidal | 0.025 | 3.75 | 0.26 | 5.00 | 0.46 |
| Alluvial silts, colloidal | 0.025 | 3.75 | 0.26 | 5.00 | 0.46 |
| Shales and hardpan | 0.025 | 6.00 | 0.67 | 6.00 | 0.67 |
| Fine gravel | 0.020 | 2.50 | 0.075 | 5.00 | 0.32 |
| Graded loam to cobbles when noncolloidal | 0.030 | 3.75 | 0.38 | 5.00 | 0.66 |
| Graded silts to cobbles when colloidal | 0.030 | 4.00 | 0.43 | 5.50 | 0.80 |
| Coarse gravel, noncolloidal | 0.025 | 4.00 | 0.30 | 6.00 | 0.67 |
| Cobbles and shingles | 0.035 | 5.00 | 0.91 | 5.50 | 1.10 |

such fine sediments, cohesion is an important factor responsible for the increase in critical velocity.

Since the permissible velocity relationship of Fig. 5.8 is restricted to a flow depth of at least 1 m (3 ft), the variation of permissible velocity with flow depth can be corrected such that the same critical unit tractive force is maintained. A permissible velocity correction factor to account for the variation in flow depth

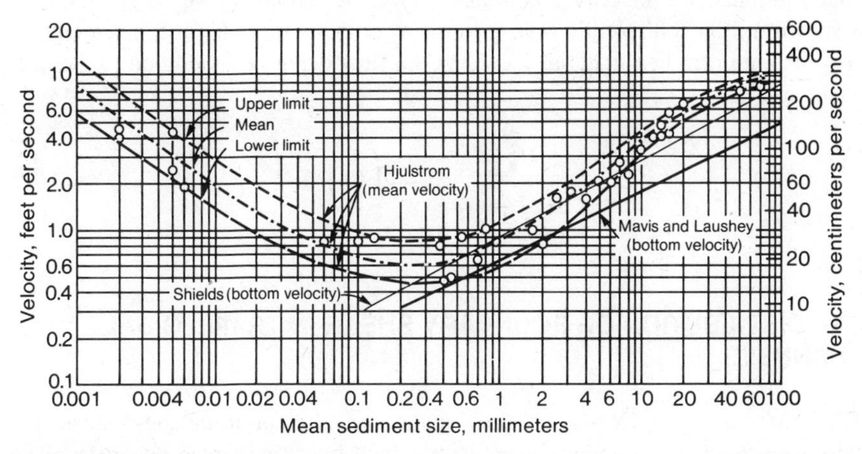


**Figure 5.8** Critical water velocities for quartz sediment as function of mean grain size (after ASCE Task Committee, 1967).

was derived by Mehrota (1983). For two flow depths $D_1$ and $D_2$, the critical unit tractive force for the same soil type is the same, that is,

$$\tau_c = \gamma R_1 S_1 = \gamma R_2 S_2 \tag{5.13}$$

where $R_1$ and $R_2$ are the respective hydraulic radii, and $S_1$ and $S_2$ are the respective channel slopes.

Now, from the Manning equation for each case, we have

$$U_1 = \frac{1}{n_1}(R_1)^{2/3}(S_1)^{1/2} \quad \text{and} \quad U_2 = \frac{1}{n_2}(R_2)^{2/3}(S_2)^{1/2} \tag{5.14}$$

and assuming that $n$ is independent of the depth and thus the same, we obtain

$$\frac{U_2}{U_1} = \left(\frac{R_2}{R_1}\right)^{2/3}\left(\frac{S_2}{S_1}\right)^{1/2} \tag{5.15}$$

The correction factor $k$ for the permissible velocity is obtained from Eqs. 5.13 and 5.15:

$$k = \frac{U_2}{U_1} = \left(\frac{R_2}{R_1}\right)^{1/6} \tag{5.16}$$

The correction factor should be applied to the permissible velocity $U_1$ if the normal depth $D_2$ is different from $D_1$. For flow depth $D_1$, the correction factor is, of course, 1.

Several formulas for critical velocity of noncohesive bed material have been proposed. For example, Mavis and Laushey (1949) and Carstens (1966) have presented formulas for the critical bottom velocity; the formula of Mavis and Laushey is shown in Fig. 5.8. Neill's (1967) formula for the permissible mean velocity $U_p$ of coarse uniform bed material considers the flow depth $D$ in addition to the sediment size $d$:

$$\frac{U_p^2}{\left(\frac{\rho_s}{\rho} - 1\right)gd} = 2.5\left(\frac{d}{D}\right)^{-0.20} \tag{5.17}$$

## 5.6 DISTRIBUTION OF BOUNDARY SHEAR IN TRAPEZOIDAL CHANNELS

The equation $\tau_0 = \gamma RS$ was obtained under the assumption of one-dimensional flow for which the boundary shear is averaged over the wetted perimeter. The shear distribution is not uniform in reality, as a result of the channel shape and curvature. The distribution of boundary shear in straight trapezoidal channels was studied by Olsen and Florey (1952) using membrane analogy. For various trapezoidal cross sections, the distributions of boundary shear around wetted perimeter were obtained, as illustrated by a sample shown in Fig. 5.9a. The results of Olsen and Florey were transformed and expanded by the Highway Research Board (1970) as shown in Figs. 5.9b and 5.9c which give the maximum shears, in terms of $\gamma RS$, on the beds and sides for various $b/D$ ratios, respectively. These curves indicate that the maximum value of the boundary shear occurs on the bed at higher values of $b/D$. As the $b/D$ ratio decreases below 2, the shear on the bed decreases but that on the sides continues to increase. This decrease in bottom shear is attributed to the side-wall effect. For trapezoidal channels of practical

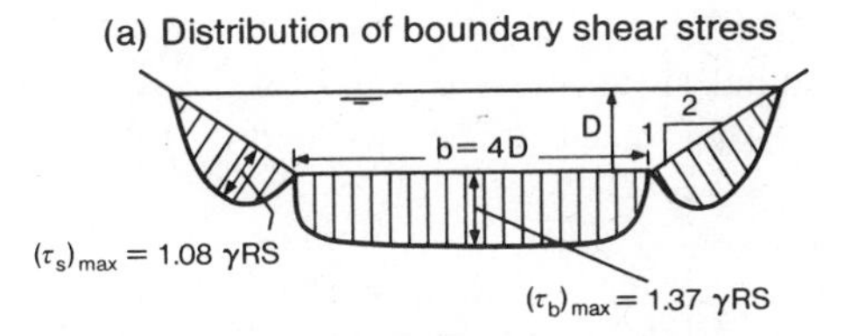


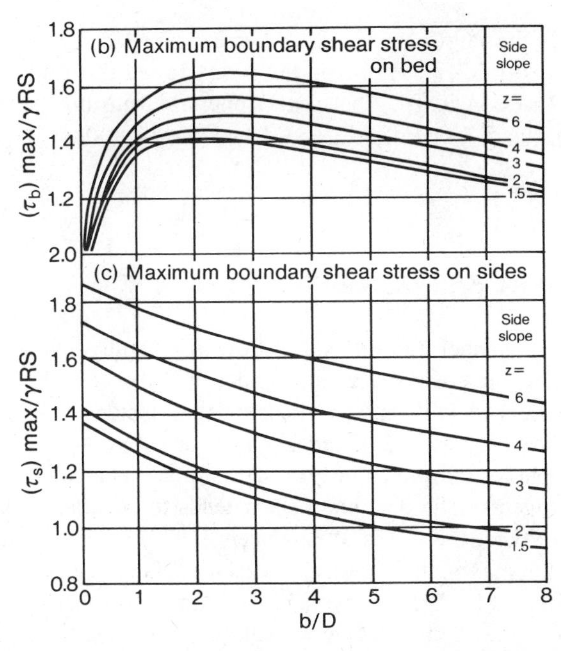


**Figure 5.9** Distributions of boundary shear stress in trapezoidal channels.

interest, the values of $b/D$ and $z$ are usually greater than 2; then the value of maximum shear can be conservatively approximated as

$$(\tau_0)_{max} = 1.5\gamma RS \tag{5.18}$$

▶ ***EXAMPLE 5.2.*** A trapezoidal channel has a depth $D = 10$ ft, bottom width $b = 30$ ft, and side slope of 2 to 1. It is constructed with riprap bank protection, and the sand bed is mobile at the design discharge. If the riprap has an angle of repose of 35°, and the channel slope is 0.002, obtain the maximum shear stress (unit tractive force) on the bed and banks. Determine the appropriate stone size for the riprap slope protection using the Highway Research Board (Federal Highway Administration) criterion.

*Solution.* For the given channel configuration, $b/D = 30/10 = 3$ and $R = 6.69$ ft. From Fig. 5.9b, $(\tau_b)_{max} = 1.42\ \gamma RS = 1.19\ \text{lb/ft}^2$. From Fig. 5.9c, $(\tau_s)_{max} = 1.15\ \gamma RS = 0.96\ \text{lb/ft}^2$.

The stone size for the riprap is selected such that the permissible shear of stone matches the actual shear stress exerted by the flow. Because of the bank slope, the permissible shear needs to be modified by the factor $K$ defined by Eq. 5.12. Now

$$K = \left(1 - \frac{\sin^2 26.6°}{\sin^2 35°}\right)^{1/2} = 0.626$$

where 26.6° is the angle for the 2 to 1 side slope. The $K$ value may also be obtained from Fig. 5.7. Then for the side slope, the required permissible $\tau_0 = 0.96/K = 1.534\ \text{lb/ft}^2$. The mean stone size with this permissible shear is obtained from Fig. 5.4 or Eq. 5.8, and $d = 1.534/4 = 0.38$ ft.

## 5.7 BOUNDARY SHEAR IN BENDS

In a curved river channel, the velocity of flow is generally higher near the outside, or concave, bank and smaller near the inside, or convex, bank. This velocity distribution is associated with greater shear on the outside of the bend. Ippen and Drinker (1962) measured the boundary shear distribution in curved trapezoidal channels. Sample results shown in Fig. 5.10 are near channel bends preceded and followed by tangents. The maximum shear tends to be near the bend exit. Using different sets of data, the Corps of Engineers (1970) compiled the relationship of Fig. 5.11 for the maximum boundary shear at channel bends. The ratio of maximum shear affected by bend to average shear in approach channel, $(\tau_0)_{max}/\bar{\tau}_0$, is given as a function of the relative curvature $r_c/B$ ($B$ is the surface width) for smooth and rough channels of equal bottom and side roughness.

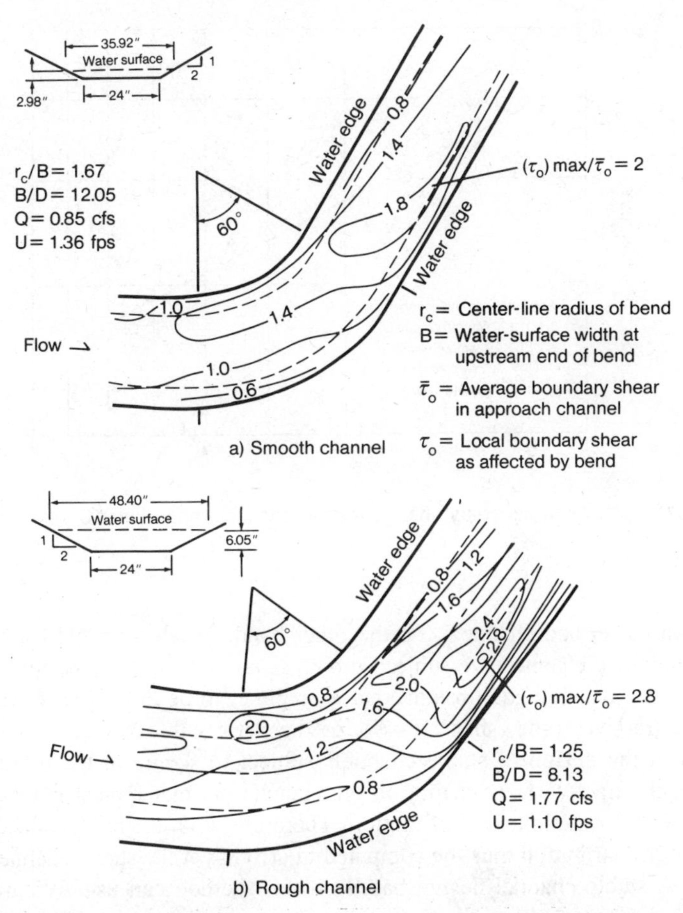


**Figure 5.10** Boundary shear distributions in curved trapezoidal channels obtained experimentally by Ippen and Drinker (1962).

## 5.8 DESIGN OF STABLE CHANNELS SUBJECT TO SCOUR BUT NOT TO SILT

The term *stable channel* refers to an unlined channel for carrying water, the banks and bed of which are not scoured by the moving water and in which objectionable deposits of sediment do not occur. Stable unlined channels are usually more economical and environmentally desirable than lined channels. However, lined channels may prove to be more feasible under certain conditions, such as steep slopes and excessive velocities.

When water is drawn from a storage reservoir, it is normally free of bed load, and the basic criterion in designing a stable channel is to prevent scour of the material from the channel boundary. In order to maintain stability at a moderate or high velocity, such channels usually have coarse bed material or are lined with

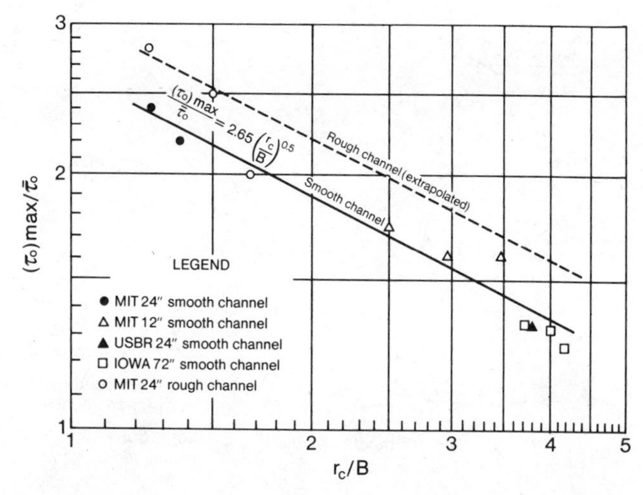


**Figure 5.11** Maximum boundary shear stress at channel bends (after Corps of Engineers, 1970).

vegetation. Sand-bed channels, on the other hand, usually transport a bed load. Such mobile bed channels are subject to both scour and silt; their stability is maintained by the dynamic equilibrium, that is, regime, to be discussed in Part III.

The tractive force (shear stress), or the permissible velocity, is the basic criterion in the design of stable channels subject to scour. In the tractive force criterion, the tractive force acting on the channel perimeter must not exceed the critical value for the initiation of motion. Therefore, both the critical shear and the actual shear distribution must be computed in arriving at the stable channel configuration. A stable channel design based on this method can usually carry a discharge slightly greater than the design discharge without serious effect on channel stability. This is because erosion caused by the greater discharge will carry away the fine particles of the bed material, leaving the coarse particles to form a protective armor layer.

In a design problem, the design discharge, bed material, and slope of the terrain are usually given; and the channel geometry is to be determined. The channel slope may be equal to the terrain slope or it may be flatter through the use of drop structures. The design procedure for channels subject only to scour is illustrated by the following examples. The procedure used in determining the channel geometry involves trial-and-error in which the channel geometry is first assumed and then adjusted based on the computed results to arrive at the final geometry.

► ***EXAMPLE 5.3.*** This example illustrates the design procedure when only the bed is subject to scour. Design a trapezoidal channel to carry the discharge of 27 cfs on a slope of 0.01 predetermined by the terrain. The channel has rigid banks with a 1.5-to-1 slope. The bed material is coarse quartz gravel with $d_{75}$ = 30 mm. Assume a Manning's coefficient of 0.025.

*Step 1*. For $d_{75} = 30$ mm, the permissible tractive force $\tau_p = 0.48$ lb/ft$^2$ is obtained from Lane's criteria in Fig. 5.3.

*Step 2*. Assume a width-depth ratio $b/D = 7$ and use Fig. 5.9b to obtain the maximum tractive force on the bed $(\tau_b)_{max} = 1.25\ \gamma RS$. For a stable channel, $\tau_p = (\tau_b)_{max}$; therefore

$$R = \frac{0.48}{1.25 \times 62.4 \times 0.01} = 0.615 \text{ ft}$$

*Step 3*. Apply the Manning formula to determine the velocity and cross-sectional geometry. Now

$$U = \frac{1.486}{n} R^{2/3} S^{1/2} = 4.30 \text{ ft/sec}$$

For the design discharge of 27 cfs, the cross-sectional area $A = Q/U = 6.28$ ft$^2$. For the $A$ value of 6.28 and the $R$ value of 0.615, the values of $b$ and $D$ are calculated to be 7.60 ft and 0.723 ft, respectively, using the geometric relationships for a trapezoid.

*Step 4*. Check the $b/D$ ratio obtained in Step 3 with that assumed in Step 2. Repeat the steps until these two values are sufficiently close. The final values so obtained are as follows: $b = 6.54$ ft and $D = 0.77$ ft.

*Step 5*. Add a free-board to the cross section.

► ***EXAMPLE 5.4*** This example illustrates the procedure for designing trapezoidal channels whose bed and banks are subject to scour. Design a stable channel to carry the discharge of 1500 cfs, laid on the slope of 0.0016. The bed material consists of noncolloidal course gravel and pebbles (very rounded material), for which $d_{50} = 1.25$ in. and $d_{75} = 1.57$ in. Obtain the design configuration of the channel.

*Step 1*. Select the angle of repose and channel roughness. From Fig. 4.8, the angle of repose for very rounded material with a mean size of 1.25 in. is 36°. The Manning coefficient may be estimated from Strickler's formula (Eq. 3.28) for straight channels with a gravel bed. In this case, the $n$ value of 0.025 is assumed.

*Step 2*. Select the side slope. For the sake of bank stability, the side slope must be less than the angle of repose. A flatter slope is generally more stable but is less efficient hydraulically. With consideration of the 36° angle of repose, the 2-to-1 side slope, which has the slope angle of 26.6°, is assumed. The $K$ value given by Eq. 5.12 and Fig. 5.7 is thus 0.647.

*Step 3*. Determine the permissible tractive forces for the bed and bank. Based on the value of $d_{75}$, the permissible tractive force for the bed $(\tau_b)_p$ is obtained from Fig. 5.3 to be 0.63 lb/ft$^2$. For the bank, $(\tau_s)_p = K\,(\tau_b)_p = 0.409$ lb/ft$^2$.

*Step 4*. Determine the maximum tractive forces based on an assumed width-depth ratio. Assume $b/D = 7$ and insert this value into Fig. 5.9 to obtain $(\tau_b)_{max} = 1.26\ \gamma RS$ and $(\tau_s)_{max} = 1.0\ \gamma RS$.

*Step 5*. Obtain the hydraulic radius. For a stable channel, the permissible tractive force must equal, or exceed, the maximum tractive force on the bed and bank. Equating the permissible and maximum tractive forces for the bed yields

$$R = \frac{(\tau_b)_p}{1.26\gamma S} = 5.01 \text{ ft}$$

Equating the forces for the bank gives

$$R = \frac{(\tau_s)_p}{1.0\gamma S} = 4.10 \text{ ft}$$

The smaller value governs the design; therefore $R = 4.10$ ft.

*Step 6*. Determine the channel geometry. The velocity is first computed from the Manning formula,

$$U = \frac{1.486}{n} R^{2/3} S^{1/2} = 6.09 \text{ ft/sec}$$

Therefore, $A = Q/U = 246.3 \text{ ft}^2$. From the values of $A$ and $R$, one has $b = 36.6$ ft and $D = 5.23$ ft.

*Step 7*. Check the $b/D$ ratio obtained with the assumed. Repeat the steps until these two values are sufficiently close.

*Step 8*. Add a free-board to the channel section.

## 5.9 LOCAL SCOUR AROUND BRIDGE PIERS

In alluvial channels, the scour around bridge piers (see Fig. 5.12), abutments, and other local obstructions is first initiated by the interference to flow and sediment transport. The erodible bed deforms until it reaches an equilibrium scour configuration for which the rate of sediment supplied to the scour area is balanced by the rate of transport out of the area, that is, $(Q_s)_{in} = (Q_s)_{out}$. Sediment transport through a scour hole is also affected by the horseshoe vortices, which, as a turbulent motion, increase the particle mobility. The sediment rate is an inverse function of the particle size. Because sediment rates flowing into and out of a scour area change with the size, at nearly the same proportion, the scour depth is not significantly affected by the sediment size which is therefore missing in most formulas for local scour.

Since the flow pattern of scour is very much complicated by the configuration of the obstruction, it is often necessary to utilize model studies in order to

**Figure 5.12** Local scour around bridge pier formed at low flow (courtesy of James Nelson).

establish the equilibrium scour depth as a function of the pertinent variables. Such model studies have been made at the University of Iowa (Laursen, 1956), Colorado State University (Shen et al. 1969), and elsewhere.

More than 10 different formulas have been developed for predicting local scour around bridge piers, based on essentially laboratory data. Despite the large number, such formulas contain a limited number of variables, namely, approach flow depth, effective pier width, Froude number, shear stress, and critical shear stress. Some of these formulas for rectangular bridge piers are described in the following paragraphs. In the case of circular piers, the scour depth is about 90% of that for rectangular piers; for sharp nosed piers, it is about 80%.

The Laursen formula (1962) contains the local scour depth as an implicit variable:

$$\frac{b}{D_0} = 5.5\frac{D_s}{D_0}\left(\frac{1}{11.5}\frac{D_s}{D_0} + 1\right)^{1.7} - 1 \tag{5.19}$$

where $D_s$ is the depth of scour below mean bed elevation, $b$ is the width of pier normal to flow, and $D_0$ is the mean depth of flow upstream of pier. In the experimental study by Laursen, the median diameter of sediment ranged between 0.46 and 2.2 mm.

The scour depth $D_s$ is implicit in Eq. 5.19. Neill (see Blench, 1969) used Laursen's 1956 design curve to obtain the following explicit formula for $D_s$:

$$\frac{D_s}{b} = 1.5\left(\frac{D_0}{b}\right)^{0.3} \tag{5.20}$$

Shen et al. (1969) used the Froude number in the scour-depth prediction:

$$\frac{D_s}{b} = 3.4(\mathbf{F}_0)^{2/3}\left(\frac{D_0}{b}\right)^{1/3} \tag{5.21}$$

where $\mathbf{F}_0 = U_0/(gD_0)^{1/2}$ is the Froude number based on the mean upstream velocity $U_o$ and depth $D_o$.

The Colorado State University (1975), or CSU, formula reported in the Federal Highway Administration manual developed as a best fit to the data available at the time is shown in Fig. 5.13. This formula has the form

$$\frac{D_s}{D_0} = 2.2\left(\frac{b}{D_0}\right)^{0.65}(\mathbf{F}_0)^{0.43} \tag{5.22}$$

Jain and Fischer (1979) studied scour around circular piers at higher Froude numbers and proposed the following formulas:

$$\frac{D_s}{b} = 2.0(\mathbf{F}_0 - \mathbf{F}_c)^{0.25}\left(\frac{D_0}{b}\right)^{0.5} \quad \text{for } \mathbf{F}_0 - \mathbf{F}_c > 0.2 \tag{5.23}$$

and

$$\frac{D_s}{b} = 1.84\left(\frac{D_0}{b}\right)^{0.3}(\mathbf{F}_0)^{0.25} \quad \text{for } \mathbf{F}_0 > \mathbf{F}_c \tag{5.24}$$

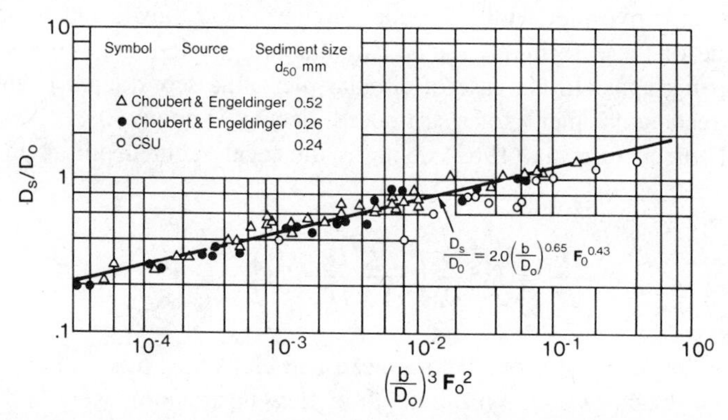


**Figure 5.13** CSU formula for scour depths of circular piers. The constant 2.0 becomes 2.2 for rectangular piers.

where $\mathbf{F}_c$ is the critical Froude number for incipient sediment motion. In the region $0 < \mathbf{F}_0 - \mathbf{F}_c < 0.2$, the larger value from both formulas is used. The procedure for computing $\mathbf{F}_c$ is as follows:

1. Determine $\tau_c$ from the Shields diagram (Fig. 5.2) or Fig. 5.3 based on the estimated median diameter of bed material.
2. Compute laminar sublayer thickness $\delta$ from Eq. 3.11 and obtain the ratio $\delta/d_{50}$.
3. Select correction factor $X$ in the logarithmic velocity distribution from Fig. 3.6.
4. Compute mean critical velocity from the equation

$$U_c = 2.5\left(\frac{\tau_c}{\rho}\right)^{1/2} \ln\left(\frac{11.02\, D_0 X}{d_{50}}\right)$$

5. Compute $\mathbf{F}_c$ from $U_c/(gD_0)^{1/2}$.

A summary and comparison of these formulas and data bases were made by Jones (1984), as shown in Fig. 5.14. Among the findings, all of the equations are for noncohesive materials. Neill's formula is a regression fit to the design curve originally presented by Laursen; therefore, Neill's and Laursen's formulas produce

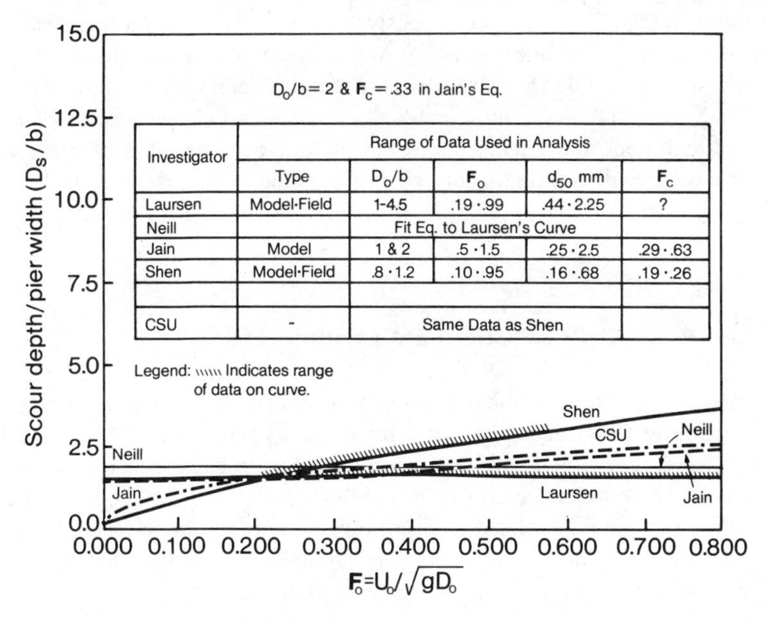


| Investigator | Range of Data Used in Analysis | | | | $F_c$ |
|---|---|---|---|---|---|
| | Type | $D_0/b$ | $F_0$ | $d_{50}$ mm | |
| Laursen | Model-Field | 1-4.5 | .19-.99 | .44-2.25 | ? |
| Neill | Fit Eq. to Laursen's Curve | | | | |
| Jain | Model | 1 & 2 | .5-1.5 | .25-2.5 | .29-.63 |
| Shen | Model-Field | .8-1.2 | .10-.95 | .16-.68 | .19-.26 |
| | | | | | |
| CSU | - | Same Data as Shen | | | |

**Figure 5.14** Summary of formulas and data used for pier scour equations from U.S. literature compiled by Jones (1984).

similar results. Shen's formula is an envelope curve that fits the uppermost scour points for all the data available at the time. The CSU equation is a best fit to many of the same points, as shown in Fig. 5.13. The Jain–Fischer formula is somewhat of a compromise between those of Laursen and Shen.

Pier alignment that is not parallel with the flow direction will create deeper scour holes because it effectively increases the dimension $b$. The obstruction to flow is affected by the pier width normal to flow: it is also related to the pier shape. The strength of turbulent motion, as well as the scour depth, can be reduced with streamlined pier shapes, as shown by Tison (1961). There have also been investigations, for example, by Shen and Schneider (1970), of the effects of pier shape on scour depth. However, a designer is not likely to be able to take advantage of the streamlined shapes because of the complex flow pattern that changes with time and stage.

Dunes affect the equilibrium depth of scour by increasing it as a trough passes the scour hole and by decreasing it as a crest passes through. Based on laboratory data, Shen et al. (1969) determined that the increase or decrease is less than one-half the dune height.

Jones (1984) cautioned us on the use of the prediction equations because of limited field measurements used and the complexity of flow conditions in a river. Perhaps the most important thing to bear in mind is that these formulas were developed based on extensive laboratory data, partly because reliable field data are difficult to obtain during the flood stage. There exist great discrepancies between field and laboratory conditions on the flow and bed-material distribution.

The equations in the above discussion are applicable to noncohesive materials with movement. The erosion of cohesive materials depends on the shear force and the strength of the cohesive bond. When shear forces are large enough to erode cohesive material, they are also sufficient to carry the sediment as suspended load. Therefore, very little material is transported as bed load. As the supply of bed sediment approaches zero, the pattern becomes that of clear-water scour. The scour will become stable when the velocity decreases with the scour hole development such that it falls below the critical velocity for the bed material.

## 5.10 LOCAL SCOUR AROUND EMBANKMENTS

Local scour occurs around embankments, spur dikes, and abutments because of the obstruction to flow caused by such structures, as illustrated in Fig. 5.15. The scour hole develops as overbank flow reenters the main channel and sets up large vortices to wash sediment away. In a mobile-bed stream, the scour reaches an equilibrium when the rates of sediment inflow and outflow are in balance. But the scour hole seldom stays constant, even in laboratory tests under a constant discharge, because the depth of scour fluctuates with time when there are dunes moving on the alluvial bed. The scour depth usually decreases when a crest of dune reaches the area, and it increases as a trough approaches. The equilibrium scour

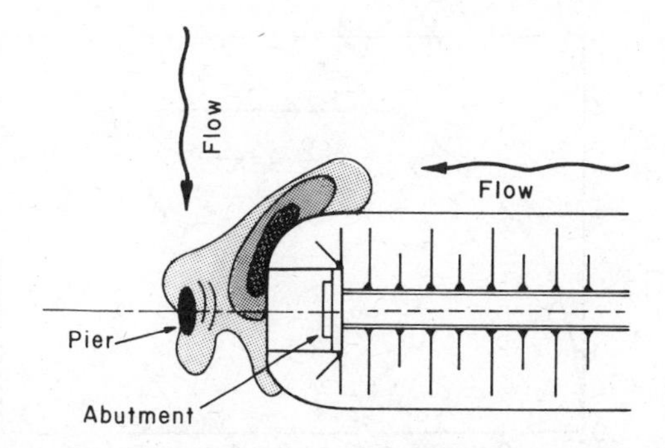


**Figure 5.15** Typical scour at an abutment (after CSU, 1975).

depth is the mean of the fluctuations. In engineering applications, the maximum scour depth may be obtained from the equilibrium scour depth plus one-half of the bed-form (dune or antidune) height.

The equilibrium scour depth should be distinguished from the "clear-water" scour depth. The latter occurs in the absence of sediment supply, and it stabilizes when the velocity becomes too small to move the material in the scour hole.

Detailed studies of abutment scour have been made mostly in laboratories (see, e.g., Laursen, 1956; Karaki, 1959; Liu et al. 1961; and Gill, 1972). Field studies are handicapped by the difficulties of making measurement during floods. Liu et al. (1961) presented an expression for the equilibrium scour depth in sand for subcritical flow:

$$\frac{D_s}{D_0} = c\left(\frac{a}{D_0}\right)^{0.40} (\mathbf{F}_0)^{0.33} \tag{5.25}$$

where $D_s$ is the equilibrium scour depth measured from the mean bed level, $a$ is the embankment length normal to the wall of a flume, $D_0$ is the approaching depth, and $\mathbf{F}_0$ is the Froude number of the approaching flow. The constant $c$ has the value of 1.1 for scour development at a spill slope. If the embankment terminates at a vertical wall and has a vertical wall on the upstream side, then the value of $c$ is 2.15 according to Liu et al. and Gill.

Using field data collected at rock dikes on the Mississippi, the Colorado State University (1975) research group presented the following equation for embankment scour.

$$\frac{D_s}{D_0} = 4(\mathbf{F}_0)^{0.33} \tag{5.26}$$

The data used to establish this formula are scattered, primarily because equilibrium depths were not measured. These depths changed significantly with the passage of dunes, which were as high as 20–30 ft.

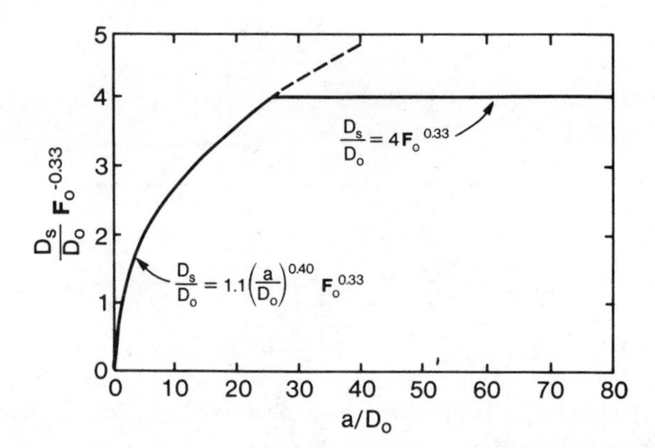


**Figure 5.16** Recommended prediction equations for embankment scour.

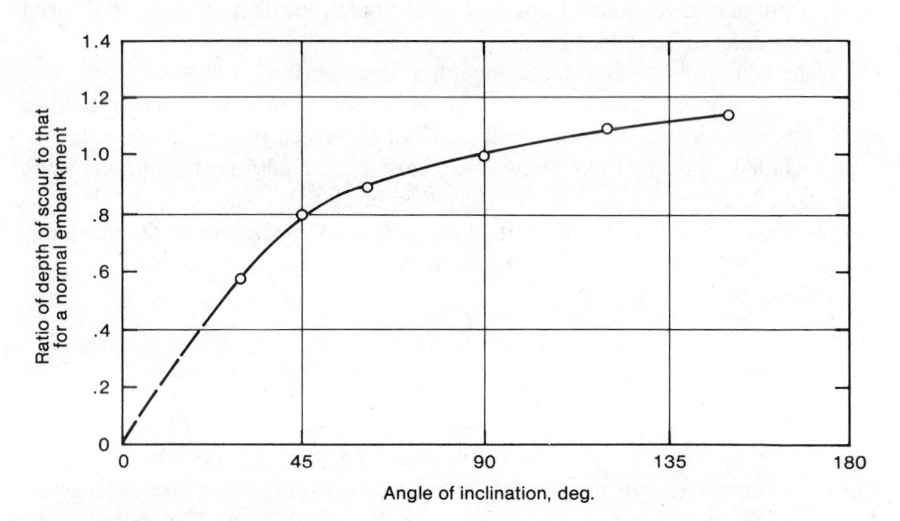


**Figure 5.17** Correction factor for embankment scour for angle of inclination.

Equations 5.25 and 5.26 are shown graphically in Fig. 5.16, together with their respective ranges of applications recommended by CSU. Equation 5.25 is recommended for embankments with $0 < a/D_0 < 25$, and Eq. 5.26 is for $a/D_0 > 25$. In applying Eq. 5.25, the embankment length is measured from the high water line at the valley bank perpendicular to the end of the embankment at the bridge. If $a/D_0 > 25$, then scour depth is independent of $a/D_0$ but depends on the approach Froude number and flow depth. A simple way to determine the maximum scour depth is to add 30% to the equilibrium scour depth. The lateral extent of scour can be estimated from the angle of repose of the material and scour depth.

The scour depth is also affected by the angle of inclination of the embankment. Those embankments directed downstream generally have smaller scour holes than those angled upstream because the scour depth is directly related to the extent of the obstruction to flow. The correction factor by Ahmad (1953) for the scour depth to account for the angle of inclination is shown in Fig. 5.17.

## REFERENCES

Ahmad, M., "Experiments on Design and Behavior of Spur Dikes," Proceedings of the IAHR, ASCE Joint Meeting, University of Minnesota, August 1953.

ASCE, closure to "Sediment Transportation Mechanics: Initiation of Motion," by the Task Committee on Preparation of Sedimentation Manual, Vito A. Vanoni, Chairman, *J. Hydraul. Div. ASCE*, **93**(HY5), pp. 297-302, September 1967.

ASCE, *Sedimentation Engineering*, Manuals and Reports on Engineering Practice, No. 54, Vito A. Vanoni, ed., 1975.

Blench, T., *Mobile Bed Fluviology*, University of Alberta Press, Edmonton, Alberta, Canada, 1969.

Carstens, M. R., "An Analytical and Experimental Study of Bed Ripples Under Water Waves," Quarterly Reports 8 and 9, School of Civil Engineering, Georgia Institute of Technology, 1966.

Colorado State University, "Highways in the River Environment: Hydraulic and Environmental Design Considerations," prepared for the Federal Highway Administration, U.S. Department of Transportation, May 1975.

Corps of Engineers, "Hydraulic Design of Flood Control Channels," EM 1110-2-1601, Department of the Army, July 1970.

Fortier, S. and Scobey, F. C., "Permissible Canal Velocities," *Trans. ASCE*, **89**, Paper No. 1588, pp. 940-984, 1926.

Gill, M. A., "Erosion of Sand Beds around Spur Dikes," *J. Hydraul. Div. ASCE*, **98** (HY9), pp. 1587-1602, September 1972.

Highway Research Board, "Tentative Design Procedure for Riprap-Lined Channels," National Academy of Sciences, National Cooperative Highway Research Program, Report 108, 1970.

Hjulstrom, F., "Studies of the Morphological Activity of Rivers as Illustrated by the River Fyris," Bulletin, Geological Institute of Upsala, Vol. XXV, Upsala, Sweden, 1935.

Ippen, A. T. and Drinker, P. A., "Boundary Shear Stress in Curved Trapezoidal Channels," *J. Hydraul. Div. ASCE*, **88**(HY5), pp. 143-180, September 1962.

Jain S. C. and Fischer E. E., "Scour Around Circular Bridge Piers at High Froude Numbers," Report FHWA-RD-79-104, Federal Highway Administration, U.S. Department of Transportation, April 1979.

Jones, J. S. "Comparison of Prediction Equations for Bridge Pier and Abutment Scour," *Transp. Res. Rec.*, **950**(2), Transportation Research Board, National Research Council, pp. 202-209, September 1984.

Karaki, S. S., "Hydraulic Model Studies of Spur Dikes for Highway Bridge Openings," Report No. CER59-SSK36, Colorado State University (also Bulletin 286, Highway Research Board, Washington), 1959.

Lane, E. W., "Design of Stable Channels," *Trans. ASCE,* **120**, Paper No. 2776, pp. 1234-1279, 1955.

Laursen E. M., "Scour Around Bridge Piers and Abutments," Bulletin 4. Iowa Highway Research Board, Iowa City, May 1956.

Laursen, E. M., "Scour at Bridge Crossings," *Trans. ASCE,* **127**(1), pp. 166-180, 1962.

Laursen, E. M., "An Analysis of Relief Bridge Scour," *J. Hydraul. Div. ASCE*, **89**(HY3), May 1969.

Liu, H. K., Chang, F. M., and Skinner, M. M., "Effect of Bridge Construction on Scour and Backwater," Department of Civil Engineering, Colorado State University, Report No. CER60-HKL22, February 1961.

Mavis, F. T. and Laushey, L. M., "Formula for Velocity at Beginning of Bed-Load Movement is Reappraised," *Civ. Eng. ASCE*, **19**(1), pp. 38-39 and 72, January 1949.

Mehrota, S. C., "Permissible Velocity Correction Factors," *J. Hydraul. Eng. ASCE*, **109**(2), pp. 305-308, February 1983.

Neill, C. R., "Mean Velocity Criterion for Scour of Coarse Uniform Bed Material," Proceedings of the Twelfth Congress, IAHR, Fort Collins, Colorado, pp. 46-54, 1967.

Olsen O. J. and Florey Q. L. (compilers), "Sedimentation Studies in Open Channels: Boundary Shear and Velocity Distribution by Membrane Analogy, Analytical and Finite-Difference Methods, reviewed by D. McHenry and R. E. Glover," U.S. Bureau of Reclamation, Laboratory Report, No. Sp-34, August 5, 1952.

Shen, H. W., Schneider, V. R., and Karaki, S. S., "Local Scour Around Bridge Piers," *J. Hydraul. Div. ASCE*, **95**(HY11), pp. 1919-1940, November 1969.

Shen H.W. and Schneider V. R., "Effect of Bridge Pier Shape on Local Scour," ASCE National Meeting on Transportation Engineering, Boston, MA, July 13–17, 1970.

Shen, H. W. and Wang, S-Y., "Analysis of Commonly Used Riprap Design Guides Based on Extended Shields Diagram," *Transp. Res. Rec.*, **950**(2), Transportation Research Board, National Research Council, pp. 217-221, September 1984.

Shields, A., "Anwendung der Aenlichkeitsmechanik und der Turbulenzforschung auf die Geschiebebewegung," Mitteilungen der Prevssischen Versuchsanstalt fur Wasserbau und Schiffbau, Berlin, Germany, translated into English by W. P. Ott and J. C. van Uchelen, California Institute of Technology, Pasadena, California 1936.

Tison, L. J., "Local Scour in Rivers," *J. Geophys. Res.*, **66**(12), pp. 4227-4232, December 1961.

Zeller, J., "Einfuhrung in den Sedimentransport offener Gerinne, Schweiz," Bauzeitung, Jgg. 81, 1963.

# 6

# ALLUVIAL BED FORMS AND FLOW RESISTANCE

The phenomenon of bed forms in alluvial rivers was perhaps first described in the classic research of Gilbert (1914). This bed feature has fascinated engineers and scientists because of its association with so many aspects of river sedimentation and river morphology. The bed forms are flow induced and directly affect the roughness or flow resistance. Therefore, computation of the river stage and flow velocity relies on the determination of bed-form roughness. In an experimental study, Brooks (1958) showed that the flow-induced roughness does not always increase with the velocity, since it may also decrease as the velocity increases; he also showed that flows with the same slope and depth and, thus, the same shear stress may have different velocities associated with separate roughness. This characteristic variation of bed-form roughness has important effects on the stage-discharge relationship during the passage of flood in the short term. It has also been confirmed by Chang (1979, 1985) that the changes in bed form are matched by responses in river morphology. The thresholds, or discontinuities, for the geometric relationships of regime rivers are described in Chapter 11 on the basis of the adjustments in alluvial bed roughness. In the following, terminology related to bed forms are defined, methods for predicting bed forms are described, and several approaches for determining the flow resistance associated with bed forms are presented.

## 6.1 BED FORMS

Many terms are used to describe bed forms. The following descriptions are based on those of the ASCE Task Force on Bed Forms in Alluvial Channels (1966). These terms are illustrated in Figs. 2.7, 6.1, 6.2, and 11.16. Except for the first two terms, they are presented in the order of increasing velocity or stream power per unit bed area $\tau_0 U$, where $\tau_0$ is the boundary shear stress and $U$ is the mean velocity.

*Bed Configuration*. This is an array of bed forms, or absence thereof, generated on the bed of an alluvial channel by the flow. Some of the synonyms of

**Figure 6.1** Ripples on natural stream bed.

this term in common use are bed geometry, forms of bed roughness, bed form, bed regime, bed irregularities, sand waves, and bed shape.

*Bars*. These are bed forms having lengths of the same order as the channel width, or greater, and having heights comparable to the mean depth of the generating flow. There are several different types of bars. Among them, *point bars* (see Fig. 2.7) are deposits of sediment that occur on the convex side or inside of channel bends. Their shape may vary with changing flow conditions, but they do not move relative to the bends. *Alternating bars* or *alternate bars* (see Fig. 11.16) tend to be distributed periodically along a channel, with bars near alternate channel banks. Their lateral extent is significantly less than the channel width. Alternating bars move slowly downstream.

*Flat Bed*. This is a bed surface devoid of bed forms. It is also known as *plane bed* or *smooth bed*.

*Ripples*. These are small bed forms with wavelengths less than approximately 1 ft and heights less than approximately 0.1 ft, as shown in Figs. 6.1 and 6.2. In longitudinal section, ripple profiles vary from approximately triangular (with long gentle upstream slopes and downstream slopes approximately equal to the angle of repose of the bed material), to symmetrical nearly sinusoidal shapes.

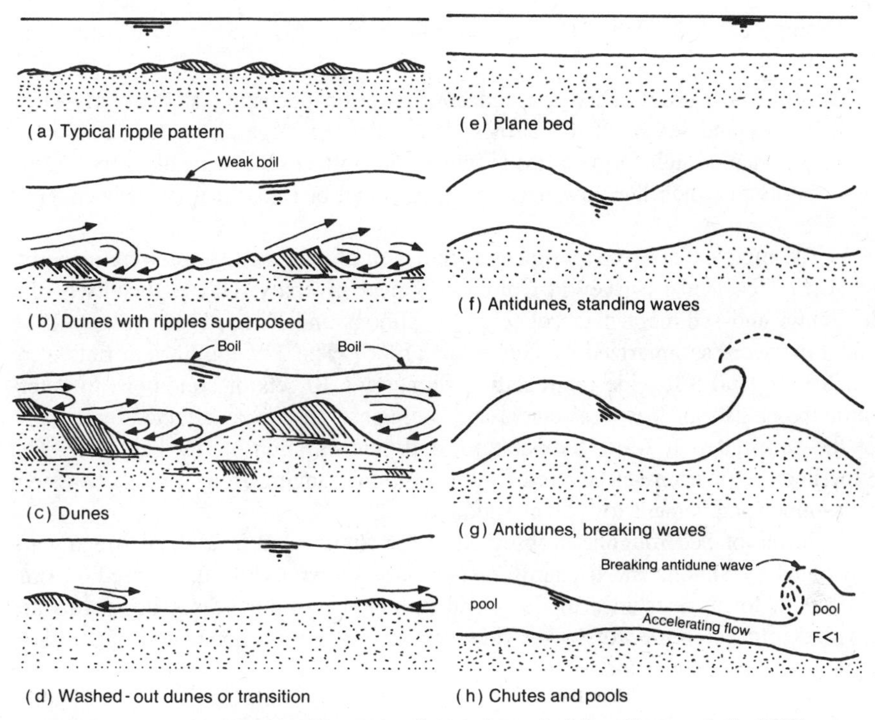


**Figure 6.2** Idealized bed forms in alluvial channels (after Simons et al., 1966).

*Dunes*. These are bed forms smaller than bars but larger than ripples that are out of phase with any water-surface gravity waves that accompany them, as shown in Fig. 6.2. Dunes generally occur at larger velocities and sediment transport rates than do ripples, which may occur on the upstream slopes of dunes. The longitudinal profiles of dunes are approximately triangular, with fairly gentle upstream slopes and downstream slopes that are approximately equal to the angle of repose of the bed material. The large lee eddies that occur in dune troughs often cause surface boils of intense turbulence.

*Transition*. This is a bed configuration consisting of a heterogeneous array of bed forms, primarily low-amplitude ripples or dunes and flat areas (see Fig. 6.2). Transition is also called *sand waves* or *washed-out dunes*; it occurs at higher flow intensity for dunes.

*Antidunes*. These are bed forms that occur in trains that are in phase with, and strongly interact with, gravity water-surface waves (see Fig. 6.2). They are also called *standing waves* or *sand waves*. The free-surface waves have larger amplitude than the sand waves. The surface waves grow with increasing velocity and Froude number, until they become unstable and break in the upstream direction. In longitudinal section, antidune profiles vary with flow and sediment properties, from approximately triangular to sinusoidal, the latter occurring at higher Froude numbers than the former. However, the sharp-crested, triangular-shaped antidunes have been observed in laboratory flumes.

*Chutes and Pools*. This is a bed configuration occurring at relatively large slopes and sediment discharges. It consists of large elongated mounds of sediment which form chutes in which the flow is supercritical, connected by pools in which the flow may be supercritical or subcritical (see Fig. 6.2).

The most extensive experimental studies of alluvial bed forms were made by the U.S. Geological Survey at Colorado State University (CSU). Results of the hydraulic and sediment data collected by Simons and Richardson between 1956 and 1961 were summarized by Guy et al. (1966). The 339 equilibrium runs were made in 2- and 8-ft-wide recirculating flumes for 10 sets of conditions to determine the bed form, flow resistance, and sediment transport as affected by the size of the bed material, water temperature, and concentration of fine sediment. The experiments for each set covered flow conditions ranging from a plane bed with no sediment movement to violent antidunes.

Forms of bed roughness observed in the flumes and in alluvial streams are illustrated by Simons and Richardson (1961), as shown in Fig. 6.2. Based on similarities in form, resistance to flow, and sediment transport, these bed forms are divided into categories of lower flow regime, transition zone, and upper flow regime in the order of increasing velocity as follows:

Lower flow regime
- Ripples
- Dunes with ripples superposed
- Dunes

Transition (Bed roughness ranges from dunes to plane bed or standing waves)

Upper flow regime
- Plane bed
- Antidunes
- Chutes and pools

The characteristics associated with these flow regimes, in terms of bed-material concentration, mode of sediment transport, type of roughness, and phase relation between bed and water surface are compared and summarized in Table 6.1. In the lower flow regime, the bed roughness is ripples or dunes or both. The bed-material load (sediment discharge composed of particle sizes found in the shifting portions of the bed) is small, and most of the movement is close to the bed over the backs of the ripples and dunes. At the crest of these roughness elements, some of the sediment particles avalanche down the faces of the crest into the trough, where they are temporarily at rest. Therefore, the movement of particles follows discrete steps while the waves move slowly downstream. The resistance to flow is large in the lower regime, with form roughness predominant.

**TABLE 6.1 Classification of Bed Forms and Their Characteristics (After Simons et al., 1965)**

| Flow Regime | Bed Form | Bed-Material Concentration (ppm) | Mode of Sediment Transport | Type of Roughness | Phase Relation Between Bed and Water Surface |
|---|---|---|---|---|---|
| Lower regime | Ripples | 10–200 | Discrete steps | Form roughness predominates | Out of phase |
| | Ripples on dunes | 100–1200 | | | |
| | Dunes | 200–2000 | | | |
| Transition zone | Washed-out dunes | 1000–3000 | — | Variable | — |
| Upper regime | Plane beds | 2000–6000 | Continuous | Grain roughness predominates | In phase |
| | Antidunes | Above 2000 | | | |
| | Chutes and pools | Above 2000 | | | |

Form roughness is attributed to the pressure difference between the upstream and downstream surfaces of the sand wave. The water-surface profile over the dune bed is mildly wavy and it is out of phase with the bed profile, that is, water surface is lower over dune crests and higher above troughs.

In the upper regime of plane bed, antidunes, and standing waves, the bed-material transport is relatively large and is nearly continuous in motion. As the particles move downstream on the surface of the sinusoidal wave, the wave itself may travel upstream. The resistance to flow is small, with grain roughness predominant, especially for the plane bed. The water-surface profile over antidunes or standing waves is in phase with the bed profile. When antidunes break, they resemble the hydraulic jump, which has an upstream supercritical flow and a downstream subcritical flow.

The change from lower flow regime into the transition occurs as the dunes are washed out to become the flat bed. This process is associated with a definite reduction in roughness, an example based on the flume data by Guy et al. (1966) is shown in Fig. 6.3. Similar changes in roughness have also been observed in natural rivers, and they have important effects on stream rating. For example, the stage–discharge curve shown in Fig. 6.4 for the Pigeon Roost Creek in Mississippi observed by Colby (1960) has a discontinuity in the relationship between the lower regime over dune beds and the upper regime for flat or antidune beds. Dawdy (1961) studied streams with discontinuous rating curves and concluded that many streams exhibit this characteristic. In studies of the Rio Grande, Nordin (1964) also observed a discontinuity of the stage–discharge relationship and associated it with the changes in bed form and roughness.

The discontinuity in the rating curve reflects short-term changes in bed roughness with the flow condition. The adjustment in regime geometry (i.e., the long-term equilibrium geometry), of rivers and particularly the thresholds separating different river types are also related to the discontinuity for flow resistance to be analyzed in Chapter 11.

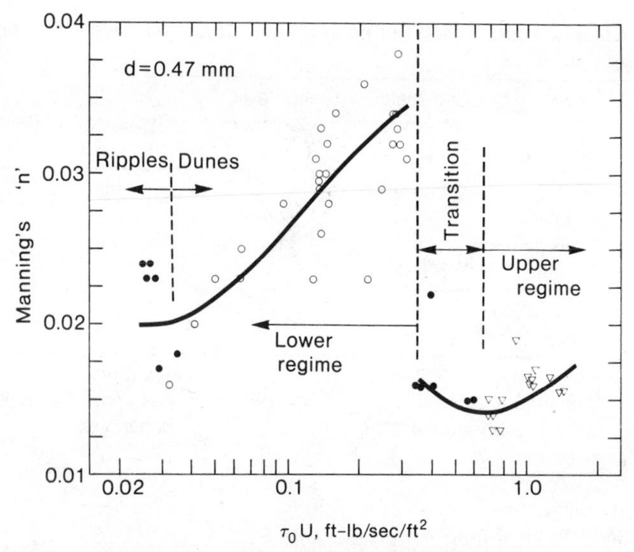


**Figure 6.3** Relationship between bed roughness and bed form based on flume data by Guy et al. (1966).

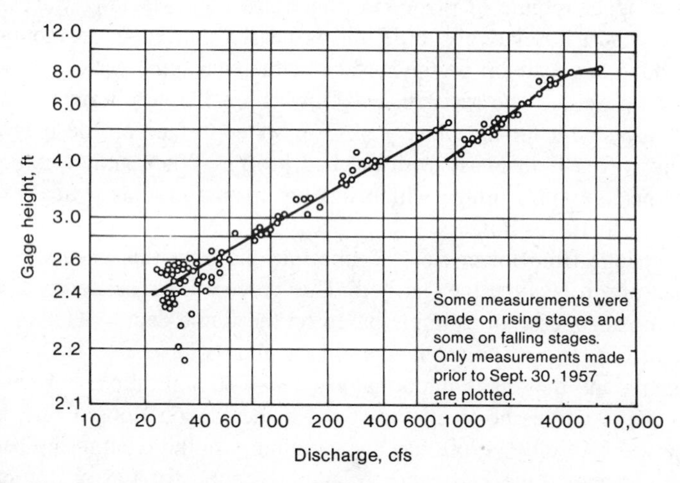


**Figure 6.4** Discontinuous stage-discharge relationship for Pigeon Roost Creek, Mississippi (Colby, 1960).

## 6.2 PREDICTION OF BED FORMS

In the pursuit for a better understanding of the interaction of the flow and bed material and the interdependence of the bed form, roughness, and sediment transport rate, rather extensive studies have been made on the prediction of bed forms following both a theoretical and an empirical approach. Because of the complexity of the physical processes, mathematical models to date have generally employed the two-dimensional flow assumption. The theoretical approach on the formation and geometry of bed forms by Kennedy (1963) was based on the stability analysis for which the instability of a perturbation was given as the cause of bed-form formation. Kennedy's stability analysis was later followed and extended by several other investigators, for example, Hayashi (1970), Engelund (1980) and Fredsoe (1982). Wave theory was also applied to study bed forms (see, e.g., Song, 1983; and Haque and Mahmood, 1985). Important contributions have been made to the prediction of bed-form occurrence and characteristic features. However, a universally acceptable analytical solution is still lacking, and the cause of bed forms is still not fully explained. In this chapter, only empirical methods are presented, including those by Simons and Richardson (1961), Athaullah (1968), Brownlie (1983) (given in Sec. 6.5), and van Rijn (1984). More comprehensive coverage of this topic has been made by ASCE (1975), Raudkivi (1976), and Simons and Senturk (1977).

## Simons and Richardson's Approach

Simons and Richardson (1961) developed a bed-form predictor as shown in Fig. 6.5, based on the extensive CSU flume data and data from several rivers and canals. Bed forms are given in terms of the median fall diameter of the bed material in the sand-sized range and the stream power, which is defined as the product of bed shear stress and mean flow velocity $\tau_0 U$. The stream power so defined is the rate of energy expenditure per unit bed area, to which the bed forms appear to be closely correlated. At low values of $\tau_0 U$ starting from the initial plane bed, the bed remains flat with no movement of the bed material. Ripples, or dunes, start to form after the beginning of motion, with ripple formation for sand sizes smaller than about 0.6 mm. The bottom line in Fig. 6.5 corresponds to the initiation of motion. Above the dune region, the line separating dunes from transition is represented by the following equations:

$$\log \tau_0 U = -0.60 + 1.05d \quad \text{for } d < 0.25 \text{ mm} \tag{6.1}$$

and

$$\log \tau_0 U = -0.442 + 0.44d \quad \text{for } d > 0.25 \text{ mm} \tag{6.2}$$

where $\tau_0 U$ is in foot-pounds per second per square foot and $d$ is the median fall diameter in millimeters.

The graphical relationship in Fig. 6.5 is only applicable for sand bed. However, it seems to show the following trend: As the fall diameter increases beyond the indicated region, the range of $\tau_0 U$ for dune formation will decrease to result in the disappearance of dunes for gravel bed.

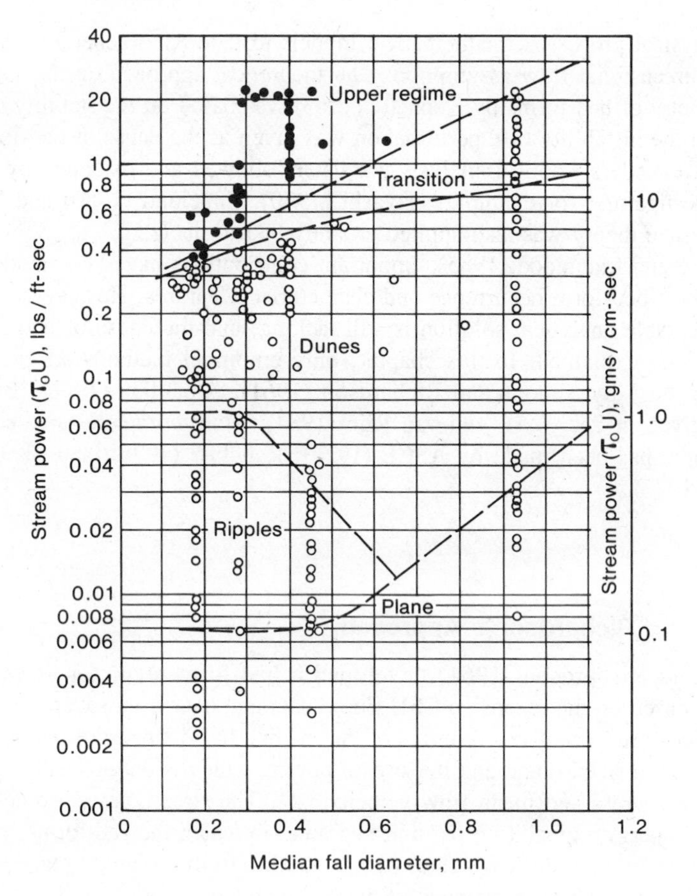


**Figure 6.5** Relationships among bed form, stream power per unit area, and median fall diameter (Simons and Richardson, 1961).

## Athaullah's Approach

Athaullah (1968), working with Simons, studied many groups of dimensionless parameters. He presented a graphical relationship, shown in Fig. 6.6 delineating different flow regimes based on the Froude number and the relative roughness $R/d$, where $R$ is the hydraulic radius and $d$ is the median bed-material size. The Froude number reflects the inertial to the gravitational effect or the channel slope at a given discharge. The relative roughness describes the dimension of alluvial channel in relation to the sediment size. The discharge of the channel, in general, is directly related to $R/d$. Figure 6.6 shows that at smaller values of $R/d$, that is, for smaller streams, lower flow regime can occur at higher Froude numbers. For very large rivers, on the other hand, the transition from lower flow regime to flat bed develops at fairly small Froude numbers. The designers of India–Pakistan canals also used the Froude number as a guide because all the canals were

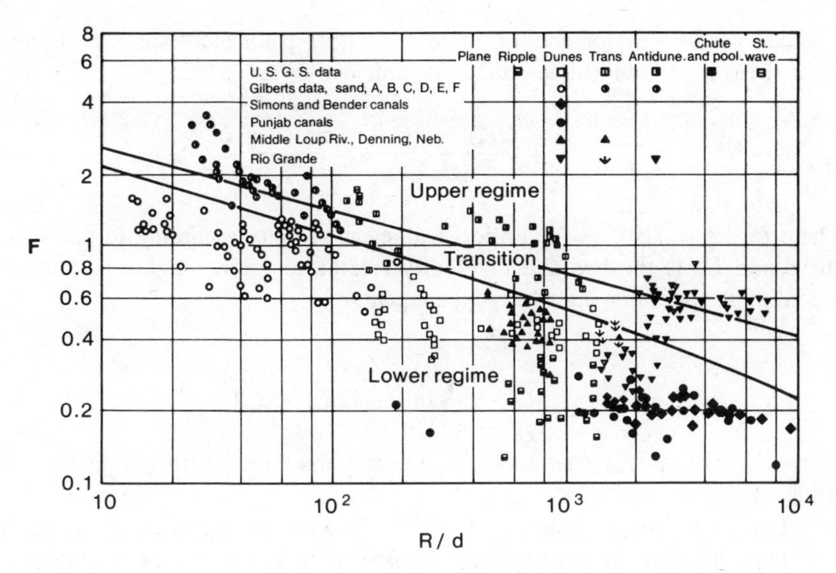


**Figure 6.6** Bed form prediction as function of Froude number and $R/d$ (Athaullah, 1968).

designed to have a Froude number less than about 0.3, within the region of lower flow regime. The demarcation line in the Athaullah–Simons relationship bears close relationship to the thresholds in river channel geometry to be described in Chapter 11.

The criticism of this predictive method is its failure to discriminate bed forms in natural streams (Simons and Senturk, 1977). The flow distribution across the width of a natural stream can be quite nonuniform such that different flow regimes often exist concurrently at the same cross section. The relationship does not include the temperature effect on bed form. A discussion of the temperature effect on bed forms is given in Sec. 6.4.

## van Rijn's Approach

In van Rijn's (1984) approach, the classification of bed form is assumed to be controlled mainly by bed-load transport, which is described by a dimensionless particle parameter, $d_*$, and a transport-stage parameter, $T$. The former is defined as

$$d_* = d\left[\frac{(\rho_s - \rho)g}{\rho \nu^2}\right]^{1/3} \tag{6.3}$$

where $d$ is the median size of bed sediment, $\rho$ is the mass density of fluid, $\rho_s$ is the mass density of sediment, $\nu$ is the kinematic viscosity, and $g$ is the gravitational acceleration. Since the particle parameter accounts for the specific gravity of sediment and for the viscosity of the fluid, it is therefore similar to the fall diameter used by Simons and Richardson. The $T$ parameter expresses the grain shear stress in relation to the critical, or Shields, stress:

$$T = \frac{(U'_*)^2 - (U_{*c})^2}{(U_{*c})^2} = \frac{\tau'_0 - \tau_c}{\tau_c} \tag{6.4}$$

where $U_{*c} = (\tau_c/\rho)^{1/2}$ is the critical bed-shear velocity according to the Shields curve and $U'_*$ is the bed-shear velocity related to grain roughness computed according to the following Chezy-type equation:

$$U'_* = \frac{g^{1/2}}{C'} U = \frac{g^{1/2}}{18 \log(12R_b)/(3d_{90})} U \tag{6.5}$$

where $C'$ is the Chezy coefficient due to grain roughness and $R_b$ is the hydraulic radius of the alluvial bed.

Using the particle parameter $d_*$ and the transport-stage parameter $T$, van Rijn developed the diagram for bed-form classification in the lower and transitional flow regime, as shown in Fig. 6.7. It shows distinct zones of ripples, dunes, and washed-out dunes based on 40 sets of flume and field data. The flume data include those by Guy et al. (1966), Delft Hydraulics Laboratory, and so on. The field data are from the Dutch rivers, Rio Parana, Japanese channels, and the Mississippi River. As shown in the figure, dune-type bed forms are present for $T < 15$. But for particles smaller than about 0.45 mm ($d_* = 10$), ripples are generated after initiation of motion but disappear for $T > 3$. The transitional flow regime with washed-out dunes is present for $15 < T < 25$. For $T > 25$, a flat bed flow will be generated. The washing-out process is described by the $T$ parameter because when $\tau'_0 >> \tau_c$, sediment particles will go into suspension and the bed forms are thus washed out. This method is perhaps more valid than the one by Simons and Richardson for the lower and transitional flow regime in field conditions because a large number of field data with small and large flow depths (up to 16 m for the Mississippi) were used. However, because the $T$ parameter is a ratio, it is sensitive to any inaccuracy in determining the Shields stress.

## 6.3 BED-FORM DIMENSIONS

Bed-form dimensions have been studied primarily because of the close relation to the hydraulic roughness. Certain flow-resistance formulas are based on bed-form dimensions. Large bed forms, such as dunes and antidunes have wave heights on the same order of magnitude as the flow depths. Therefore, their dimensions must affect navigation and should also be considered in determining the scour depth along bank protection and at bridge piers, and so on. Ordinarily, one-half of the bed-form height is added to the scour depth.

Methods for determining bed-form dimensions have been proposed by Kennedy (1963), Yalin (1964), Ranga Raju and Soni (1976), Fredsoe (1982), and

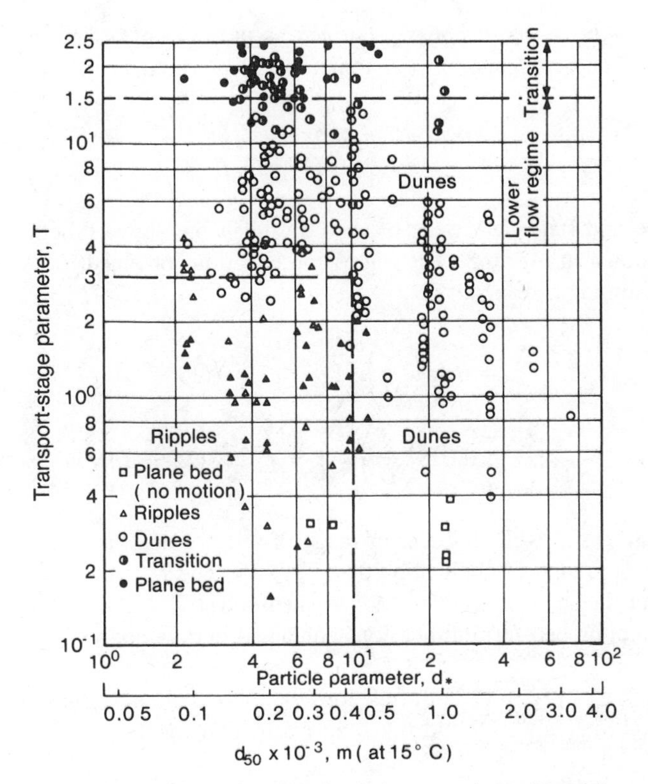


**Figure 6.7** Diagram for bed-form classification in lower and transitional flow regimes (van Rijn, 1984).

van Rijn (1984), among others. Yalin (1964) argues that dunes are reflective of large-scale eddies because the flow and bed form have to reach a state of mutual compatibility. Since the full depth of flow is perturbed, the wavelength of the bed form $\lambda$, as well as the eddy size, should be proportional to the depth of flow $D$, say,

$$\lambda = 2\pi D \tag{6.6}$$

Van Rijn's (1984) method for bed-form dimensions is restricted to dune-type bed forms in the lower and transitional flow regime. The ripple bed form, which is supposed to be independent of the flow depth, is not included in this method. From basically dimensional reasoning, the bed-form height $\Delta$ is related to other parameters as follows:

$$\frac{\Delta}{D} = F\left(\frac{d}{D}, d_*, T\right) \tag{6.7}$$

Likewise, the *bed-form steepness*, defined as the ratio of bed-form height to its length $\Delta/\lambda$, can be expressed as a similar functional relationship:

$$\frac{\Delta}{\lambda} = G\left(\frac{d}{D}, d_*, T\right) \tag{6.8}$$

Flume and field data were used to establish the above functional relationships, as shown in Fig. 6.8. The regression equations obtained for $\Delta/D$ and $\Delta/\lambda$ by curve-fitting are, respectively,

$$\frac{\Delta}{D} = 0.11\left(\frac{d}{D}\right)^{0.3}(1 - e^{-0.5T})(25 - T) \tag{6.9}$$

$$\frac{\Delta}{\lambda} = 0.015\left(\frac{d}{D}\right)^{0.3}(1 - e^{-0.5T})(25 - T) \tag{6.10}$$

These equations, as well as the error range of a factor of 2, are shown in the figures within the range of application $0 < T < 25$. The $T = 0$ line corresponds to the bed-load threshold, and dunes are washed out at $T > 25$. Note that the temperature effect on bed-form dimensions contained in $d_*$ is not reflected in Eqs. 6.9 and 6.10.

As shown in Fig. 6.8, both functions exhibit maximum values at the $T$ value of about 5. From Eqs. 6.9 and 6.10, an expression for the bed-form length can also be derived:

$$\lambda = 7.3D \tag{6.11}$$

This relation, similar to Eq. 6.6, indicates that the dune length is related only to the mean flow depth. Therefore, as the bed-form height is reduced during the increasing stage of flow, the bed-form length remains essentially unchanged. To use van Rijn's method, the values of $\Delta$ and $\lambda$ may be computed if the depth, mean flow velocity, and particle size are known.

Kennedy (1963) proposed the following relation for the antidune wavelength:

$$\lambda = 2\pi\frac{U^2}{2g} \tag{6.12}$$

which compares with observed wavelengths reasonably well. The surface waves over antidunes broke at high velocities. The steepness (ratio of wave height to wavelength) at incipient breaking was found by Kennedy to be about 0.14.

## 6.4 EFFECT OF WATER TEMPERATURE

The effect of water temperature on bed forms is included in some of the relations described so far and it is contained among the stage–discharge predictors pre-

| | Source | Flow velocity U, m/s | Flow depth D, m | Particle size d, μm | Temperature °C |
|---|---|---|---|---|---|
| Flume data | Guy et al | 0.34-1.17 | 0.16-0.32 | 190 | 8-34 |
| | Guy et al | 0.41-0.65 | 0.14-0.34 | 270 | 8-34 |
| | Guy et al | 0.47-1.15 | 0.16-0.32 | 280 | 8-34 |
| | Guy et al | 0.77-0.98 | 0.16 | 330 | 8-34 |
| | Guy et al | 0.48-1.00 | 0.10-0.25 | 450 | 8-34 |
| | Guy et al | 0.53-1.15 | 0.12-0.34 | 930 | 8-34 |
| | Williams | 0.54-1.06 | 0.15-0.22 | 1350 | 25-28 |
| | Delft Hydr. Lab. | 0.45-0.87 | 0.26-0.49 | 790 | 12-18 |
| | Stein | 0.52-0.95 | 0.24-0.31 | 400 | 20-26 |
| | Znamenskaya | 0.53-0.80 | 0.11-0.21 | 800 | – |
| Field data | Dutch Rivers | 0.85-1.55 | 4.4-9.5 | 490-3600 | 5-20 |
| | Rio Parana | 1.0 | 12.7 | 400 | – |
| | Japanese Channels | 0.53-0.89 | 0.25-0.88 | 1100-2300 | – |
| | Mississippi River | 1.35-1.45 | 6-16 | 350-550 | – |

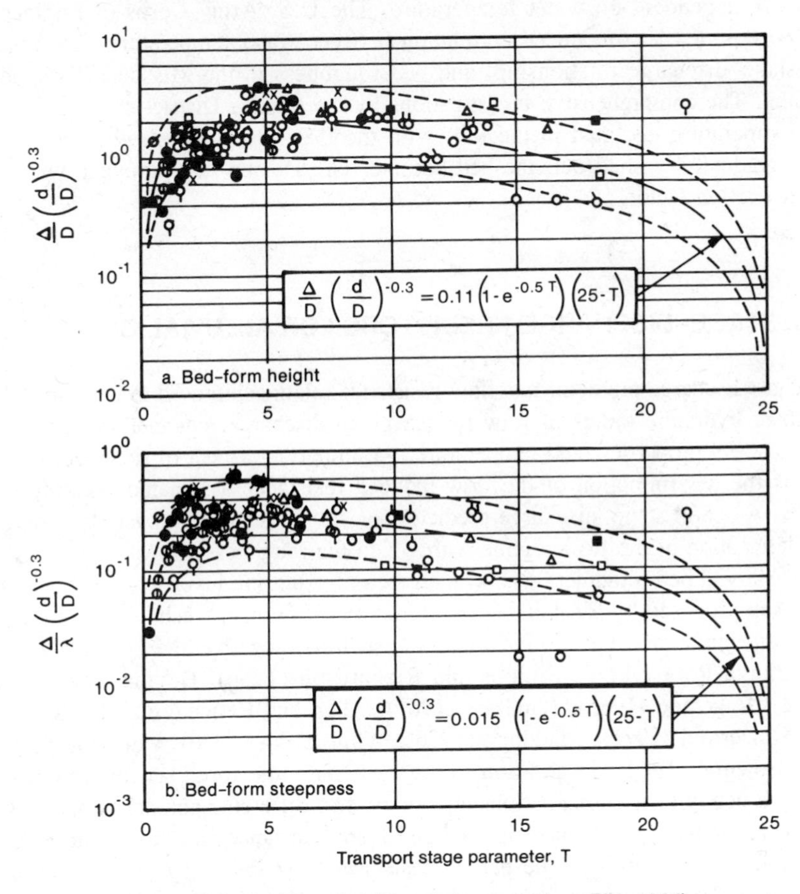


**Figure 6.8** Bed-form height and steepness (van Rijn, 1984).

sented in Sec. 6.5. In this section, studies made specifically to determine temperature effects are discussed. The temperature effect on bed form is simpler than on sediment discharge.

The viscosity of water decreases with a rise in temperature; therefore, a particle has a greater fall velocity and larger fall diameter at a higher temperature. According to Fig. 6.5, a change in the fall diameter with temperature can cause a dramatic change in bed-form resistance, provided a point is close to the threshold between the lower flow regime and the transition.

In the case of natural rivers, Colby and Scott (1965) found rather spectacular effects of water temperature on the Middle Loup River, in that bed forms were more pronounced in summer than in winter and thus the friction factor was strongly dependent on water temperature. The U.S. Army Corps of Engineers (1969) reported a pronounced correlation between water temperature, changes in the stage–discharge relationship, and bed roughness in the Missouri River near Omaha. The comprehensive investigations by the Omaha District showed that as the temperature declined in the fall, with the flow discharge steady, the dunes were gradually washed out, the bed became flat, and the stage dropped with the lower friction factor.

## 6.5 STAGE–DISCHARGE PREDICTORS FOR ALLUVIAL CHANNELS

A stage–discharge predictor is a flow-resistance relationship used to determine the depth or hydraulic radius of flow for the given discharge, channel shape, slope, bed-material properties, and temperature. An important aspect of river sedimentation is the determination of the flow-induced resistance associated with the bed forms. A useful stage–discharge predictor for alluvial channels must also provide the delineation of the flow regime with which the roughness varies.

Alluvial bed roughness has been an area of extensive investigation. Since the approach proposed by Einstein and Barbarossa (1952), many techniques for stage–discharge prediction have been proposed, such as those by Shen (1962), Garde and Ranga Raju (1966), Simons and Richardson (1966), Haynie and Simons (1968), Engelund (1966), Raudkivi (1967), Alam and Kennedy (1969), Mostafa and McDermid (1971), Maddock (1976), White et al. (1980), van Rijn (1982), and Brownlie (1983), in addition to the earlier regime formulas for flow resistance. Existing resistance relationships follow two different approaches: those that divide resistance into grain resistance and form resistance and those that do not. *Grain resistance* refers to the part of resistance contributed by the surface drag (tangential force) whereas *form resistance* or *form drag* is caused by the pressure difference between the front and back surfaces of the bed forms.

The divided resistance approach can be expressed in terms of the energy gradient as

$$S = S' + S'' \tag{6.13}$$

or it can be for the hydraulic radius,

$$R = R' + R'' \tag{6.14}$$

where $S'$ and $R'$ are the energy gradient and hydraulic radius resulting from grain roughness; $S''$ and $R''$ are the same variables associated with form roughness. The values of $S'$ and $R'$ can be determined from one of the fixed-bed relations given in Sec. 3.5. Multiplying Eq. 6.13 by $\gamma R$ yields the relation for shear stress:

$$\tau_0 = \tau_0' + \tau_0'' \tag{6.15}$$

Dividing Eq. 6.15 by $\rho U^2$ gives the relation for friction factor:

$$f = f' + f'' \tag{6.16}$$

where $\tau_0'$ and $f'$ are counterparts to $S'$, and $\tau_0''$ and $f''$ are to $S''$.

Selected methods for stage–discharge prediction by Einstein and Barbarossa (1952), Engelund (1966), van Rijn (1982), and Brownlie (1983) are given in the following.

## Einstein and Barbarossa's Method

The division of total resistance into grain resistance and form resistance was introduced by Einstein and Barbarossa (1952). In the technical approach, the hydraulic radius (or the cross section) is divided according to Eq. 6.14.

Under fully rough conditions, $R'$ is obtained from the Manning-Strickler formula in the form (see Eq. 3.38)

$$\frac{U}{U_*'} = 7.66\left(\frac{R'}{d_{65}}\right)^{1/6} \tag{6.17}$$

where $U_*' = (gR'S)^{1/2}$ is the shear velocity related to grain roughness. Note that $d_{65}$ is used as the roughness height $k_s$. For those cases where grain roughness does not produce fully rough conditions, $R'$ is obtained from Eq. 3.36, the logarithmic law for the mean velocity, which is

$$\frac{U}{U_*'} = 5.75 \log\left(12.27\frac{R'}{k_s}X\right) \tag{6.18}$$

Equation 6.18 may be replaced by Eqs. 3.32 and 3.33, which do not rely on graphically determined values.

The form roughness is assumed to be related to the sediment transport rate along the channel bed because flow resistance due to bed forms is a function of flow to which the sediment rate may be related. A functional relation was thus suggested for the lower flow regime:

$$\frac{U}{U''_*} = F(\Psi') \tag{6.19}$$

where $\Psi'$ is the intensity of shear on representative particles and is given by

$$\Psi'_{35} = \frac{\rho_s - \rho}{\rho} \frac{d_{35}}{R'S} \tag{6.20}$$

Note that this parameter represents the ratio of submerged weight to the shear stress and is thus reversely related to the Shields stress.

The functional relationship for Eq. 6.19 was developed based on the field data shown in Fig. 6.9. Since $U/U''_*$ may be replaced by $(8/f'')^{1/2}$ and $\Psi'_{35}$ is inversely related to $\tau'_0$, this relation shows that the friction factor due to form roughness decreases with the grain shear in the lower flow regime in which its application is limited.

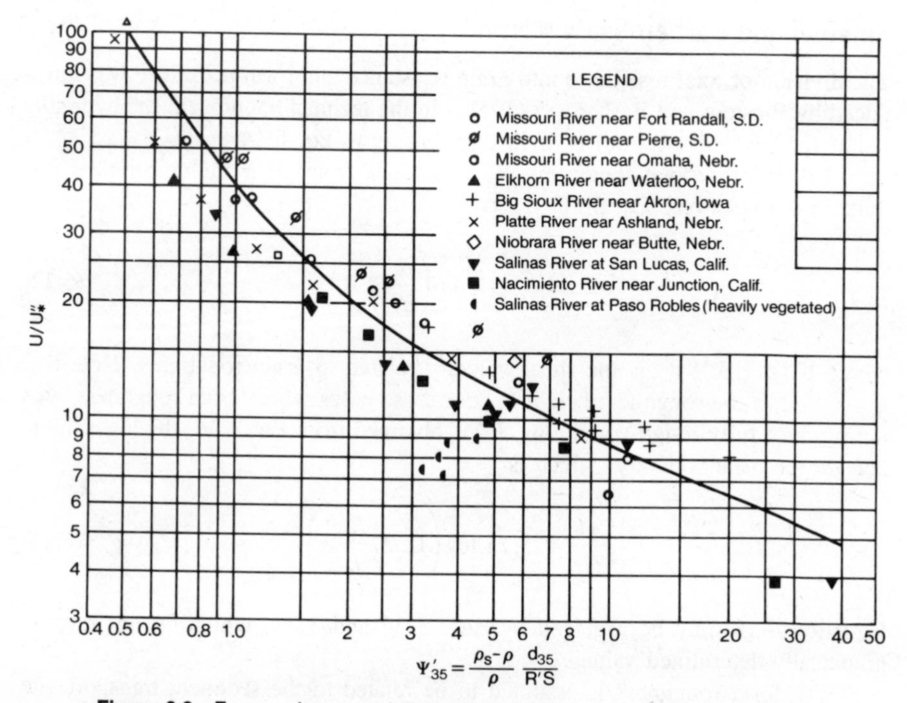


**Figure 6.9** Form resistance relationship by Einstein and Barbarossa (1952).

Good agreement of this relation with field data from the Rio Grande was reported by Nordin (1964); that regarding the Missouri and Mississippi was reported by Harrison and Mellema (1967). However, poor agreement was reported by Garde and Ranga Raju (1966) with flume and field data. The fallacy of this method, according to Yalin (1976), is attributed to the fact that sediment transport (or really the friction factor) is treated as a function of a single dimensionless variable.

## Engelund's Method

This method by Engelund (1966) employs the divided slope approach by assuming that $S = S' + S''$, where $S'$ is due to skin friction and $S''$ is due primarily to expansion losses associated with flow separation downstream of the dune crest. The magnitude of the expansion head loss $\Delta H''$ may be estimated from the formula

$$\Delta H'' = \alpha \frac{(U_1 - U_2)^2}{2g} \tag{6.21}$$

where $\alpha$ is the loss coefficient, $U_1$ is the mean velocity above the crest, and $U_2$ is the mean velocity over the trough. When the dune height is denoted $h$ and the mean depth is denoted $D$, then Eq. 6.21 becomes

$$\Delta H'' = \frac{\alpha}{2g}\left[\frac{q}{D - \frac{1}{2}h} - \frac{q}{D + \frac{1}{2}h}\right]^2 \simeq \alpha \frac{U^2}{2g}\left(\frac{h}{D}\right)^2$$

where $q$ is the discharge per unit channel width and $U = q/D$ is the mean velocity.

The energy gradient $S''$ is the head loss $\Delta H''$ divided by the distance of one wavelength $\lambda$, that is,

$$S'' = \frac{\Delta H''}{\lambda} = \frac{\alpha}{2}\frac{h^2}{\lambda D}\mathbf{F}^2 \tag{6.22}$$

Substituting Eq. 6.22 into Eq. 6.13 yields

$$S = S' + \frac{\alpha}{2}\frac{h^2}{\lambda D}\mathbf{F}^2$$

The depth $D$ is replaced by the hydraulic radius $R$. Multiplying both sides by $\gamma R/(\gamma_s - \gamma)d$ gives

$$\frac{\gamma RS}{(\gamma_s - \gamma)d} = \frac{\gamma RS'}{(\gamma_s - \gamma)d} + \frac{\alpha}{2}\frac{\gamma h^2}{(\gamma_s - \gamma)\lambda d}\mathbf{F}^2 \tag{6.23}$$

Engelund assumed that $\tau_0' = \gamma RS' = \gamma R'S$. Because of this assumption, the divided slope approach becomes the divided hydraulic radius approach. Let

$$\tau_* = \frac{\gamma RS}{(\gamma_s - \gamma)d}$$

$$\tau_*' = \frac{\gamma R'S}{(\gamma_s - \gamma)d}$$

and

$$\tau_*'' = \frac{\alpha}{2} \frac{\gamma h^2}{(\gamma_s - \gamma)\lambda d} \mathbf{F}^2$$

then Eq. 6.23 becomes

$$\tau_* = \tau_*' + \tau_*'' \tag{6.24}$$

where $\tau_*$, $\tau_*'$, and $\tau_*''$ are the dimensionless total shear stress, shear stress due to grain roughness, and shear stress due to bed-form roughness, respectively.

The grain roughness is based on the following logarithmic resistance formula (see Eq. 3.39):

$$\frac{U}{(gR'S)^{1/2}} = 6 + 2.5 \ln \frac{R'}{2.5d} \tag{6.25}$$

which agrees closely with the fully rough Nikuradse data and Eq. 3.34. Using the flume data by Guy et al.(1966), Engelund and Hansen (1967) obtained the following relationship for lower flow regime with a ripple or dune bed:

$$\tau_*' = 0.06 + 0.4\tau_*^2, \quad \text{or } \tau_* = 1.581(\tau_*' - 0.06)^{1/2}, \quad \text{for } \tau_*' < 0.55 \tag{6.26}$$

For the upper flow regime with $0.55 < \tau_*' < 1$, we have

$$\tau_*' = \tau_* \tag{6.27}$$

These relations are shown in Fig. 6.10 with a discontinuity between lower and upper flow regimes occurring at about $\tau_*' = 0.55$. Equation 6.26 was extended by Brownlie (1983) from the curve in Fig. 6.10 into higher regions of the upper flow regime for $\tau_*' > 1$:

$$\tau_*' = [0.702(\tau_*)^{-1.8} + 0.298]^{-1/1.8}$$

or

$$\tau_* = [1.425(\tau_*')^{-1.8} - 0.425]^{-1/1.8} \tag{6.28}$$

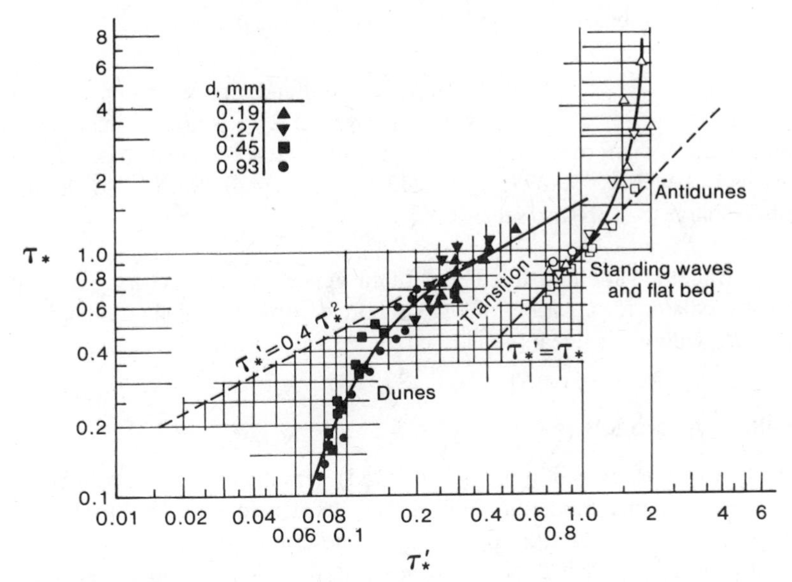


**Figure 6.10** Engelund's universal relationship between dimensionless grain shear stress $\tau_*'$ and total shear stress $\tau_*$.

For a given set of $Q$, $d$, $S$ and flow and sediment characteristics, the depth may be determined using the Engelund method. The procedure is demonstrated by the following example.

► ***EXAMPLE 6.1.*** A wide alluvial channel carries a unit discharge $q$ of 2.21 $m^3$/sec/m. If the channel slope is 0.0003 and median sediment size is 0.4 mm, determine the flow depth.

*Step 1*. Assume a value for $D'$ or $R'$. The trial value of 1.0 m for $D'$ is first assumed in this case.

*Step 2*. Compute $\tau_*'$ and the mean velocity $U$:

$$\tau_*' = \frac{\gamma D'S}{(\gamma_s - \gamma)d} = \frac{1 \times 0.0003}{(2.65 - 1)0.0004} = 0.455$$

From Eq. 6.25, we obtain

$$U = (gd'S)^{1/2}\left(6 + 2.5 \ln \frac{D'}{2.5d}\right) = 1.26 \text{ m/sec}$$

*Step 3*. Compute $\tau_*$. Since $\tau_*' < 0.55$, the alluvial bed is in the lower flow regime. Therefore, $\tau_*$ is computed using Eq. 6.26, and its value so obtained is 0.993.

*Step 4*. Obtain the depth $D$. From $\tau_* = \gamma DS/(\gamma_s - \gamma)d = 0.993$, $D$ is computed to be 2.18 m.

*Step 5*. Compute $q$ based on $U$ and $D$. In this case, the unit discharge $q = UD = 1.26 \times 2.18 = 2.75\ \text{m}^3/\text{sec/m}$.

*Step 6*. Compare the computed $q$ with the given $q$ and repeat the steps until these two $q$'s match. Since the computed discharge of 2.75 is greater than the given value of 2.50, this procedure is repeated. The final results that match the given $q$ are as follows: $\tau'_* = 0.377$, $U = 1.13$ m/sec, $\tau_* = 0.890$, $D = 1.96$ m, and unit discharge $= 2.21\ \text{m}^3/\text{sec/m}$.

The Engelund method has been found generally satisfactory as a depth–discharge predictor for a variety of cases reported. However, it should not be used for channels with a very coarse sand bed.

## Brownlie's Approach

This method was developed by Brownlie (1983) with the objective that it be easily adaptable to computer modeling applications. Its development, determination of flow regime, and procedure of application are described below.

Brownlie's approach deviates from those that divide flow resistance into grain roughness and form roughness in that bed forms are considered but treated as if they were large-scale grains or equivalent roughness. Different concepts for equivalent roughness were used by previous investigators, for example, Simons and Richardson (1966) and van Rijn (1982). In Brownlie's approach, sediment properties are described by the median size $d$, specific weight $\rho_s g$, and the geometric standard deviation $\sigma_g$ under the assumption of log-normal distribution. Dimensionless groups in the approach are given as

$$\frac{RS}{d} = \frac{(\rho_s - \rho)}{\rho}\tau_* = F\left(\frac{q}{(gd^3)^{1/2}}, S, \sigma_g\right) \tag{6.29}$$

In this functional relation, the shear stress $RS$, that is, flow resistance, is related to the unit discharge $q$, channel slope $S$, and sediment properties. Note that the Reynolds number is not included for fully rough flow.

For flow over a dune bed, the friction factor may be defined by the power-law equation of Eq. 3.29, but with roughness height replaced by a measure of the dune height $k_d$, that is,

$$\frac{U}{(gRS)^{1/2}} = a\left(\frac{R}{k_d}\right)^{1/6} \tag{6.30}$$

where $a$ is a coefficient of proportionality. After considerable rearrangement, it becomes

$$\left(\frac{\rho_s - \rho}{\rho}\right)\tau_* = a^{-0.6}\left(\frac{k_d S}{d}\right)^{0.1}(q_* S)^{0.6} \tag{6.31}$$

where $q_* = q/(gd^3)^{1/2}$. It shows that the flow resistance is related to $k_d$ raised to the power of 0.1. Therefore, the shear stress is not strongly dependent on $k_d$ and an exact definition is not a critical factor in the prediction of $\tau_*$. It is further assumed that $k_d/d$ is proportional to the product of undetermined powers of $q_*$ and $S$, since the dune height in relation to sediment size is a direct function of the power expenditure represented by $q_*S$. The effect of nonuniform bed materials is considered by using the variable $\sigma_g$ raised to an unknown power. Substituting these assumptions into Eq. 6.31 yields

$$\left(\frac{\rho_s - \rho}{\rho}\right)\tau_* = w(q_* S)^x S^y \sigma_g^z \tag{6.32}$$

where $w$, $x$, $y$, and $z$ are constants to be fitted empirically. By taking logarithms of both sides of Eq. 6.32, these coefficients were determined by multiple regression. The data and the best fit line for the lower flow regime are shown in Fig. 6.11, and those for the upper flow regime are shown in Fig. 6.12.

The regression equation for the lower flow regime is

$$\frac{R}{d} = 0.3724(q_*)^{0.6539} S^{-0.2542} (\sigma_g)^{0.1050} \tag{6.33}$$

and for the upper regime, it is

$$\frac{R}{d} = 0.2836(q_*)^{0.6248} S^{-0.2877} (\sigma_g)^{0.08013} \tag{6.34}$$

From each equation, the hydraulic radius and flow depth may be easily computed based on the given variables. This method is preferred because of its large data base and its good correlation among the variables. For data used in the analysis, the median particle size ranges from 0.088 to 2.8 mm, the unit discharge is from 0.012 to 40 $m^3/sec/m$, the slope is from $3.0 \times 10^{-6}$ to $3.7 \times 10^{-2}$, the hydraulic radius is between 0.025 and 17 m, and the temperature ranges from 0 to 63°C.

For a given set of independent variables, there are two possible depths (or hydraulic radii): one for the lower flow regime and the other for the upper regime. Determination of the flow regime in Brownlie's approach is based on the forces on bed-sediment particles to which bed deformation is related. Dimensionless parameters that are considered to be indicative of flow regime are selected as $F_g$, $d/\delta$ and $S$, where

$$F_g = \frac{U\rho^{1/2}}{[(\rho_s - \rho)gd]^{1/2}} \tag{6.35}$$

is the grain Froude number representing the square root of the ratio of drag force on a particle to its immersed weight. The parameter $d/\delta$ is the ratio of the mean

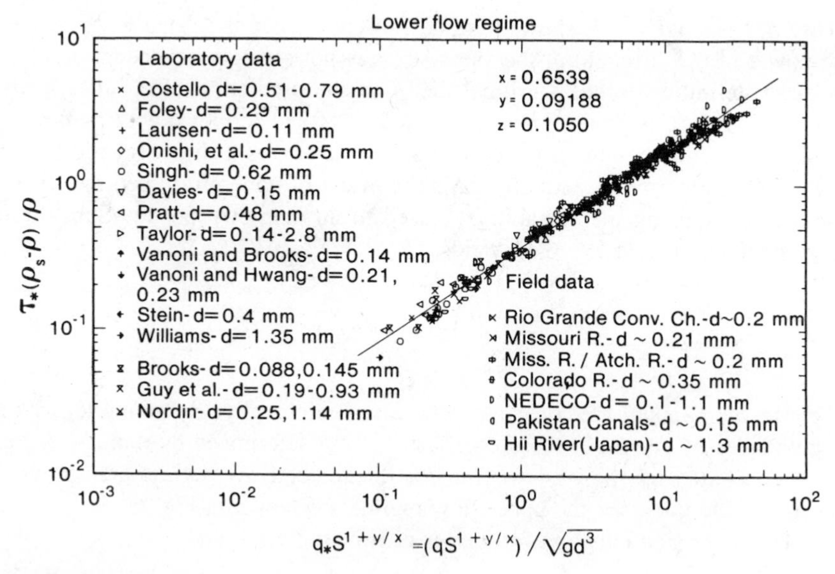


**Figure 6.11** Relationships among dimensionless shear stress $\tau_*$ and $q_*$ and $S$ for lower flow regime (Brownlie, 1983).

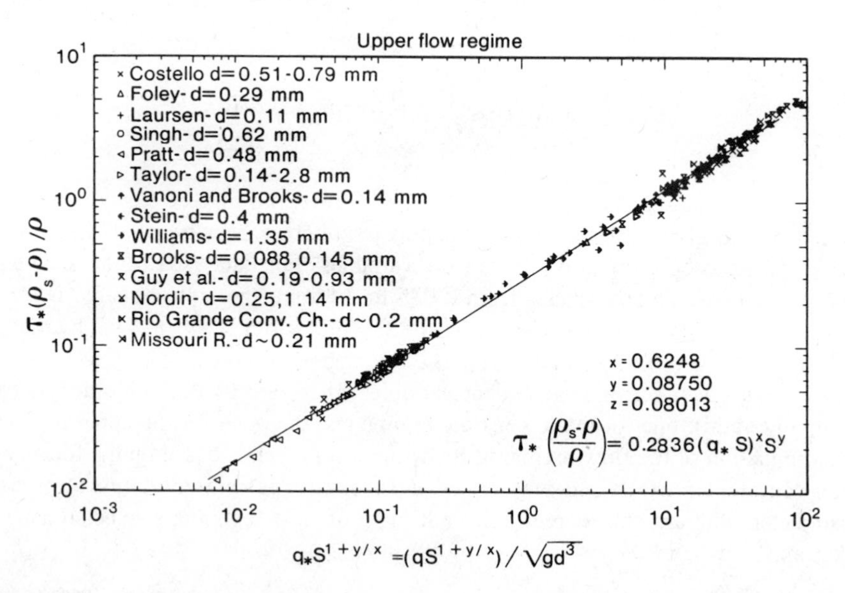


**Figure 6.12** Relationship among $\tau_*$ and $q_*$ and $S$ for upper flow regime and flat beds prior to initiation of motion (Brownlie, 1983).

grain size to the thickness of the laminar sublayer, where $\delta = 11.6\nu/U'_*$. The variable $U'_*$ is the shear velocity obtained from Eq. 6.34 for the upper flow regime; that is, it is the shear velocity that would occur when no dunes are present.

Flow regimes delineated on the basis of $F_g$ and $S$ are illustrated in Fig. 6.13. It shows that beyond a slope of 0.006, only the upper regime exists. For lower values of $S$, an approximate dividing line is given by

$$F_g = F'_g = 1.74S^{-1/3} \tag{6.36}$$

where $F'_g$ is the $F_g$ value along this line. However, the overlap of both flow regimes along this line indicates a transition that is refined by the ratios $F_g/F'_g$ and $d/\delta$ which are shown in Fig. 6.14. By neglecting viscous effects, the maximum velocity of the lower regime can be determined from $F_g = 0.8\ F'_g$, and the minimum velocity of the upper regime from $F_g = 1.25\ F'_g$. When temperature effects are important, the transition values of $F_g$ is also a function of $d/\delta$, as shown in Fig. 6.14. The transition region can be defined by the equations for the upper limit of the lower flow regime,

$$\log \frac{Fg}{F'_g} = \begin{cases} -0.2026 + 0.07026 \log \dfrac{d}{\delta} + 0.9330\left(\log \dfrac{d}{\delta}\right)^2 & \text{for } \dfrac{d}{\delta} < 2 \\ \log 0.8 \quad \text{for } \dfrac{d}{\delta} \geq 2 \end{cases} \tag{6.37}$$

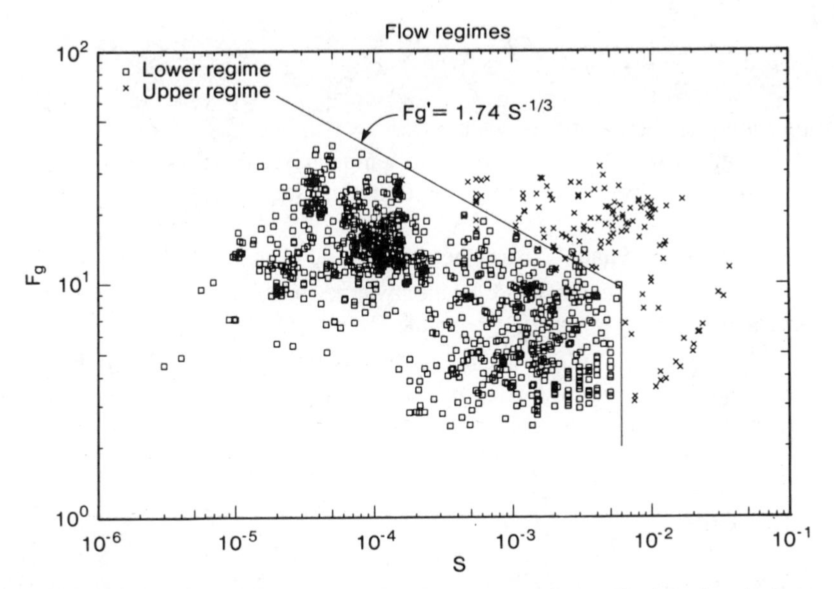


**Figure 6.13** Determination of flow regimes: grain Froude number $F_g$ plotted against slope $S$ (Brownlie, 1983).

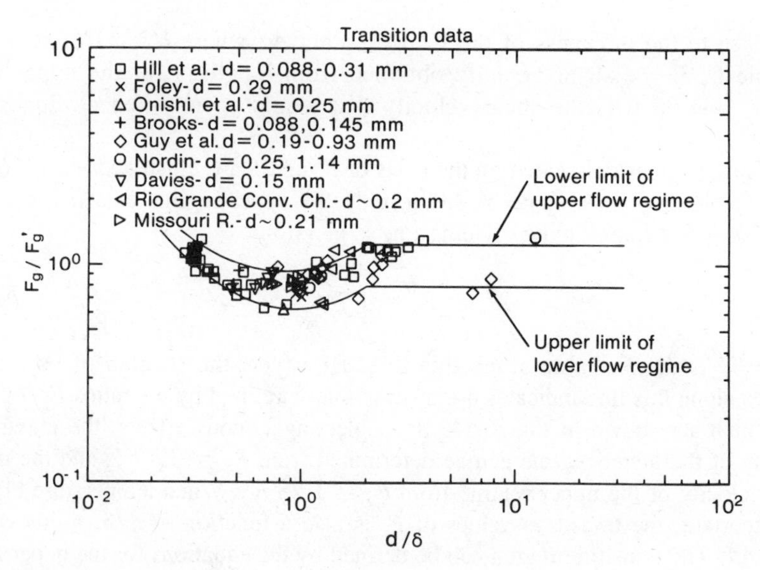


**Figure 6.14** Viscous effects on the transition from lower flow regime to upper flow regime (Brownlie, 1983).

and by the equations for the lower limit of the upper flow regime,

$$\log \frac{Fg}{F_g'} = \begin{cases} -0.02469 + 0.1517 \log \dfrac{d}{\delta} + 0.8381\left(\log \dfrac{d}{\delta}\right)^2 & \text{for } \dfrac{d}{\delta} < 2 \\ \log 1.25 \quad \text{for } \dfrac{d}{\delta} \geq 2 \end{cases} \tag{6.38}$$

The transition lies between these values.

For a given set of $q$, $S$, $d$, $\sigma_g$, and temperature, the flow depth and flow regime may be obtained according to the following steps.

*Step 1*. Compute $d/\delta$ and $F_g'$ from the equation $F_g' = 1.74\ S^{-1/3}$.

*Step 2*. Compute $R$ from Eq. 6.33 for the lower regime and from Eq. 6.34 for the upper regime. With $U = q/R$, compute $F_g$ from Eq. 6.35 for each case.

*Step 3*. Use $F_g/F_g'$ and $d/\delta$ to locate two points in Fig. 6.14, one for the upper regime and one for the lower regime. There may be one valid solution or, for rare cases, two valid solutions.

## REFERENCES

Alam, A. M. Z. and Kennedy, J. K. "Friction Factors for Flow in Sand Bed Channels," *J. Hydraul. Div. ASCE*, **95**(HY6), pp. 1973-1992, November 1969.

ASCE, *Sedimentation Engineering*, Manual and Reports on Engineering Practice, No. 54, V. A. Vanoni, ed. 1975.

ASCE Task Force on Bed Forms in Alluvial Channels, "Nomenclature for Bed Forms in Alluvial Channels," *J. Hydraul. Div. ASCE*, **92**(HY3), pp. 51-64, May 1966.

Athaullah, M., "Prediction of Bed Forms in Erodible Channels," Ph.D. Thesis, Department of Civil Engineering, Colorado State University, Fort Collins, CO. 1968.

Brooks, N. H., "Mechanics of Streams with Moveable Beds of Fine Sand," *Trans. ASCE*, **123**, pp. 526-594, 1958.

Brownlie, W. R., "Flow Depth in Sand-Bed Channels," *J. Hydraul. Eng. ASCE*, **109**(7), pp. 959-990, July 1983.

Chang, H. H., "Geometry of Rivers in Regime," *J. Hydraul. Div. ASCE*, **105**(HY6), pp. 671-706, June 1979.

Chang, H. H., "River Morphology and Thresholds," *J. Hydraul. Eng. ASCE*, **111**(3), pp. 503-519, March 1985.

Colby, B. R., "Discontinuous Rating Curves for Pigeon Roost and Cuffawa Creeks in Northern Mississippi," Report ARS-41-36, Agricultural Research Service, April 1960.

Colby, B. R. and Scott, C. H., "Effects of Water Temperature on the Discharge of Bed Material," *USGS Professional Paper 462-G*, 1965.

Dawdy, D. R., "Depth-Discharge Relations of Alluvial Streams—Discontinuous Rating Curves," *USGS Water Supply Paper 1948-C*, 1961.

Einstein, H. A. and Barbarossa, N., "River Channel Roughness," *Trans. ASCE*, **117**, pp. 1121-1146, 1952.

Engelund, F., "Hydraulic Resistance of Alluvial Streams," *J. Hydraul. Div. ASCE*, **92**(HY2), pp. 315-326, March 1966.

Engelund, F., "Instability of Erodible Beds," *J. Fluid Mech.*, **42**, pp. 225-240, 1980.

Engelund, F. and Hansen, E., *A Monograph on Sediment Transport in Alluvial Streams*, Teknisk Vorlag, Copenhagen, Denmark, 1967.

Fredsoe, J., "Shape and Dimensions of Stationary Dunes in Rivers," *J. Hydraul. Div. ASCE*, **108**(HY8), pp. 932-947, August 1982.

Garde, R. J. and Ranga Raju, K. G., "Resistance Relationships for Alluvial Channel Flow," *J. Hydraul. Div. ASCE*, **92**(HY4), pp. 77-100, July 1966.

Gilbert, G. K., "Transportation of Debris by Running Water," *USGS Professional Paper 86*, 1914.

Guy, H. P., Simons, D. B., and Richardson, E. V., "Summary of Alluvial Channel Data from Flume Experiments, 1956–61," *USGS Professional Paper 462-I*, 1966, 96 pp.

Haque, M. I. and Mahmood, K., "Geometry of Ripples and Dunes," *J. Hydraul. Eng. ASCE,* **111**(1), pp. 48-63, January 1985.

Harrison, A. S. and Mellema, W. J., "Movable Bed Model for Alluvial Channel Studies," Proceedings of the Twelfth Congress, IAHR, Fort Collins, Colorado, 1967.

Hayashi, T., "Formation of Dunes and Antidunes in Open Channels," *J. Hydraul. Div. ASCE*, **96**(HY2), pp. 431-439, February 1970.

Haynie, R. B. and Simons, D. B., "Design of Stable Channels in Alluvial Materials," *J. Hydraul. Div. ASCE*, **94**(HY6), pp. 1399-1420, November 1968.

Kennedy, J. F., "The Mechanics of Dunes and Antidunes in Erodible-Bed Channels," *J. Fluid Mech.*, **16**(4), pp. 521-544, August 1963.

Maddock, T. Jr., " Equations for Resistance to Flow and Sediment Transport in Alluvial Channels," *Water Resour. Res.*, **12**(1), pp. 11-21, February 1976.

Mostafa, M. G. and McDermid, R. M., discussion of "Sediment Transportation Mechanics: F. Hydraulic Relations for Alluvial Streams," by the Task Committee for Preparation of Sedimentation Manual, Committee on Sedimentation of the Hydraulics Division, Vito A. Vanoni, Chairman, *J. Hydraul. Div. ASCE*, **97**(HY10), pp. 1777-1780, October 1971.

Nordin, C. F. Jr., "Aspects of Flow Resistance and Sediment Transport, Rio Grande near Bernalillo, New Mexico,' *USGS Water Supply Paper 1498-H*, 1964.

Ranga Raju, K. G. and Soni, J. P., "Geometry of Ripples and Dunes in Alluvial Channels," *J. Hydraul. Res., IAHR*, **14**(3), pp. 77-100, 1976.

Raudkivi, A. J., *Loose Boundary Hydraulics*, 2nd ed., Pergamon Press, Oxford, England, 1976.

Raudkivi, A. J., "Analysis of Resistance in Fluvial Channels," *J. Hydraul. Div. ASCE*, **93**(HY5), pp. 73-84, May 1967.

Shen, H. W., "Development of Bed Roughness in Alluvial Channels," *J. Hydraul. Div. ASCE*, **88**(HY3), pp. 45-58, 1962.

Simons, D. B. and Richardson, E. V., "Forms of Bed Roughness in Alluvial Channels," *J. Hydraul. Div. ASCE*, **87**(HY3), pp. 87-105, 1961.

Simons, D. B. and Richardson, E. V., "Resistance to Flow in Alluvial Channels," *USGS Professional Paper 422-J*, 1966.

Simons, D. B., Richardson, E. V., and Nordin, C. F., "Sedimentary Structures Generated by Flow in Alluvial Channels," Am. Assoc. Petrol. Geologists, Special Publ. No. 12, 1965.

Simons, D. B. and Senturk, F., *Sediment Transport Technology*, Water Resources Publications, P.O. Box 2841, Littleton, Colorado, 1977.

Song, C. C. S., "Modified Kinematic Model: Application to Bed Forms," *J. Hydraul. Eng. ASCE*, **109**(8), pp. 1133-1151, August 1983.

U.S. Army Corps of Engineers, "Missouri River Channel Regime Studies, Omaha District," *MRD Sediment Series*, No. 13B, November 1969.

van Rijn, L. C., "Equivalent Roughness of Alluvial Bed," *J. Hydraul. Div. ASCE*, **108**(HY10), pp. 1215-1218, October 1982.

van Rijn, L. C., "Sediment Transport, Part III: Bed Forms and Alluvial Roughness," *J. Hydraul. Eng. ASCE*, **110**(12), pp. 1733-1754, December 1984.

White, W. R., Paris, E., and Bettess, R., "The Frictional Characteristics of Alluvial Streams: A New Approach," *Proc. Inst. Civ. Eng.*, **69**(1), pp. 737-750, September 1980.

Yalin, M. S., "Geometrical Properties of Sand Waves," *J. Hydraul. Div. ASCE*, **90**(HY5), pp. 105-119, 1964.

Yalin, M. S., *Mechanics of Sediment Transport*, 2nd ed., Pergamon Press, Oxford, England, 1976.

# 7

# SEDIMENT MOVEMENT IN RIVERS

An important aspect of fluvial processes is the movement of sediment in rivers, to which river morphology and river channel changes are closely related. The term *load*, as used in sediment transport, may refer to the sediment that is in motion in a stream. It is also used to denote the rate at which sediment is moved, for example, cubic feet per second or tons per day. The latter usage is preferred in river morphology.

There are two common classifications of the load in a stream. The first divides the load into bed load and suspended load; the second separates the load into wash load and bed-material load (or bed-sediment load). *Bed load* is defined as that part of the load moving on, or near, the bed by rolling, saltation, or sliding. Suspended load, by definition, moves in suspension.

*Wash load* refers to the finest portion of sediment, generally silt and clay, that is washed through the channel, with an insignificant amount of it being found in the bed. Bed-material load or bed-sediment load, on the other hand, consists of particles that are generally found in the bed material. An alluvial stream bed, such as those shown in Fig. 7.1, is formed during the fluvial process of sorting, through which clay and silt are removed as wash load. The discharge of wash load depends primarily on the rate of supply; it is generally not correlated with the flow characteristics. Bed-material load, on the other hand, is usually correlated with water discharge.

Formulas for predicting sediment discharge can be found in abundance. Such formulas have been developed for noncohesive sediment in steady uniform flow. Therefore, they are not applicable to cohesive sediment, which normally constitutes the wash load and is independent of the flow characteristics. Because of the complexity of the sediment transport processes, prediction of the transport rate has not been accomplished following purely theoretical pursuit. All existing sediment transport formulas have been established relying on calibration using flume and field data supposedly under the steady uniform flow condition.

Sediment transport formulas presented herein were developed based on three different approaches, which are given as follows, together with their respective formulas:

**Figure 7.1** Alluvial beds of a desert wash (top) and a small stream (bottom). The bed material is primarily sand, with clay and silt removed through fluvial process.

1. Shear Stress Approach: DuBoys formula, Shields formula, Einstein bed-load function, Meyer-Peter–Muller formula, Einstein–Brown formula, and Parker et al. formula for gravel.
2. Power Approach: Engelund–Hansen formula, Ackers–White formula, and Yang formula.
3. Parametric Approach: Colby relations.

Sediment transport formulas are also classified according to their applicabilities into bed-load formulas, suspended load formulas, and bed-material load formulas as follows:

1. Bed-Load Formulas: DuBoys formula, Shields formula, Meyer-Peter–Muller formula, Einstein bed-load function, Einstein–Brown formula, and Parker et al. formula.
2. Suspended Load Formulas: Einstein suspended-load method.
3. Bed-material Load Formulas: Colby relations, Engelund–Hansen formula, Ackers–White formula, and Yang formula.

In the development of river sedimentation, there has been perhaps more work done on the prediction of sediment discharge than any other aspects. Such predictive methods are published frequently. Only a limited number of sediment transport relations are included in this chapter. Other commonly used formulas have been developed by Laursen (1958), Rottner (1959), Bishop et al. (1965), Bagnold (1966), Chang et al. (1967), Toffaleti (1968), Shen and Hung (1972), Graf (1971), Ranga Raju et al. (1981), Brownlie (1982), van Rijn (1984a, 1984b), among others. Extensive coverage of this subject can be found in the respective books by Graf (1971), Raudkivi (1976), Bogardi (1974), Yalin (1977), ASCE (1975), Garde and Ranga Raju (1985), Simons and Senturk (1977), and Chien and Wan (1983).

## 7.1 BED-LOAD FORMULAS

The development of predictive methods for bed-load transport started in western Europe, where rather coarse bed materials in streams are carried mainly as bed load. It should be noted that the shear stress on the channel boundary is used in most of the predictive formulas. The shear stress is the tractive force per unit area applied to the channel boundary on which bed load moves by rolling, sliding, and sometimes saltating. Several bed-load formulas are described as follows.

### DuBoys Formula

The bed-load formula by DuBoys (1879) assumes that uniform sediment grains move as a series of superimposed layers with each thickness $d'$ of the same magnitude as the grain diameter. The velocity of the layers is assumed to vary linearly toward the surface, as shown in Fig. 7.2. If there are $(n - 1)$ layers in motion, then the surface layer will have a velocity $(n - 1)\Delta u$, where $\Delta u$ is the velocity increment between two adjacent layers. The volume rate of bed load is then obtained from the average velocity $(n - 1)\,\Delta u/2$ and the total thickness, that is,

$$q_b = \tfrac{1}{2} n d'(n - 1)\,\Delta u \tag{7.1}$$

where $q_b$ is the bed-load discharge per unit channel width.

The shear stress exerted on the bed, $\tau_0$, is balanced by the frictional forces between the successive layers, that is,

$$\tau_0 = f_s(\gamma_s - \gamma)\, n d' \tag{7.2}$$

where the friction coefficient $f_s$ is assumed constant.

The threshold for bed load is when the top layer is at the incipient motion, given by $n = 1$; thus,

$$\tau_c = f_s(\gamma_s - \gamma) d' \tag{7.3}$$

From Eqs. 7.2 and 7.3, one has

$$n = \frac{\tau_0}{\tau_c} \tag{7.4}$$

Substituting the value of $n$ into Eq. 7.1 yields

$$q_b = \frac{d'\Delta u}{2\tau_c^2}\tau_0(\tau_0 - \tau_c)$$

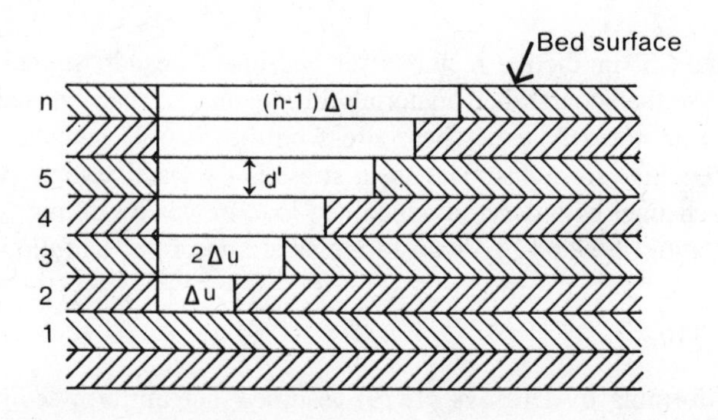


**Figure 7.2** Moving layers of sediment as conceived by DuBoys (1879).

or

$$q_b = C_d \tau_0 (\tau_0 - \tau_c) \tag{7.5}$$

where $C_d$ is the characteristic sediment coefficient. Equation 7.5 is DuBoys' bed-load formula; it relates the bed-load discharge per unit channel width $q_b$ to the excess shear stress $\tau_0 - \tau_c$. This concept by DuBoys has influenced many other investigators.

Relations for $C_d$ and $\tau_c$ in Eq. 7.5 were obtained by Straub (1935) based upon experiments in small laboratory flumes with a sand bed. In metric units, he gave

$$C_d = \frac{0.17}{d^{3/4}} \qquad (\mathrm{m^3/kg/sec}) \tag{7.6}$$

$$\tau_c = 0.061 + 0.093d \qquad (\mathrm{kg/m^2}) \tag{7.7}$$

where $d$ is the diameter of uniform sand in millimeters. Their respective relations in English units are

$$C_d = \frac{0.173}{d^{3/4}} \qquad (\mathrm{ft^6/lb^2/sec}) \tag{7.8}$$

$$\tau_c = 0.0125 + 0.019d \qquad (\mathrm{lb/ft^2}) \tag{7.9}$$

The unit for $d$ is still in millimeters. The critical shear stress $\tau_c$ in the DuBoys formula is generally somewhat different in value from the corresponding Shields stress given in Sec. 5.2.

### Shields Formula

A dimensionless formula also based on the excess of shear stress was proposed by Shields (1936):

$$\frac{q_b \gamma_s}{q S \gamma} = 10 \frac{\tau_0 - \tau_c}{(\gamma_s - \gamma) d} \tag{7.10}$$

where $q$ is the water discharge per unit width. This equation is dimensionally homogeneous and can thus be used in any system of units. It was not Shields' intention to establish a universal equation but to represent bed-load transport as a function of the major factors. Physically, the left-hand side of Eq. 7.10 may be interpreted as being the dimensionless bed-load discharge, which is influenced by two opposing forces represented on the right-hand side: the excess shear stress and the submerged weight of sediment particles embodied in $(\gamma_s - \gamma)d$. The excess shear stress is considered to be the applied force responsible for the motion of particles; this movement is resisted by the submerged weight of transported sediment.

This concept of opposing forces for sediment motion is also included in the Shields diagram for incipient sediment motion and many other sediment transport formulas.

## Meyer-Peter–Muller Formula

The dimensionless Meyer-Peter–Muller formula (1948) and the physical meanings for its respective terms normalized by $(\gamma_s - \gamma)d_m$ are given by

$$\underbrace{\overset{\text{I}}{\left[\frac{q_b(\gamma_s - \gamma)}{\gamma_s}\right]^{2/3}\left(\frac{\gamma}{g}\right)^{1/3}\frac{0.25}{(\gamma_s - \gamma)d_m}}}_{\text{Bed-load discharge}} = \underbrace{\overset{\text{II}}{\frac{(k/k')^{3/2}\gamma RS}{(\gamma_s - \gamma)d_m}}}_{\text{Effective shear}} - \underbrace{\overset{\text{III}}{0.047}}_{\text{Critical shear}} \tag{7.11}$$

In this basically empirical equation, the bed-load discharge $q_b$ is in weight per unit time and unit channel width. Being dimensionally homogeneous, it may be used under any consistent set of units. It is applicable to graded sediments, for which the effective diameter $d_m$ of the sediment mixture is defined as

$$d_m = \sum_i p_i d_i \tag{7.12}$$

where $i$ is the size fraction index, $d_i$ is the mean size of a fraction of the bed material, and $p_i$ is its fraction by weight. The quantities $k$ and $k'$, which are reciprocals of Manning's roughness coefficient, are given by

$$U = kR^{2/3}S^{1/2} \tag{7.13}$$

$$U = k'R^{2/3}(S')^{1/2} \tag{7.14}$$

where $U$ is the cross-sectionally averaged velocity, $R$ is the hydraulic radius, $S$ is the total energy gradient, and $S'$ is the energy gradient caused by grain roughness. The value of $k'$ can be obtained from Strickler's formula for grain roughness (see Sec. 3.5), that is,

$$k' = \frac{26}{(d_{90})^{1/6}} \tag{7.15}$$

where $d_{90}$ is the grain size of the bed material for which 90% is finer, in meters. Note that this formula is valid only if $d_{90}$ is in meters and time is in seconds.

Term I in Eq. 7.11 represents the bed-load discharge per unit channel width measured in submerged weight and normalized by $(\gamma_s - \gamma)d_m$; it is related to the shear stress caused by grain roughness (term II) subtracted by the critical shear stress (term III). The grain shear stress is considered directly responsible in moving the particles. The form roughness also affects the shear stress because of its influence on the depth. The ratio $k/k'$ is used to provide the grain shear stress as a portion of the total (grain plus form) shear stress. The value of $k/k'$ varies between 0.5 and 1; it is 0.5 for strong bedforms and 1 in the absence of bedforms. Bedforms such as dunes and ripples are usually characteristic to the sand bed and are usually poorly developed in coarse sediments for which the total roughness is essentially caused by grain roughness. Term III as the dimensionless critical shear is similar to the critical Shields stress.

The experiments in developing the formula were made in laboratory flumes with widths ranging between 15 cm and 2 m, water depth between 1 and 120 cm, effective diameter of sediments between 6.4 and 30 mm, and specific gravity for sediments from 1.25 to over 4. This formula is therefore more applicable to coarse sediments with little suspended load. It has enjoyed considerable popularity in Europe.

### Einstein Bed-Load Function

Einstein (1942, 1950) presented the most extensive analysis on bed-load transport, based on fluid mechanics and probability. Over the years his analysis has been a focus of research that resulted in improvements and introduction of modifications. In bed-load transport, Einstein (1942) obtained two dimensionless parameters: the bed-load intensity $\Phi$ and the flow intensity $\Psi$, given, respectively, as

$$\Phi = \frac{q_b}{\gamma_s}\left(\frac{\gamma}{\gamma_s - \gamma}\frac{1}{gd^3}\right)^{1/2} \tag{7.16}$$

$$\Psi = \frac{(\gamma_s - \gamma)d}{\gamma R'S} \tag{7.17}$$

where $R'$ is the hydraulic radius with respect to grain roughness described in Sec. 6.5. The relationship for $\Phi = F(\Psi)$ was determined based on empirical data. Such a relationship states that the bed-load discharge normalized by its submerged weight depends on the shear stress related to the grain resistance.

Recognizing that bed-load transport of grains in a mixture is affected by other grains because of the difference in size and the hiding effect, Einstein divided the bed material into several size fractions, each of which may be represented by the geometric mean. Sediment discharge of each fraction is computed separately, and the total sediment discharge is obtained from

$$q_b = \sum_{i=1}^{n} p_i q_{bi} \tag{7.18}$$

and

$$Q_b = q_b b \tag{7.19}$$

where $n$ is the number of size fractions, $q_{bi}$ is the sediment discharge of a size fraction per unit width, and $Q_b$ is the sediment discharge of the section with a bed width $b$.

Later, Einstein (1950) presented the more sophisticated dimensionless parameters of bed-load intensity and flow intensity for each size fraction, respectively:

$$\Phi_{*i} = \frac{q_{bi}}{p_i \gamma_s} \left[ \left( \frac{\gamma}{\gamma_s - \gamma} \right) \frac{1}{(gd_i)^3} \right]^{1/2} \tag{7.20}$$

and

$$\Psi_{*i} = \overbrace{\xi_i}^{\text{I}} \overbrace{Y \left[ \frac{\log 10.6}{\log(10.6 d_x / \Delta)} \right]^2}^{\text{II}} \overbrace{\frac{(\gamma_s - \gamma) d_i}{\gamma R' S}}^{\text{III}} \tag{7.21}$$

The parameter $\Psi_{*i}$ has three parts as designated. Part III is the reciprocal of the grain shear stress. Part I consists of the hiding factor $\xi_i$, which is a function of $d_i/d_x$ as shown in Fig. 7.3, where $d_x$ is the characteristic grain diameter of the

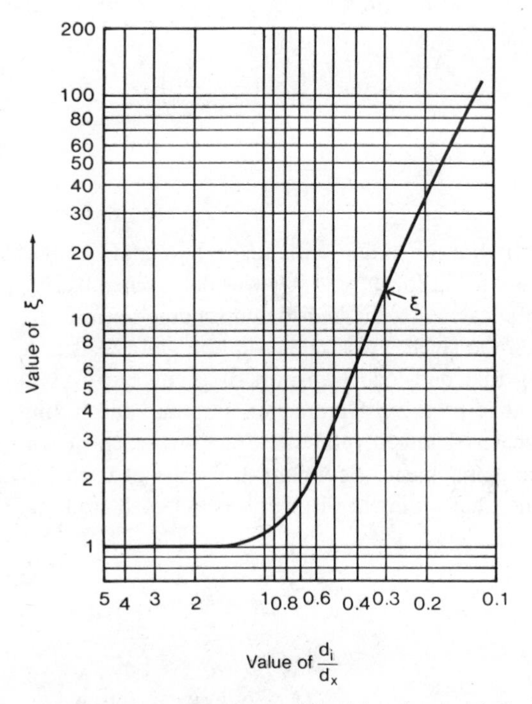


**Figure 7.3** Hiding factor in Einstein's bed-load function in relation to $d_i/d_x$.

mixture. For uniform grains, the hiding factor is unity, but its value increases in relation to $d_i/d_x$. In other words, the hiding effect increases for smaller particles in a mixture. The value of $d_x$ is a function of sediment size for rough bed,

$$d_x = 0.77\Delta \quad \text{if } \Delta/\delta > 1.8 \tag{7.22}$$

and a function of viscosity for smooth bed,

$$d_x = 1.39\,\delta \quad \text{if } \Delta/\delta < 1.8 \tag{7.23}$$

where $\Delta$ is defined in Fig. 3.6 and $\delta = 11.6\,\nu/U'_*$ is the laminar sublayer thickness.

Part II in Eq. 7.21 accounts for the hydrodynamic lift. The correction factor $Y$ describes the change of the lift coefficient in mixtures with various roughness as a function of $d_{65}/\delta$, as shown in Fig. 7.4. The factors $\xi_i$ and $Y$ were determined experimentally, and they have been modified subsequently, for example, by Shen and Hung (1983). The lift is related to the square of the near bed velocity. From experiments, it was found that the velocity acting on particles of a mixture must be measured at a distance $0.35d_x$ above the theoretical bed. At this point, the velocity can be obtained from Eq. 3.35:

$$u = u'_* 5.75 \log\left(10.6\,\frac{d_x}{\Delta}\right) \tag{7.24}$$

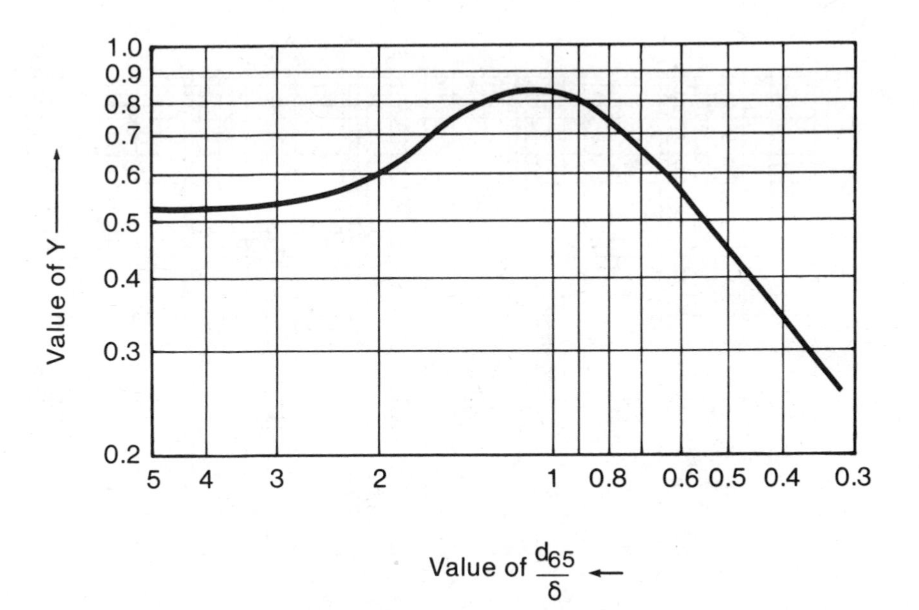


**Figure 7.4** Factor Y in Einstein's bed-load function in relation to $d_{65}/\delta$.

The factor $[\log 10.6/\log(10.6d_x/\Delta)]^2$ in Eq. 7.21 results from this velocity distribution for smooth and rough beds.

After fitting the analytical relation to flume data, the following equation relating $\Psi_*$ and $\Phi_*$ was obtained

$$1 - \frac{1}{\pi^{1/2}} \int_{-(1/7)\Psi_{*i}-2}^{(1/7)\Psi_{*i}-2} e^{-t^2}\, dt = \frac{43.5\Phi_{*i}}{1 + 43.5\Phi_{*i}} \tag{7.25}$$

This equation is the Einstein bed-load function, which is also represented graphically in Fig. 7.5. The bed-load discharge $q_{bi}$ is contained in the parameter $\Phi_{*i}$ which is related to $\Psi_{*i}$. In applying this relationship, the parameter $\Psi_{*i}$ is computed from the given characteristics of bed sediment and flow conditions for each size fraction. The corresponding value of $\Phi_{*i}$ is then obtained from Fig. 7.5 or Eq. 7.25. Then the bed-load discharge per unit width is obtained from the value of $\Phi_{*i}$ according to Eq. 7.20. Total sediment discharge is obtained from Eqs. 7.18 and 7.19.

## Einstein–Brown Formula

This formula is a modification of the 1942 Einstein formula by Rouse, Boyer, and Laursen described in a chapter authored by Brown (1950). This formula employs

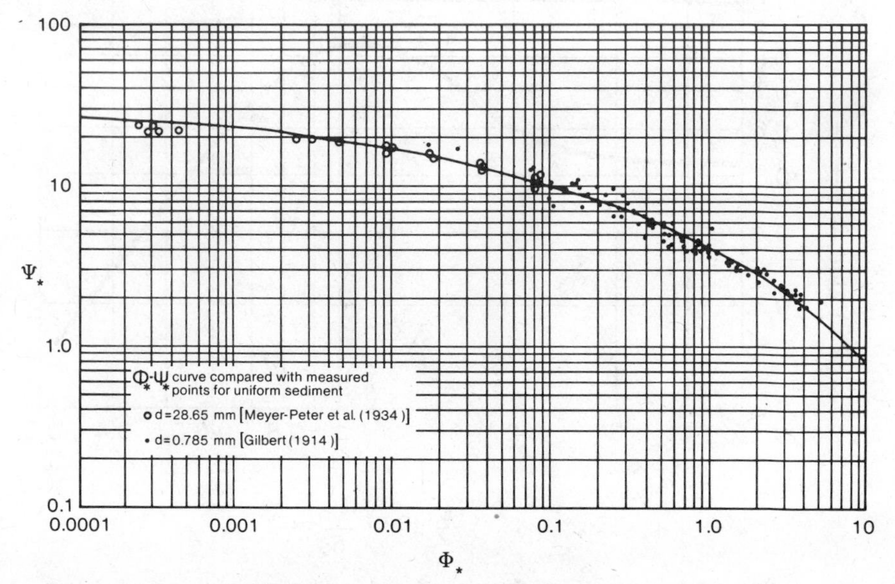


**Figure 7.5** Einstein's bed-load function.

the parameters $\Phi$ and $\Psi$; their relationship, shown in Fig. 7.6, is represented by the following equations:

$$\Phi = 40\left(\frac{1}{\Psi}\right)^3 = 40\tau_*^3 \quad \text{where } \Psi \leq 5.5 \quad (\tau_* \geq 0.182) \tag{7.26}$$

$$0.465\,\Phi = e^{-0.391\Psi} \quad \text{where } \Psi > 5.5 \tag{7.27}$$

in which

$$\Phi = \frac{q_b}{\gamma_s F[g(s-1)d^3]^{1/2}} \tag{7.28}$$

$$\Psi = \frac{(\gamma_s - \gamma)d}{\tau_0} = \frac{1}{\tau_*} \tag{7.29}$$

$$F = \left[\frac{2}{3} + \frac{36\nu^2}{gd^3(s-1)}\right]^{1/2} - \left[\frac{36\nu^2}{gd^2(s-1)}\right]^{1/2} \tag{7.30}$$

where $\tau_*$ is the Shields stress, $\nu$ is the kinematic viscosity, and $s(=\gamma_s/\gamma)$ is the specific gravity of sediment. This dimensionless relationship given by Eqs. 7.26 and 7.27 may be used for any consistent set of units, with $d$ usually taken as the median size $d_{50}$. Equation 7.26 is for more active bed-load transport, for which the dimensionless bed-load discharge $\Phi$ can be written as a function of the Shields stress. The parameter $F$ in the Einstein–Brown formula appears in the Rubey (1933) formula for fall velocity and was introduced to account for the effects of the fall velocity.

Different from many other bed-load formulas, the Einstein–Brown formula does not use the critical shear stress. This relationship was investigated by Gill (1968) using Gilbert's and Simons–Richardson's data. He observed a departure of the Einstein–Brown formula from the data in the zone of low shear stress when $\tau_0$

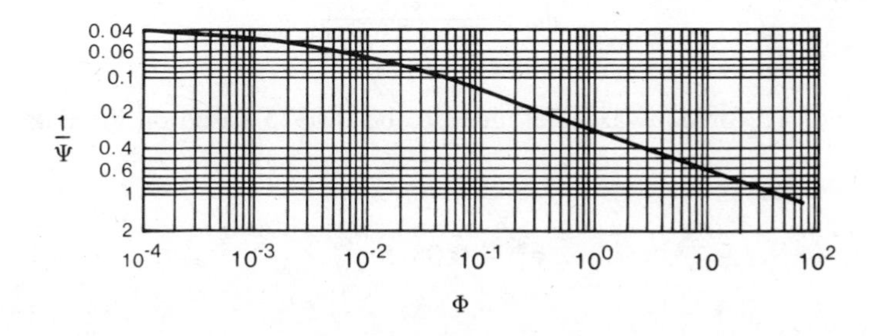


**Figure 7.6** Function $\Phi$ versus $1/\Psi$ for Einstein-Brown formula.

and $\tau_c$ are of the same order of magnitude. To correct the departure, he suggested the following modification:

$$\Phi = 40\left(\frac{\tau_0}{\tau_c} - 1\right)^3 \tag{7.31}$$

At higher values of shear stress, this equation reduces to the Einstein–Brown formula. At low shear stress, this equation bears resemblance to the DuBoys formula.

## Parker et al. Formula

The bed-load equation developed by Parker et al. (1982) is for streams of mostly gravel and coarser bed materials. Such streams usually possess a surface layer markedly coarser than the substrate. This layer, referred to as the *pavement*, is different from the immobile armor. In paved gravel-bed streams, bed motion is considered as a normal event in that the bed is active for infrequent periods of flood. The coarsest pavement grains are often mobile, whereas the armored bed is immobile.

Parker et al. (1982) used field data to study the size distribution of bed load in paved gravel-bed streams. The concept of similarity transformation was employed to combine different curves pertaining to different size fractions into a single universal curve. In this approach, individual bed-load relations were developed empirically for each of 10 grain-size ranges in Oak Creek (Oregon). By choosing proper parameters, these individual bed-load relations collapse, at least approximately, into a single curve, as shown in Fig. 7.7. The chosen parameters include the dimensionless bed load $W_i^*$ for a size fraction:

$$W_i^* = \frac{(s - 1)q_{bi}}{p_i(gDS)^{1/2}DS} \tag{7.32}$$

The parameter $\phi_i$ is the dimensionless shear stress in relation to the reference value $\tau_{ri}^*$:

$$\phi_i = \frac{DS}{(s - 1)\,d_i\tau_{ri}^*} \tag{7.33}$$

The value of $\tau_r^*$ for $d_{50}$ was determined to be 0.0875 based on the data and for other sizes:

$$\tau_{ri}^* = 0.0875\,\frac{d_{50}}{d_i} \tag{7.34}$$

From similarity analysis, they deduced that all grain size ranges are of approximately equal mobility when the critical condition for breaking the pave-

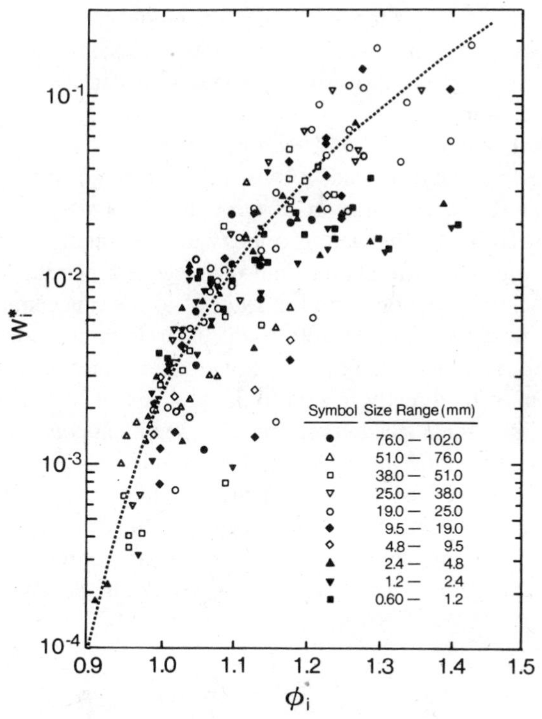


**Figure 7.7** Similarity plot of $W_i^*$ with indicated size ranges (Parker et al., 1982).

ment is exceeded. Because of the approximate equal mobility of all sizes, only one grain size—subpavement $d_{50}$—is used to characterize bed-load discharge as a function of the dimensionless shear:

$$W^* = 0.0025 \exp[14.2(\phi_{50} - 1) - 9.28(\phi_{50} - 1)^2] \quad \text{where } 0.95 < \phi_{50} < 1.65 \tag{7.35}$$

$$W^* = 11.2\left(1 - \frac{0.822}{\phi_{50}}\right)^{4.5} \quad \text{where } \phi_{50} > 1.65 \tag{7.36}$$

where $\phi_{50}$ is based on the subpavement $d_{50}$. The above equations were empirically fitted using field data from several streams, with median grain sizes ranging from 18 to 28 mm.

## 7.2 TURBULENT DIFFUSION AND DIFFUSION EQUATION

Diffusion of suspended sediment occurs in a stream flow through two essential mechanisms. The first is the transport of sediment by the velocity fluctuations in turbulent flow; the second involves the mixing of sediment grains with the surrounding fluid. It is easier to observe the diffusion of a small amount of dye after it is introduced into a clear stream flow. At the point of introduction, the volume of the dye is small, but this volume grows with diffusion while it moves downstream with the flow velocity. For the two-dimensional steady uniform flow under consideration, the main velocity is in the downstream direction and the time average of the vertical velocity is zero. Therefore, the increase in volume is not due to the time-averaged velocity; instead, it must be attributed to the more complicated diffusion process. Outward transport of the dye occurs when a small amount of dyed fluid is moved by the fluctuating velocity to a region of lower dye concentration and mixes with the surrounding fluid of lower concentration. During inward transport, turbulent fluctuations bring in fluid of lower concentration to mix with that of a higher concentration. Therefore, diffusion is in the direction of decreasing concentration, and the tendency is to equalize the concentration. The transport rate of the dye must be directly related to the gradient of its concentration.

In a two-dimensional steady uniform flow, sediment concentration varies and diffusion occurs only along the $z$ (vertical) direction. The sediment concentration does not change, and diffusion does not occur in the longitudinal and horizontal directions. The instantaneous turbulent velocity fluctuates; it can be written as the sum of the time-average velocity and a fluctuating velocity, as illustrated in Fig. 3.3. The same representation can be applied to the sediment concentration, that is,

$$\hat{C} = C + C' \tag{7.37}$$

where $\hat{C}$ is the instantaneous local concentration, $C$ is its time average, and $C'$ is the fluctuation from the mean value. The mean velocity is in the $x$ (downstream) direction, but there exists a vertical velocity fluctuation $w'$ even though the time average of the vertical velocity is zero in this two-dimensional flow. Through a horizontal surface area $dxdy$, the vertical fluctuations contribute to a vertical flow rate $w'\,dxdy$, which is also associated with an instantaneous sediment transport $w'\hat{C}\,dxdy$. The instantaneous transport per unit area is $w'\hat{C}$; the time average of this quantity is

$$q_1 = \overline{w'\hat{C}} \tag{7.38}$$

where the overbar denotes time average. Substituting Eq. 7.37 into Eq. 7.38 gives

$$q_1 = \overline{w'(C + C')} = \overline{w'C} + \overline{w'C'} = \overline{w'}\;\overline{C} + \overline{w'C'} \tag{7.39}$$

The term $\overline{w'}\overline{C}$ vanishes because the time average of $w'$ is zero. The quantity $\overline{w'C'}$ is zero in a flow with uniform concentration. Because of the gravitational force on sediment grains, the concentration usually decreases in the upward direction. With this gradient for sediment concentration, the upward fluctuation transports more sediment than does the downward fluctuation. Although the individual time averages $\overline{w'}$ and $\overline{C'}$ are zero, the time average of their product is not zero because there is a preponderance of positive (upward) $w'$ associated with positive $C'$ and of negative $w'$ associated with negative $C'$. In a steady flow, the upward diffusion effect on sediment is balanced by the downward gravitational force.

The transport rate per unit area due to diffusion is assumed to be proportional to the gradient of concentration, that is,

$$q_1 \propto -\frac{dC}{dz}$$

The negative sign denotes that the transport is in the direction of decreasing concentration. This relationship is analogous to thermometric conductivity in heat flow. Written as an equation, it becomes

$$q_1 = \varepsilon_s \frac{dC}{dz} \tag{7.40}$$

where $\varepsilon_s$ is the diffusion coefficient.

The diffusion equation for unsteady nonuniform distribution of sediment in a two-dimensional flow is derived from the continuity of sediment transport into and out of an elemental volume. Figure 7.8 shows the elemental volume $\Delta x \Delta z$ for which the dimension normal to the $xz$ plane is taken to be unity. For a time increment $\Delta t$, sediment flows contributed by water flow and diffusion are shown in the figure for the respective $x$ and $z$ directions. The local velocity components in the $x$ and $z$ directions are $u$ and $w$, and the diffusion coefficients are $\varepsilon_x$ and $\varepsilon_z$, respectively.

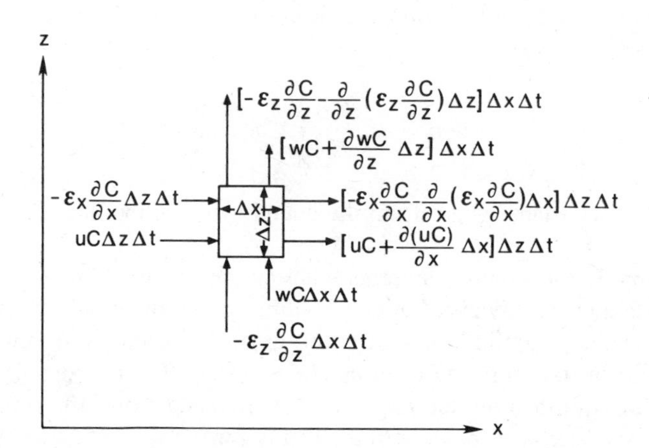


**Figure 7.8** Elements of sediment inflow and outflow for an elemental volume in two-dimensional flow.

From the continuity for sediment, the inflow sediment minus the outflow sediment is equal to the change of sediment within the volume, that is,

$$\left[-\frac{\partial uC}{\partial x}+\frac{\partial}{\partial x}\left(\varepsilon_x\frac{\partial C}{\partial x}\right)-\frac{\partial wC}{\partial z}+\frac{\partial}{\partial z}\left(\varepsilon_z\frac{\partial C}{\partial z}\right)\right]\Delta z\,\Delta x\,\Delta t=\frac{\partial C}{\partial t}\Delta x\,\Delta z\,\Delta t \tag{7.41}$$

After simplification and introduction of the flow continuity $\partial u/\partial x + \partial w/\partial z = 0$, Eq. 7.41 becomes

$$\frac{\partial C}{\partial t}+u\frac{\partial C}{\partial x}+w\frac{\partial C}{\partial z}=\frac{\partial}{\partial x}\left(\varepsilon_x\frac{\partial C}{\partial x}\right)+\frac{\partial}{\partial z}\left(\varepsilon_z\frac{\partial C}{\partial z}\right) \tag{7.42}$$

This is the two-dimensional diffusion equation for suspended sediment.

## 7.3 SUSPENDED-SEDIMENT DISCHARGE

For the two-dimensional steady uniform flow under consideration, the sediment concentration varies and diffusion occurs only along the $z$ (vertical) direction. The sediment concentration does not change, and diffusion does not occur in the longitudinal and transverse directions. Under this assumption, the diffusion equation (Eq. 7.42) can be reduced to the form

$$w\frac{\partial C}{\partial z}=\frac{\partial}{\partial z}\left(\varepsilon_z\frac{\partial C}{\partial z}\right) \tag{7.43}$$

Now, the vertical velocity component $w$ is replaced by the downward fall velocity for sediment $-w_s$ and the diffusion coefficient $\varepsilon_z$ is denoted by $\varepsilon_s$. After integration, Eq. 7.43 becomes

$$w_sC+\varepsilon_s\frac{dC}{dz}=\text{Constant} \tag{7.44}$$

The integration constant is zero from the boundary condition that $dC/dz = 0$ for $C = 0$.

Equation 7.44 contains two terms representing two opposing tendencies that jointly maintain the steady distribution of sediment concentration. The first term is the rate of sediment settling through a unit area; the second term represents the upward sediment transport due to turbulent diffusion. The two opposing tendencies reach an equilibrium condition under the steady flow, for which the net exchange across any layer parallel to the bed is zero.

The vertical variation of sediment concentration can be obtained from Eq. 7.44. For this purpose, the diffusion coefficient for mass transfer needs to be expressed in terms of appropriate variables. Von Karman (1934) pointed out that the diffusion coefficient for mass transfer was related to that for momentum transfer in turbulent flow according to the Reynolds analogy. First, for the momentum transfer.

$$\text{Momentum flux} = \rho(\nu + \varepsilon)\frac{du}{dz} \simeq \rho\varepsilon\frac{du}{dz} = \tau \tag{7.45}$$

where the molecular viscosity $\nu$ is small in relation to the diffusion coefficient $\varepsilon$, or eddy viscosity, and hence ignored. For the mass transfer, the molecular diffusion coefficient is also negligible as compared with the turbulent diffusion coefficient $\varepsilon_s$, thus

$$\text{Mass flux} \simeq \rho\varepsilon_s\frac{dC}{dz} \tag{7.46}$$

The Reynolds analogy is valid if the mechanisms controlling both transfers—momentum and mass transfer—are identical; then the two coefficients are equal, that is,

$$\varepsilon_s = \varepsilon \tag{7.47}$$

The validity of Eq. 7.47 has been tested by several investigators. Brush et al. (1962), Majumdar and Carstens (1967), and Jobson and Sayre (1970) have provided experimental evidence to show that Eq. 7.47 is valid for fine particles, but for coarse particles $\varepsilon_s < \varepsilon$. This inequality may be considered by the equation

$$\varepsilon_s = \beta\varepsilon \tag{7.48}$$

A method for determining $\beta$ is described later in this section.

For a two-dimensional steady uniform flow, the shear stress $\tau$ at distance $z$ above the bed (see Eq. 3.2) is given by

$$\tau = \gamma(D - z)S = \tau_0\left(1 - \frac{z}{D}\right) \tag{7.49}$$

and

$$\tau = \rho\varepsilon\frac{du}{dz}, \quad \text{or } \varepsilon = \frac{\tau/\rho}{du/dz} \tag{7.50}$$

If the logarithmic vertical velocity profile given by Eq. 3.23 is used, the velocity gradient $du/dz$ can be written as

$$\frac{du}{dz} = \frac{d}{dz}\left(\frac{U_*}{\kappa}\ln\frac{z}{k_s}\right) = \frac{U_*}{\kappa}\frac{1}{z} \tag{7.51}$$

where $\kappa$ is von Karman's constant, which has a mean value of 0.4 for clear fluids. From Eqs. 7.49–7.51, it follows that

$$\varepsilon = \kappa U_* \frac{z}{D}(D - z) \tag{7.52}$$

Substituting Eq. 7.52 into Eq. 7.44 and separating variables $C$ and $z$ yields

$$\frac{dC}{C} + \frac{w_s}{\kappa U_*} \frac{D\,dz}{z(D - z)} = 0 \tag{7.53}$$

Let

$$z_* = \frac{w_s}{\kappa U_*} \tag{7.54}$$

and then integrating Eq. 7.53 yields

$$[\ln C]_a^z = \left[\ln\left(\frac{D - z}{z}\right)^{z_*}\right]_a^z \tag{7.55}$$

and

$$\frac{C}{C_a} = \left(\frac{D - z}{z} \frac{a}{D - a}\right)^{z_*} \tag{7.56}$$

The quantity $C_a$ denotes the concentration of sediment with fall velocity $w_s$ at the level $z = a$. Equation 7.56 was developed by Rouse (1937) and is often referred to as the *Rouse equation*. This equation gives the concentration with reference to $C_a$ at any level $z$ from the bed. Graphical results of the Rouse equation for different $z_*$ values are depicted in Fig. 7.9. At any $z_*$ value, the concentration is higher near the bed and it decreases in the direction toward the water surface. The vertical distribution of concentration is more uniform for small values of $z_*$, that is, for finer sediments or more turbulent flow or both. Coarse grains with a larger value of $z_*$ are more concentrated near the bed.

Einstein and Chien (1954) quantified $\beta$ in Eq. 7.48 by the relation

$$z_* = z'_* \beta = \frac{w_s}{\kappa U_*} \tag{7.57}$$

The value of $\beta$ was obtained by comparing the measured suspended-load distribution with Eq. 7.56, which is also shown graphically in Fig. 7.9. The relationship between $z_*$ and $z_*'$ so obtained is shown in Fig. 7.10; it is used at present until further knowledge becomes available.

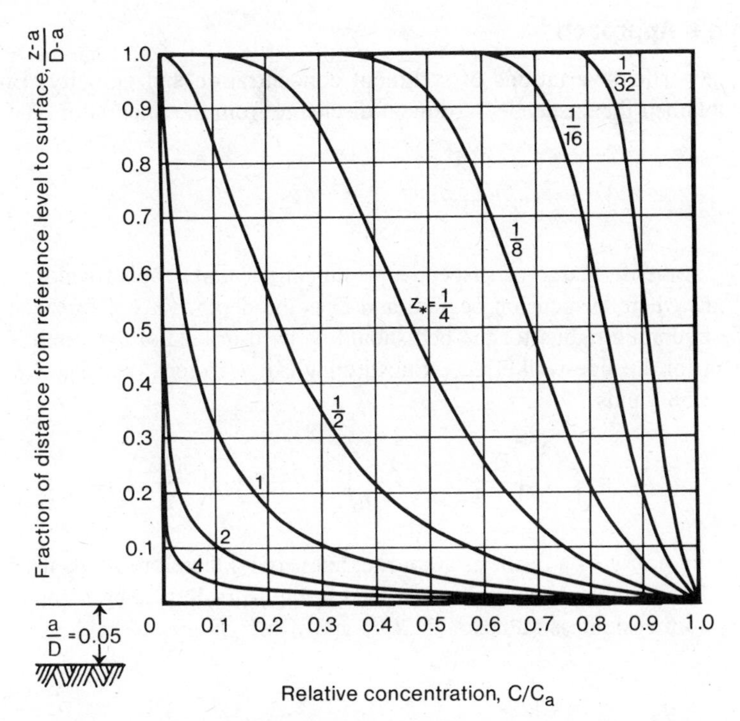


**Figure 7.9** Graph of suspended sediment distribution equation for a/D = 0.05 and several values of exponent $z_*$ (after Vanoni, 1941).

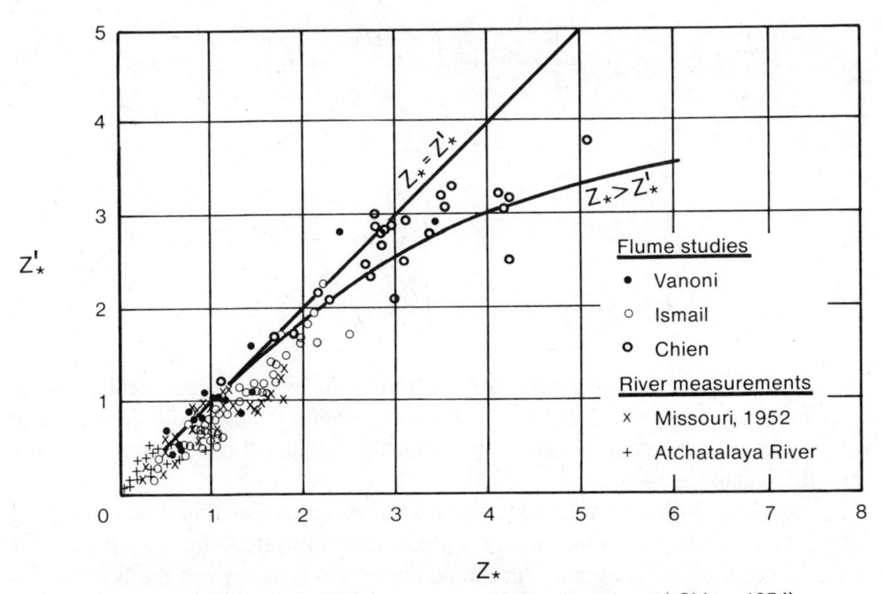


**Figure 7.10** Relationship between $z_*$ and $z_*'$ (after Einstein and Chien, 1954).

## Einstein's Approach

From the vertical variations of sediment concentration and velocity, Einstein (1950) obtained the suspended-sediment discharge from the integration

$$q_{ss} = \int_a^D Cu\,dz \tag{7.58}$$

where $q_{ss}$ is the discharge of suspended sediment per unit channel width, $a$ is the lower limit where suspension begins, and $D$ is the depth. Except for wide channels, the hydraulic radius for the bed should be used instead of the depth in order to account for the side-wall effects. Substituting Eq. 7.56 for $C$ and Eq. 3.35 for $u$ into Eq. 7.58 yields

$$q_{ss} = \int_a^D C_a\left(\frac{D-z}{z}\frac{a}{D-a}\right)^{z_*} 5.75U_*' \log\left(\frac{30.2z}{\Delta}\right) dz \tag{7.59}$$

The velocity profile is a result of grain roughness only; therefore, $U_*'$ is used in the formula. The variable $\Delta$ is $d_{65}/X$ as defined in Fig. 3.6. Replacing $a$ and $z$ by the respective dimensionless values $A = a/D$ and $\eta = z/D$ gives

$$\begin{aligned} q_{ss} &= \int_A^1 DCu\,d\eta \\ &= DU_*'C_a\left(\frac{A}{1-A}\right)^{z_*} 5.75\int_A^1\left(\frac{1-\eta}{\eta}\right)^{z_*}\log\frac{30.2\eta}{\Delta/D}\,d\eta \\ &= 11.6C_aU_*'a\left[2.303\log\left(\frac{30.2D}{\Delta}\right)I_1 + I_2\right] \end{aligned} \tag{7.60}$$

where

$$I_1 = 0.216\frac{A^{z_*-1}}{(1-A)^{z_*}}\int_A^1\left(\frac{1-\eta}{\eta}\right)^{z_*} d\eta \tag{7.61}$$

$$I_2 = 0.216\frac{A^{z_*-1}}{(1-A)^{z_*}}\int_A^1\left(\frac{1-\eta}{\eta}\right)^{z_*} \ln\eta\,d\eta \tag{7.62}$$

The values of $I_1$ and $I_2$ as functions of $A$ and $z_*$ are given in graphical forms in Figs. 7.11 and 7.12, which are used if the computation is made without the use of a computer. These integrals can be conveniently evaluated using numerical integration (Nakato, 1984).

Equation 7.60 is employed to obtain the suspended-sediment discharge for different size fractions. For each size fraction, Einstein selected the value $a = 2d_i$ as the lower limit of integration in the equation. He also defined the bed load as the sediment moving in the bed layer, which has a thickness of $2d_i$. The material

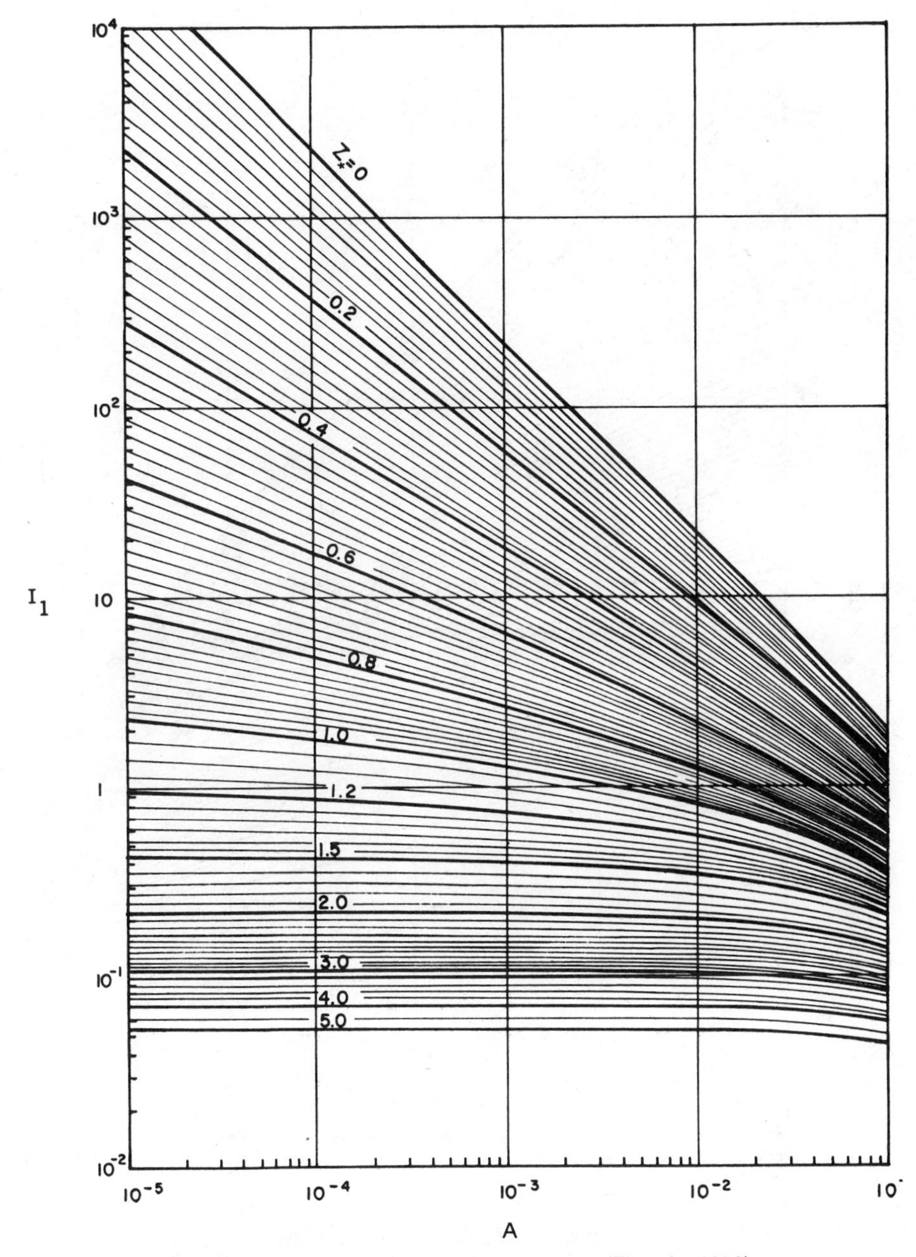


**Figure 7.11** Function $I_1$ in terms of A and $z_*$ (Einstein, 1950).

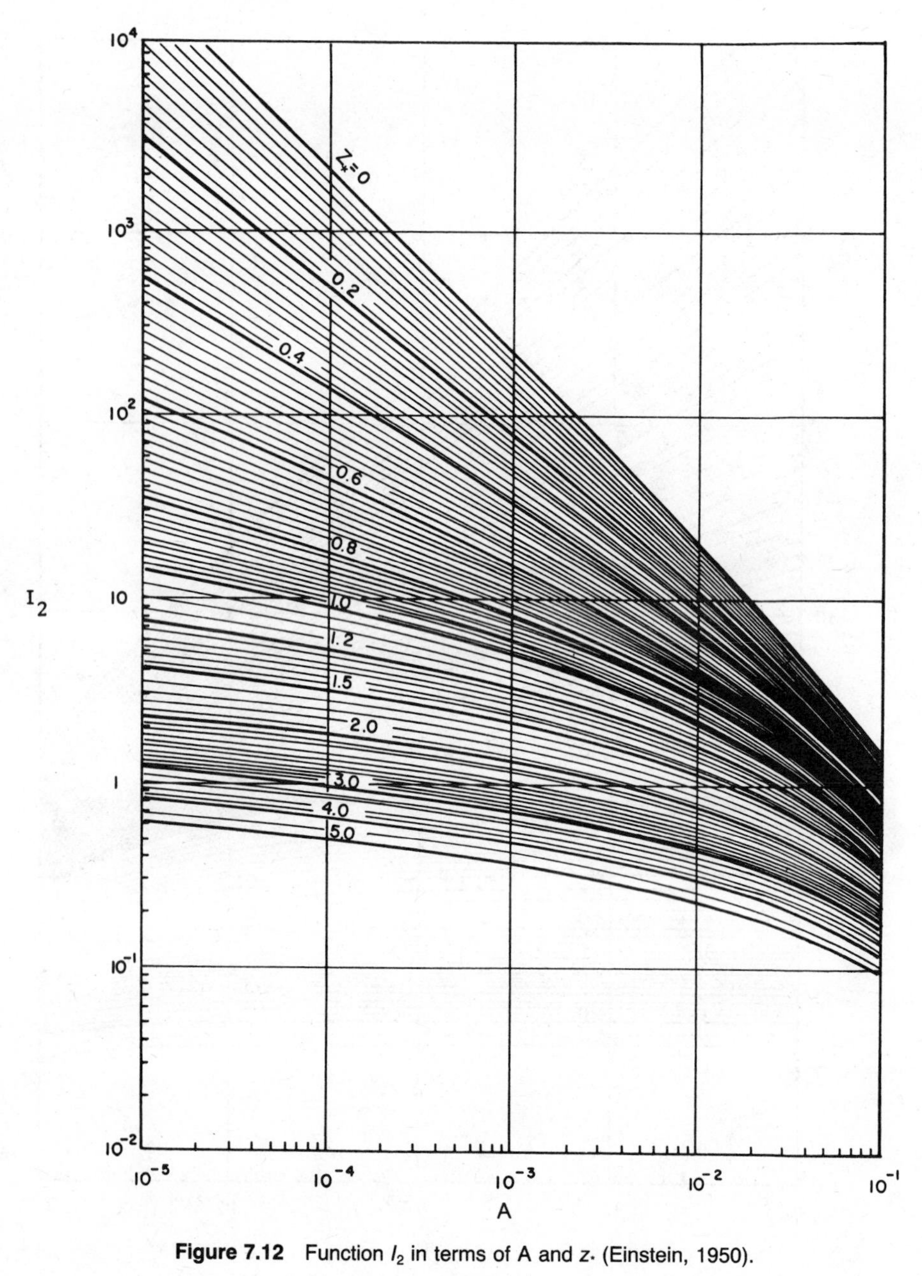


**Figure 7.12** Function $I_2$ in terms of A and $z_*$ (Einstein, 1950).

within the bed layer becomes the source of the suspended load. The concentration for a certain size fraction $C_{ai}$ was assumed to be related to the bed-load discharge per unit width $q_b$

$$C_{ai} = \frac{p_i q_b}{a u_b} \tag{7.63}$$

where $u_b$ is the unknown bed-layer velocity. Led by experimental results, Einstein further assumed that $u_b = 11.6U'_*$, which is recognized as the velocity at the edge of the laminar sublayer. Substituting Eq. 7.63 into Eq. 7.60 yields

$$q_{ssi} = p_i q_b \left[ 2.303 \log\left(\frac{30.2D}{\Delta}\right) I_1 + I_2 \right] \tag{7.64}$$

This equation is employed to compute the suspended-sediment discharge of a size fraction. Being dimensionally homogeneous, it may be used in any consistent set of units.

The bed-material discharge for a size fraction $q_{si}$ is the summation of bed-load discharge and suspended-load discharge, that is,

$$\begin{aligned} q_{si} &= q_{bi} + q_{ssi} \\ &= q_{bi}(1 + P_E I_1 + I_2) \end{aligned} \tag{7.65}$$

where $P_E = 2.303 \log(30.2D/\Delta)$ is the transport parameter.

▶ ***EXAMPLE 7.1.*** A wide alluvial channel has a depth of 10 ft and a slope of 0.0001. The bed material is primarily fine gravel for which $d_{65}$ is 0.01 ft. Assuming bed forms are missing and water temperature is 60°F, find: (a) an equation for the velocity profile along the depth and (b) the suspended load per unit channel width of a very fine sediment with its concentration given by

$$C = 1000 - 10z^2$$

where $C$ is in parts per million by weight and $z$ is in feet.

*Solution:* (a) The vertical variation of horizontal velocity may be obtained from Eq. 3.35, in which the equivalent roughness $k_s$ is given by $d_{65}/X$. In the absence of bed forms, the shear velocity resulting from grain roughness is given by $U'_* = (gDS)^{1/2} = 0.179$ ft/sec. The thickness of the laminar sublayer is given by $\delta = 11.6\nu/U'_* = 11.6 \times 1.21 \times 10^{-5}/0.179 = 0.00078$ ft. Then, we obtain the ratio $k_s/\delta = d_{65}/\delta = 0.01/0.00078 = 12.8$. Based on the value of this ratio, the value of $X$ is obtained as 1 from Fig. 3.6 and thus $k_s = d_{65} = 0.01$ ft. Substituting the obtained values of $U'_*$ and $k_s$ into Eq. 3.35 yields the velocity profile

$$u = 1.03 \log 3020z$$

(b) Based on the profiles of velocity and sediment concentration, the suspended load may now be obtained by integration illustrated by Eq. 7.58, that is,

$$\begin{aligned} q_{ss} &= 10^{-6}\gamma \int_0^{10} Cu\,dz = 10^{-6}\gamma \int_0^{10} (1000 - 10z^2)0.447 \ln 3020z\,dz \\ &= 4.48 \times 10^{-7}\gamma \left\{ \frac{1000}{3020} [3020z(\ln 3020z - 3020z)]_0^{10} \right. \\ &\quad \left. - \frac{10}{3}[z^3(\ln 3020z - \tfrac{1}{3})]_0^{10} \right\} \\ &= 1.67 \text{ lb/sec/ft} \end{aligned}$$

## 7.4 BED-MATERIAL LOAD FORMULAS

Bed-material load is the summation of the bed load and the suspended load excluding the wash load, that is,

$$q_s = q_b + q_{ss} \tag{7.66}$$

or

$$Q_s = Q_b + Q_{ss} \tag{7.67}$$

where $q_s$ and $Q_s$ are the bed-material discharges per unit channel width and for the cross section, respectively; $q_b$ and $Q_b$ are the corresponding quantities for bed load; and $q_{ss}$ and $Q_{ss}$ are for suspended load. Bed-material load is also referred to as the *total load*. Several total load formulas are described in this section.

### The Colby Relations

The bed-material discharge per unit channel width is given by Colby (1964) as a function of mean flow velocity, depth, mean sediment size, water temperature, and concentration of fine sediment (silt and clay). These variables encompass effects on sediment transport due to stream power and shear stress. These variables are included in four graphical relations shown in Figs. 7.13–7.15. The range of application is given by the limits of each variable in these figures. The first figure gives the uncorrected bed-material discharge $q_{s1}$ in terms of mean velocity for six median sizes of bed material, four depths of flow, and water temperature of

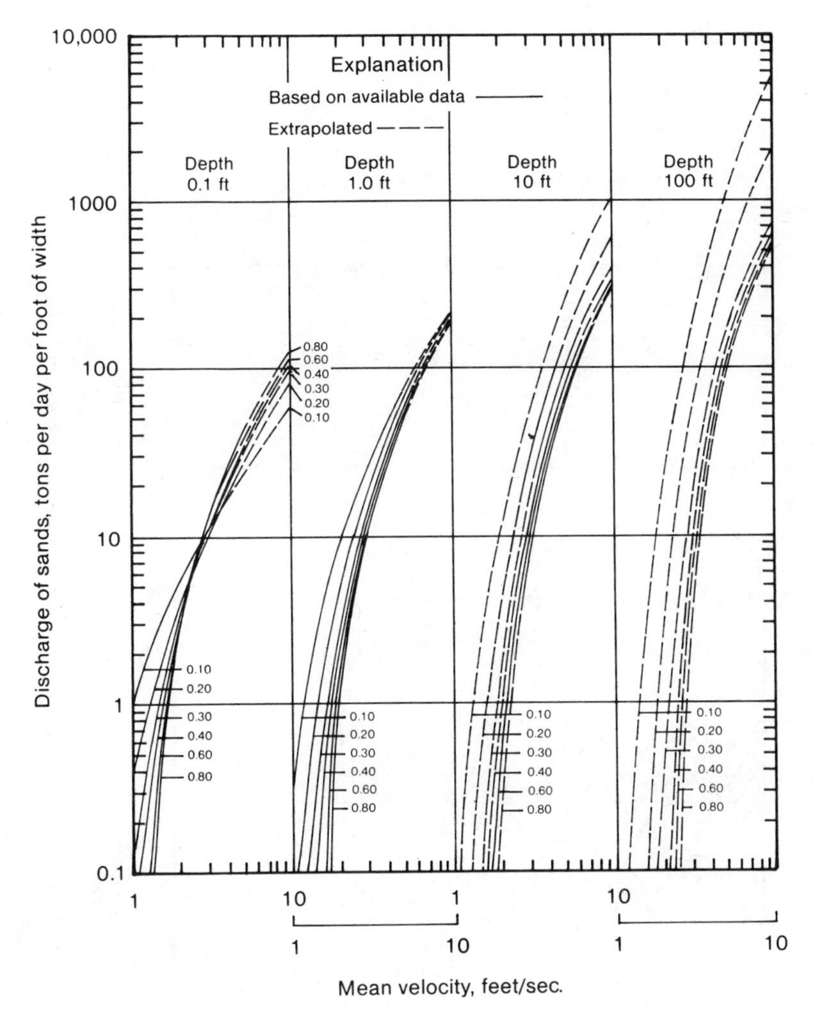


**Figure 7.13** Relationship of discharge of sands to mean velocity for six median sizes of bed sands, four depths of flow, and a water temperature of 60° F (Colby, 1964).

60°F. In using this graphical relationship, if the actual flow depth is different from the four depths, then the values of $q_{s1}$ should be read for the actual velocity at two depths that bracket the actual depth. Then interpolation is done on a logarithmic graph of $q_{s1}$ versus $D$ to obtain the $q_{s1}$ value for the actual $D$ and $U$.

The uncorrected sediment discharge so obtained is correct only for median sediment size between 0.2 and 0.3 mm, temperature of 60°F, and negligible fine sediment concentration. If the actual conditions are different, then the correction factors $k_1$ for water temperature and $k_2$ for the effect of concentration of fine sediment are obtained from Fig. 7.14. The correction factor for the median size of bed

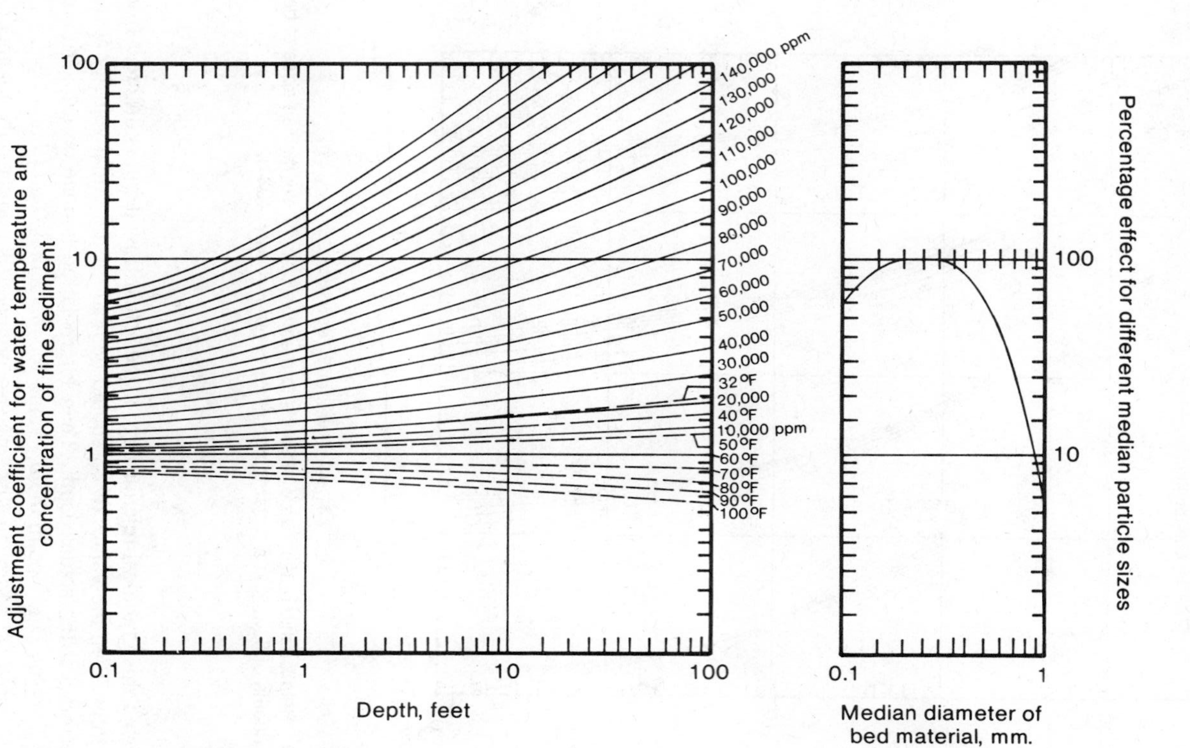


**Figure 7.14** Approximate effect of water temperature and concentration of fine sediment on the relationship of discharge of sands to mean velocity (Colby, 1964).

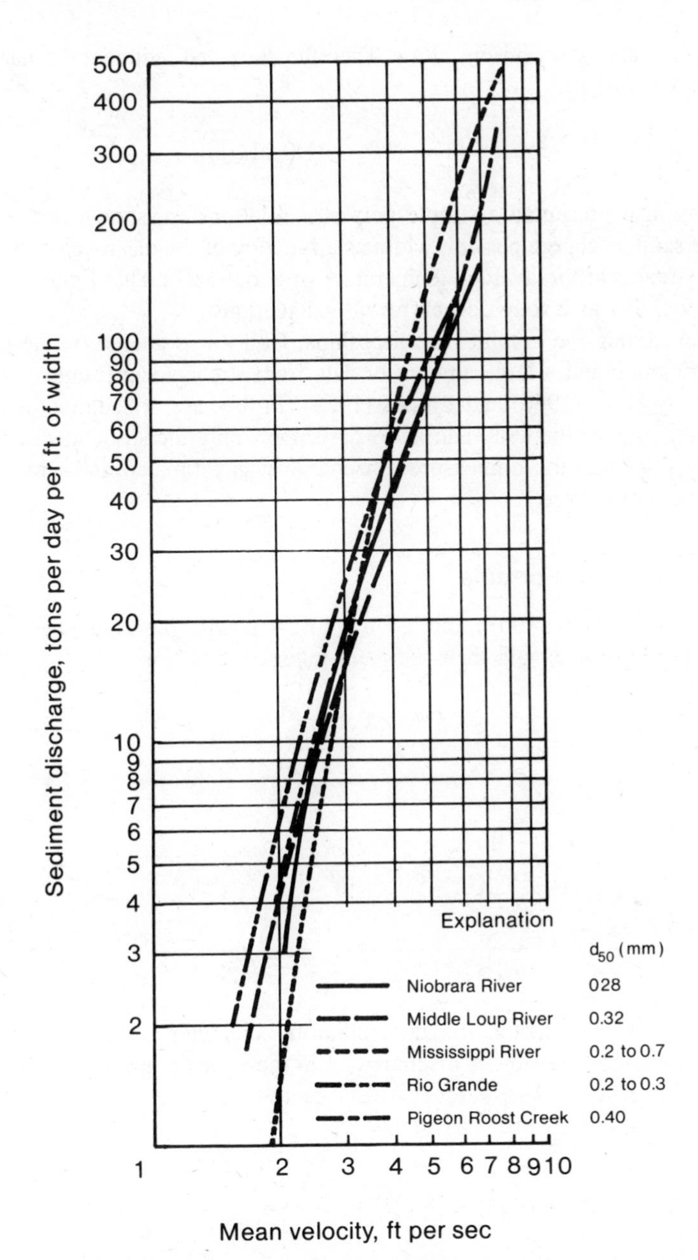


**Figure 7.15** Relationship between observed discharge of sands and mean velocity for five sand-bed streams at an average temperature of about 60° F (Colby, 1964).

sediment $k_3$ is also given in Fig. 7.14. Then the corrected sediment discharge $q_s$ is computed according to

$$q_s = [1 + (k_1 k_2 - 1)0.01 k_3] q_{s1} \tag{7.68}$$

Colby also presented a simple graphical relationship, shown in Fig. 7.15, giving the sand discharge per unit width as a function of the mean velocity for five sand-bed streams at the average temperature of about 60°F. This figure is recommended by Colby as a rough check for all calculations.

In developing the graphical relationships, Colby was guided by the Einstein bed-load function and a large amount of data from streams and flumes, including those by Guy et al. (1966) and Brooks (1958). In the case of sediment data from large rivers, such as the Colorado and Mississippi, only the suspended-load samplers were used and the unmeasured discharge was estimated and added to the total sediment discharge.

## Engelund–Hansen Formula

Engelund and Hansen (1967) applied Bagnold's stream power concept and the similarity principle to obtain their sediment transport equation:

$$f'\phi = 0.1(\tau_*)^{5/2} \tag{7.69}$$

with

$$f' = \frac{2gRS}{U^2} \tag{7.70}$$

$$\phi = \frac{q_s}{\gamma_s[(s-1)gd^3]^{1/2}}, \quad \tau_* = \frac{\tau_0}{(\gamma_s - \gamma)d} \tag{7.71}$$

where $f'$ is the friction factor, $d$ is the median fall diameter of the bed material, $\phi$ is the dimensionless sediment discharge, $s$ is the specific gravity of sediment, and $\tau_*$ is the dimensionless shear stress or the Shields stress. Substituting Eqs. 7.70 and 7.71 into Eq. 7.69 yields

$$C_s = 0.05\left(\frac{s}{s-1}\right)\frac{US}{[(s-1)gd]^{1/2}}\,\frac{RS}{(s-1)d} \tag{7.72}$$

where $C_s$ $(=Q_s/Q)$ is the sediment concentration by weight. This equation relates sediment concentration to the $U$–$S$ product (which is the rate of energy expenditure per unit weight of water) and the $R$–$S$ product (which is the shear stress). Strictly speaking, the Engelund–Hansen formula should be applied to streams with a dune bed in accordance with the similarity principle. However, it can be applied to upper flow regime with particle size greater than 0.15 mm without serious error.

## Ackers–White Formula

Based on Bagnold's stream power concept, Ackers and White (1973) related the concentration of bed-material load as a function of the mobility number $F_g$:

$$C_s = cs\frac{d}{R}\left(\frac{U}{U_*}\right)^n\left(\frac{F_g}{A} - 1\right)^m \tag{7.73}$$

where $n$, $c$, $A$, and $m$ are coefficients. The mobility number $F_g$ is given by

$$F_g = \frac{U_*^n}{[gd(s-1)]^{1/2}}\left[\frac{U}{(32)^{1/2}\log(10R/d)}\right]^{1-n} \tag{7.74}$$

They also expressed the sediment size by a dimensionless grain diameter $d_g$:

$$d_g = d\left[\frac{g(s-1)}{\nu^2}\right]^{1/3} \tag{7.75}$$

where $\nu$ is the kinematic viscosity of water.

In deriving the mobility factor for sediment transport, they distinguished bed load and suspended load. The transport of coarse sediments in the form of bed load is attributed to the stream power that generates the grain shear stress, $\tau_0'U$, which is reflected in the second part of $F_g$ in Eq. 7.74. For fine sediments, which travel mainly in suspension, the turbulent intensity that sustains the suspension is assumed to be a function of the total bed shear; thus the stream power is $\tau_0 U$. The first part of $F_g$ reflects the power expenditure associated with turbulent intensity of the flow. The coefficient $n$ is the transition exponent, which depends on the sediment size; it is used when both modes of transport are present, and it is zero for coarse sediments with bed load only. The coefficient $A$ may be interpreted as the critical value for $F_g$.

The coefficients are determined from best-fit curves of almost 1000 sets of laboratory data with sediment size greater than 0.04 mm and Froude number less than 0.8. Values of these coefficients are listed as follows:

| Coefficient | $d_g > 60$ | $60 \geq d_g > 1$ |
|---|---|---|
| $c$ | 0.025 | $\log c = 2.86 \log d_g - (\log d_g)^2 - 3.53$ |
| $n$ | 0.0 | $1 - 0.56 \log d_g$ |
| $A$ | 0.17 | $0.23/(d_g)^{1/2} + 0.14$ |
| $m$ | 1.50 | $9.66/d_g + 1.34$ |

## Yang's Unit Stream Power Equation

Yang (1972) related the bed-material load to the rate of energy dissipation of the flow as an agent for sediment transport. For steady uniform flow, there is no change in kinetic energy, and the rate of energy dissipation is due to the change in potential energy. The rate of dissipation for potential energy per unit weight of water over a reach length $x$ with a total drop of $z$ is

$$\frac{dz}{dt} = \frac{dx}{dt}\frac{dz}{dx} = US \tag{7.76}$$

The velocity–slope product is referred to as the unit stream power. Since sediment transport mainly occurs under turbulent flow condition, total sediment concentration must be directly related to the unit stream power. Yang and Molinas (1982) made a derivation to show that this direct relationship can be obtained from basic turbulent flow theories. The basic form of Yang's unit stream power equation is

$$\log C_s = M + N \log \frac{US}{w_s} \tag{7.77}$$

where $M$ and $N$ are dimensionless parameters related to flow and sediment characteristics, and $w_s$ is the fall velocity of sediment. The coefficients for $M$ and $N$ were determined based on multiple regression analysis of laboratory data. Thus Yang's 1973 equation for sand transport is

$$\log C_s = 5.435 - 0.286 \log \frac{w_s d}{\nu} - 0.457 \log \frac{U_*}{w_s} + \left(1.799 - 0.409 \log \frac{w_s d}{\nu} - 0.314 \log \frac{U_*}{w_s}\right) \log\left(\frac{US}{w_s} - \frac{U_c S}{w_s}\right) \tag{7.78}$$

where $C_s$ is in parts per million by weight. Yang (1984) later developed the following equation for gravel transport:

$$\log C_s = 6.681 - 0.633 \log \frac{w_s d}{\nu} - 4.816 \log \frac{U_*}{w_s} + \left(2.784 - 0.305 \log \frac{w_s d}{\nu} - 0.282 \log \frac{U_*}{w_s}\right) \log\left(\frac{US}{w_s} - \frac{U_c S}{w_s}\right) \tag{7.79}$$

The average flow velocity for incipient sediment motion $U_c$ in Eqs. 7.78 and 7.79 is normalized by the fall velocity $w_s$. The ratio of these two opposing velocities is related, by Yang (1973), to the shear Reynolds number. For the smooth and transition regions where the shear Reynolds number is between 1.2 and 70, the ratio is represented by the equation

$$\frac{U_c}{w_s} = \frac{2.5}{\log (U_* d/\nu) - 0.06} + 0.66, \qquad 1.2 < \frac{U_* d}{\nu} < 70 \tag{7.80}$$

In the complete rough region where the shear Reynolds number is greater than 70, the ratio is a constant, independent of the shear Reynolds number:

$$\frac{U_c}{w_s} = 2.05, \qquad 70 \le \frac{U_* d}{\nu} \tag{7.81}$$

These criteria for incipient motion are supported by experimental studies performed by Talapatra and Ghosh (1983).

The sand and gravel equations (Eqs. 7.78 and 7.79) are identical in form but have different numerical values for the coefficients, which reflect the variation in particle size between sand and gravel. The equation for sand is used if the sediment diameter is less than 2 mm; otherwise, the gravel equation should be used. However, application of the gravel equation should be limited to the maximum size of about 10 mm. If size fractions are used for graded sediment, the total bed-material concentration can be computed by

$$C_s = \sum_{i=1}^{n} p_i C_{si} \tag{7.82}$$

► ***EXAMPLE 7.2.*** Compute the bed-material load in an alluvial canal using the respective methods by Engelund and Hansen, Ackers and White, and Yang. The canal carries a discharge of 105 $m^3$/sec with the water temperature of 15°C. The channel has a slope of 0.00027, an alluvial bed width of 46 m, a flow depth of 2.32 m, and a side slope of 2 to 1. The bed material (specific gravity = 2.65) has the following composition:

| Fraction Diameter (mm) | Geometric Mean (mm) | Fraction by Weight ($p$) |
|---|---|---|
| 0.062–0.125 | 0.088 | 0.04 |
| 0.125–0.25 | 0.177 | 0.23 |
| 0.25 –0.50 | 0.354 | 0.37 |
| 0.50 –1.0 | 0.707 | 0.27 |
| 1.0 –2.0 | 1.414 | 0.09 |

*Solution*. The kinematic viscosity $\nu$ for the given temperature is $1.139 \times 10^{-6}$ $m^2$/sec. For the given width and depth, the cross-sectional area of flow is computed to be 117.5 $m^2$, the hydraulic radius $R$ = 2.08 m, the mean velocity $U$ = 0.893 m/sec, and the shear velocity $U_* = (gRS)^{1/2}$ = 0.0742 m/sec.

*Engelund–Hansen Method*. Substituting the necessary values into Eq. 7.72 and multiplying it by $10^6$ yields

$$C_s = \frac{0.00164}{d^{3/2}} \text{ ppm} \tag{7.83}$$

The variable values are tabulated and explained as follows:

| $d$ ($10^{-3}$ m) (1) | $p$ (2) | $C_s$ (ppm) (3) | $pC_s$ (ppm) (4) |
|---|---|---|---|
| 0.088 | 0.04 | 1985 | 79 |
| 0.177 | 0.23 | 696 | 160 |
| 0.354 | 0.37 | 246 | 91 |
| 0.707 | 0.27 | 87 | 23 |
| 1.414 | 0.09 | 31 | 3 |
| | | Total | 356 |

Col. 1. Geometric mean diameter.
Col. 2. Fraction of material represented by the geometric mean diameter.
Col. 3. Sediment concentration computed from Eq. 7.83.
Col. 4. Product of $p$ and $C_s$.

*Ackers–White Method*. Values of variables for this method are tabulated and explained as follows:

| $d$ ($10^{-3}$ m) (1) | $d_g$ (2) | $c$ (3) | $n$ (4) | $A$ (5) | $m$ (6) | $F_g$ (7) | $C_s$ (ppm) (8) | $p$ (9) | $pC_s$ (ppm) (10) |
|---|---|---|---|---|---|---|---|---|---|
| 0.088 | 2.02 | 0.00178 | 0.829 | 0.302 | 6.12 | 1.678 | 16800 | 0.04 | 672 |
| 0.177 | 4.07 | 0.00695 | 0.659 | 0.254 | 3.71 | 1.032 | 511 | 0.23 | 118 |
| 0.354 | 8.14 | 0.0136 | 0.490 | 0.221 | 2.53 | 0.650 | 144 | 0.37 | 53 |
| 0.707 | 16.3 | 0.0293 | 0.322 | 0.197 | 1.93 | 0.420 | 74 | 0.27 | 20 |
| 1.414 | 32.5 | 0.0322 | 0.153 | 0.180 | 1.64 | 0.278 | 31 | 0.09 | 3 |
| | | | | | | Total | | | 866 |

Col. 1. Geometric mean diameter.
Col. 2. Dimensionless grain diameter defined in Eq. 7.75.
Cols. 3–6. Coefficients in the Ackers–White formula.
Col. 7. Mobility number given by Eq. 7.74.
Col. 8. Sediment concentration obtained from the Ackers–White formula (Eq. 7.73) and multiplied by $10^6$.
Col. 9. Fraction of material represented by the geometric mean diameter.
Col. 10. Product of $p$ in Col. 8 and $C_s$ in Col. 9.

*YANG'S METHOD*. Values of variables and their explanations for this method are given as follows:

| $d$ ($10^{-3}$ m) (1) | $w_s$ (m/sec) (2) | $U_*d/\nu$ (3) | $U_c/w_s$ (4) | $C_s$ (ppm) (5) | $p$ (6) | $pC_s$ (ppm) (7) |
|---|---|---|---|---|---|---|
| 0.088 | 0.0068 | 0.522 | 4.24 | 525 | 0.04 | 21 |
| 0.177 | 0.0196 | 3.04 | 3.15 | 190 | 0.23 | 44 |
| 0.354 | 0.0457 | 14.8 | 2.58 | 110 | 0.37 | 41 |
| 0.707 | 0.0954 | 59.3 | 2.22 | 93 | 0.27 | 25 |
| 1.414 | 0.165 | 204 | 2.05 | 101 | 0.09 | 9 |

Col. 1. Geometric mean diameter.
Col. 2. Fall velocity obtained from Fig. 4.5 for the assumed shape factor of 0.7.
Col. 3. Shear Reynolds number.
Col. 4. Critical velocity–fall velocity ratio obtained from Eq. 7.80 or 7.81.
Col. 5. Sediment concentration computed from Yang's formula (Eq. 7.78 or 7.79.)
Col. 6. Fraction of material represented by the geometric mean diameter.
Col. 7. Product of Cols. 5 and 6.

The results obtained using these methods illustrate the variation of sediment rate with the predictive method. Most of the discrepancy in this case is attributed to the variation of fine sediment close to the wash-load size range for which the sediment rate may not be predicted accurately. If the sediment discharge is computed on the basis of $d_{50}$ instead of the size fractions, then results for these three methods are closer in value.

## 7.5 EVALUATION OF FORMULAS

Because of the large number of sediment transport formulas now in existence, a user faces the problem of selecting a reliable formula suitable for the physical conditions. The suitability of a formula must be judged by the physical foundation, generality of the basic assumptions used, and most of all, by the comparison of sediment discharge prediction with the measurement. One apparent way to judge the accuracy of a formula is by graphical comparison. In the example shown in Fig. 7.16, the ratio of concentration calculated by the Ackers–White formula to observed concentration is plotted as a function of the observed concentration for field data (Brownlie, 1981). The median line for the ratio is shown together with those for the 16th and 84th percentile.

Several comparisons of accuracy for different formulas have been made, such as those by White et al. (1973), ASCE (1975), Alonso (1980), Brownlie (1981), and Yang (1986). The comparison of 14 formulas by Brownlie (1981) is shown in Fig. 7.17. The bars show the 16th and 84th percentile of the values of

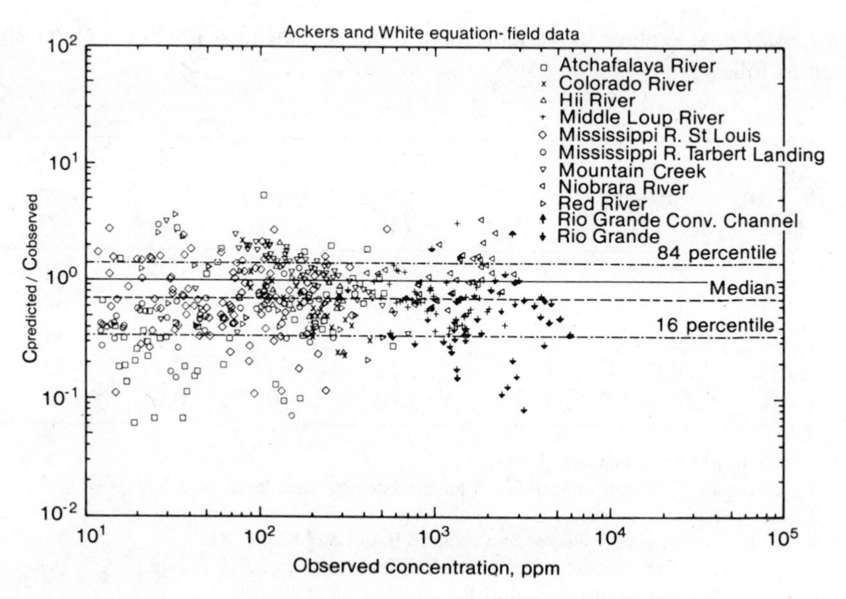


**Figure 7.16** Ratio of concentration predicted by the Ackers-White formula to observed concentration as a function of observed concentration, for field data (Brownlie, 1981).

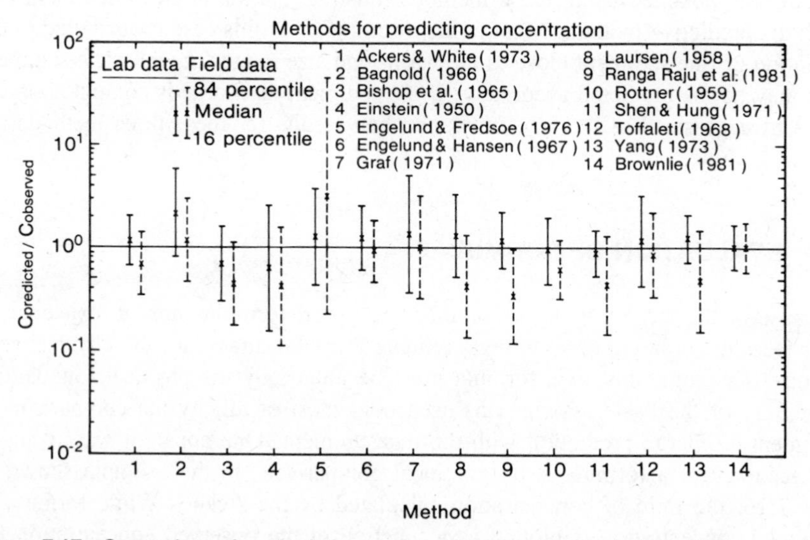


**Figure 7.17** Comparison of methods for predicting sediment concentration. Median and 16th and 84th percentile values are based on the approximation of a log-normal distribution of errors (Brownlie, 1981).

the predicted-concentration–measured-concentration ratio for flume data (solid lines) and field data (dashed lines). The median value is indicated by ×. Although Brownlie's formula rates very well in the comparison, it is cautioned that the coefficients in this formula were determined from the data used in the figure. One of the major problems on the rating of sediment transport formulas stems from the fact that similar data sets were used by different workers in developing their respective formulas.

Alonso (1980) studied and compared the accuracy of eight sediment transport formulas. The comparison for sand transport is given in Table 7.1, where MPME

**TABLE 7.1 Analysis of Discrepancy Ratio Distribution of Different Transport Formulas (Alonso, 1980)**

| Formula | Number of Tests | Ratio Between Predicted and Measured Load: Mean | 95% Confidence Limits of the Mean | | Standard Deviation | Percentage of Tests with Ratio Between $\frac{1}{2}$ and 2 |
|---|---|---|---|---|---|---|
| *Field Data* | | | | | | |
| Ackers–White | 40 | 1.27 | 1.05 | 1.48 | 0.68 | 87.8 |
| Engelund–Hansen | 40 | 1.46 | 1.28 | 1.64 | 0.56 | 82.9 |
| Laursen | 40 | 0.65 | 0.49 | 0.80 | 0.48 | 56.1 |
| MPME | 40 | 0.83 | 0.50 | 1.15 | 1.02 | 58.5 |
| Yang | 40 | 1.01 | 0.89 | 1.13 | 0.39 | 92.7 |
| Bagnold | 40 | 0.39 | 0.31 | 0.47 | 0.26 | 32.0 |
| Meyer-Peter–Muller | 40 | 0.24 | 0.22 | 0.27 | 0.09 | 0 |
| Yalin | 40 | 2.59 | 2.08 | 3.11 | 1.62 | 46.3 |
| *Flume Data with $D/d \geq 70$* | | | | | | |
| Ackers–White | 177 | 1.34 | 1.24 | 1.54 | 1.29 | 73.0 |
| Engelund–Hansen | 177 | 0.73 | 0.63 | 0.83 | 0.68 | 51.1 |
| Laursen | 177 | 0.81 | 0.73 | 0.88 | 0.51 | 71.4 |
| MPME | 177 | 3.11 | 2.95 | 3.52 | 2.75 | 42.1 |
| Yang | 177 | 0.99 | 0.93 | 1.08 | 0.60 | 79.8 |
| Bagnold | 177 | 0.85 | 0.81 | 1.22 | 2.50 | 20.8 |
| Meyer-Peter–Muller | 177 | 0.40 | 0.39 | 0.47 | 0.49 | 18.5 |
| Yalin | 177 | 1.62 | 1.38 | 2.23 | 4.08 | 32.6 |
| *Flume Data with $D/d < 70$* | | | | | | |
| Ackers–White | 48 | 1.12 | 0.93 | 1.28 | 0.52 | 89.6 |
| Engelund–Hansen | 48 | 0.75 | 0.59 | 0.90 | 0.50 | 66.7 |
| Laursen | 48 | 1.04 | 0.76 | 1.32 | 0.99 | 79.2 |
| MPME | 48 | 1.34 | 1.04 | 1.64 | 1.04 | 66.7 |
| Yang | 48 | 0.90 | 0.79 | 1.05 | 0.51 | 85.4 |
| Bagnold | 48 | 1.53 | 1.46 | 1.87 | 1.14 | 45.8 |
| Meyer-Peter–Muller | 48 | 1.03 | 1.00 | 1.27 | 0.83 | 72.9 |
| Yalin | 48 | 1.92 | 1.45 | 2.41 | 1.65 | 64.6 |

refers to Meyer-Peter–Muller formula (1948) for bed load and Einstein (1950) formula for suspended load. Alonso limited his comparisons of field data to those where the bed-material load can be measured by special facilities to avoid the uncertainty of unmeasured load. The poor correlation of the Meyer-Peter–Muller formula may be attributed to the lack of reliable measuring device for bed load in the field. Also, the effective diameter for the sediment mixture should be used in this formula but size fractions used in the study should not be used.

Sediment discharge in natural rivers is related to the variables employed in various formulas. Formulations based on variables that are more pertinent to sediment transport should generally be more accurate. Nearly all formulas contain coefficients that were established based on actual data. Such data should not be used to evaluate the accuracy of the formula. In addition to the theoretical basis, sediment transport is complicated by enormous uncertainties that are generally not incorporated in formulas, such as erratic hydrological phenomena, geological heterogeneities, and constraints. Therefore, it is very difficult, if not impossible, to recommend one equation for universal application. The user is required to look into the theoretical and empirical foundation on which each equation was developed. Basic assumptions and physical limitations must be clearly understood. Test and calibration of the equation with field data at the site of application are highly desirable.

## 7.6 EFFECT OF WATER TEMPERATURE

Water temperature affects sediment discharge through two primary mechanisms. The first is the temperature effect on the viscosity of water and, hence, on the Reynolds number and the fall velocity. Because the viscosity decreases with a rise in water temperature, a sediment particle has a higher fall velocity in summer, which tends to reduce the concentration of suspended sediment. When only this mechanism is considered, a rise in temperature leads to less sediment discharge. However, the temperature also influences sediment discharge through its effect on the resistance to flow described in Sec. 6.4. This mechanism may have an important effect on sediment discharge if the bed form exists in the vicinity of the transition from dune bed to flat bed across which pronounced changes in friction factor occurs. In addition to the above mechanisms, water temperature also affects sediment discharge because of the change in the condition for the initiation of sediment motion.

The effect of water temperature on sediment discharge was reported by Lane et al. (1949). They observed that the sediment discharge in the Lower Colorado River in winter, with water temperature of 40°F, was 2.5 to 3 times that in summer when the water temperature was 80°F. The Missouri River was observed by Straub (1955) to carry more suspended sediment in winter than in summer for comparable water discharges.

The experiments by Hong et al. (1984) using fine sand ($d_{50}$ = 0.11 mm) covered a wide range of temperatures, from about 0 to 30°C. At higher velocities with Froude numbers of 0.5 and 0.8, temperature reductions produced (1) major increases in sediment concentration in the bed layer, (2) more uniform distributions of suspended sediment along the vertical, and (3) smaller, but still significant, increases in friction factor. But at low velocity (Froude number = 0.3) flows, which were just above the threshold of sediment motion, the measured quantities for sediment were not significantly affected by water temperature. While these experiments answered important questions of the temperature effect on sediment discharge, they were handicapped by the small flow depth of about 0.24 ft, which does not permit the formation of dunes. The transition between dune bed and flat bed induced by the variation in temperature, which may have very important effects on sediment transport, was not investigated.

## 7.7 EFFECT OF SUSPENDED SEDIMENT ON FLOW CHARACTERISTICS

It was observed by Vanoni (1946, 1953) and Brooks (1954) in flume experiments that the velocity profile is affected by the suspended-sediment concentration, as exemplified in Fig. 7.18. Under the same flow depth and channel slope in a 33-in.-wide flume, the measured velocity profiles in flow with clear water (mean concentration $\overline{C} = 0$) and with heavy concentration of suspended sediment ($\overline{C} = 15.8$ gm/liter) of well-sorted 0.1-mm sand are shown. The profile with a heavy concentration of suspended sediment has a higher velocity, as shown in Fig. 7.18a and a corresponding steeper velocity gradient, as shown in Fig. 7.18b. From the von Karman–Prandtl logarithmic velocity distribution law (see Eq. 3.20), we have

$$\frac{u}{U_*} = \frac{2.303}{\kappa} \log \frac{z}{D} \tag{7.84}$$

where $U_*$ is the shear velocity given by $(\tau_0/\rho)^{1/2}$. The value of von Karman's constant $\kappa$ may be obtained from the slope of the straight line with $u/U_*$ plotted against $\log(z/D)$. Of course, this requires the data of the velocity profile and boundary shear stress $\tau_0$ which are computed from $\gamma DS$. For small slopes, it is difficult to measure the slope precisely. This has resulted in the inaccuracy of $\tau_0$ and the scatter in the $\kappa$ value.

The slope of the velocity profile, namely the velocity gradient, can be obtained from Eq. 7.84 by differentiation. It can be seen that an increase in velocity gradient $du/dz$ for a given $U_*$ is associated with a decrease in von Karman's $\kappa$. Fowler (1953) also obtained, from analysis of data, the decrease in $\kappa$ value with sediment concentration from the sediment-laden Missouri River and clear streams of the St. Clair and St. Mary's rivers.

It can also be seen, from $\tau = \rho\varepsilon\,(du/dz)$ and $\tau = \tau_0(D - z)/D$ (Eq. 3.2), that for a given $\tau_0$, $D$, and $z$, an increase in $du/dz$ will result in a decrease in $\varepsilon$

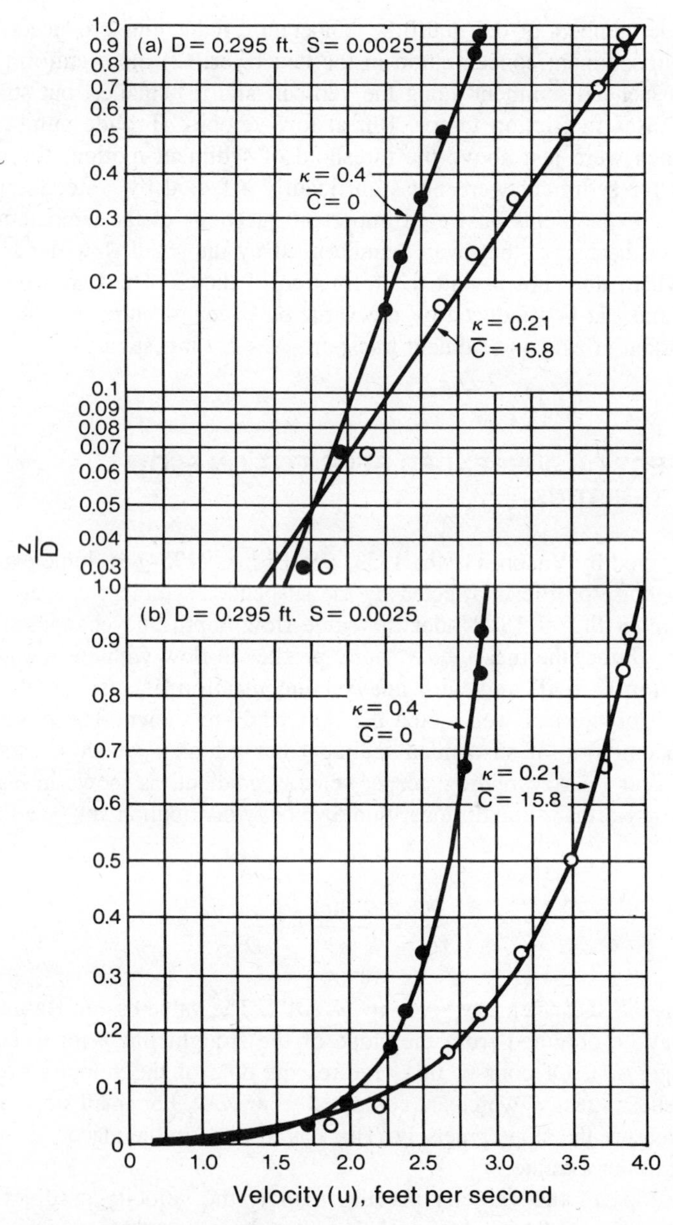


**Figure 7.18** Semi-logarithmic and linear graphs of velocity profiles with clear water and with a heavy suspended load of 0.1-mm sand (after Vanoni, 1953).

because $\tau$ remains constant. In other words, the higher velocity (and, hence, higher velocity gradient) under the same shear stress is attributed to a lesser turbulent shear or smaller value of the momentum transfer coefficient. To explain the observed decrease in $\kappa$ and $\varepsilon$ with sediment concentration, Vanoni (1946) has hypothesized that the turbulent intensity is damped by the suspended sediment, since it may be reasoned that the energy to keep the sediment in suspension must come from the turbulence.

The variation of $\kappa$ with sediment concentration was obtained by Einstein and Chien (1954) following the energy consideration described in the following text. The rate of energy, or power, expenditure on a sediment grain in maintaining its suspension is the submerged weight times its fall velocity. For a given size of sediment in a column of fluid of unit cross-sectional area, the power is

$$p_i = \left(1 - \frac{\rho}{\rho_s}\right) g w_{si} \int_0^D C_i \, dz \tag{7.85}$$

where $i$ is the size index, and $C$ is the sediment concentration at the level $z$ expressed in mass per unit volume. If the depth-averaged concentration $\overline{C}$ is used, then the integral in the foregoing equation may be replaced by $\overline{C}D$. The total power, $p_s$, to maintain the suspension of sediment of several size fractions in the column of water is given by

$$p_s = \sum p_i = \left(1 - \frac{\rho}{\rho_s}\right) gD \sum_{i=1}^{n} \overline{C}_i w_{si} \tag{7.86}$$

where $n$ is the number of size fractions. Now, the power expenditure of the water column in overcoming the flow resistance is given by

$$p_f = \gamma DUS \tag{7.87}$$

Therefore,

$$\frac{p_s}{p_f} = \left(1 - \frac{\rho}{\rho_s}\right) \frac{\Sigma \overline{C}_i w_{si}}{\rho US} \tag{7.88}$$

The von Karman constant $\kappa$ is expressed as a function of the ratio $p_s/p_f$ based on the data of several investigators, as shown in Fig. 7.19. This relation may be used to estimate $\kappa$ for the given sediment concentration.

The velocity profile in turbid water may be approximated by Eq. 7.84, with corrections made for $\rho$ and $\kappa$ to reflect the effects of suspended sediment. The density of sediment-laden water may be obtained from the equation

$$\rho = \rho_f + (\rho_s - \rho_f)C \tag{7.89}$$

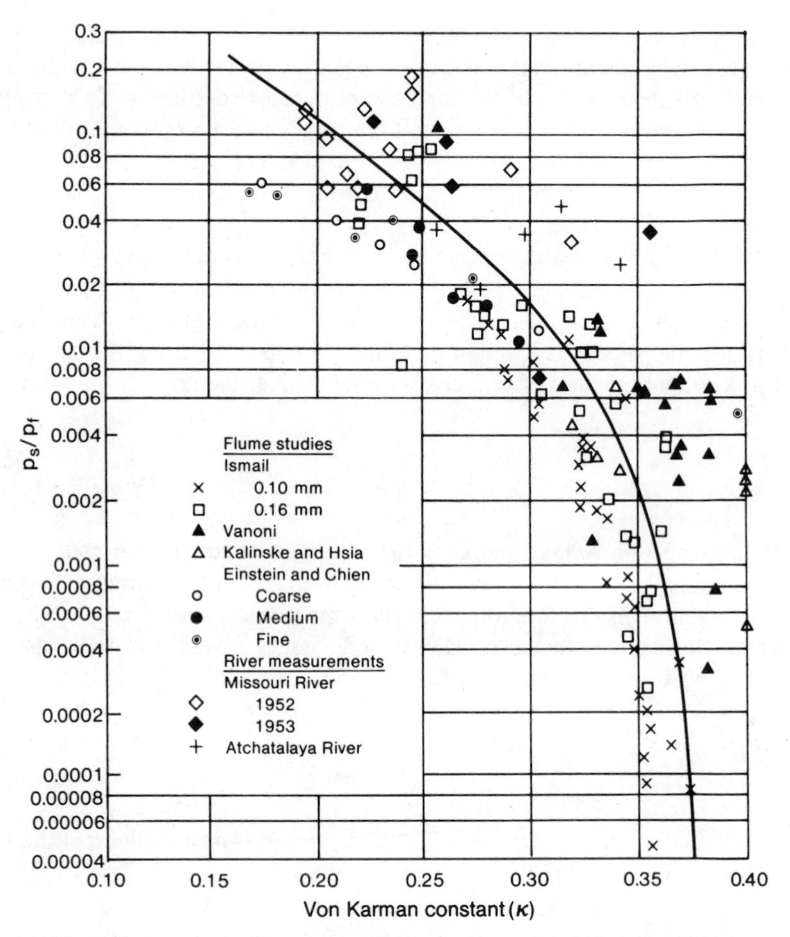


**Figure 7.19** Variation of von Karman's constant $\kappa$ with suspended-sediment concentration (after Einstein and Chien, 1954).

where $\rho_f$ is the density of water, $\rho_s$ is the density of sediment, and $C$ is the average concentration of sediment by volume at distance $z$ above the channel bed. The dynamic viscosity increases with the concentration, and according to Graf (1971) it may be obtained from

$$\mu = \frac{\mu_f}{1 - 2.5C} \tag{7.90}$$

where $\mu$ is the dynamic viscosity of the water-sediment mixture, and $\mu_f$ is the dynamic viscosity of clear water. The kinematic viscosity $\nu$ of a water-sediment mixture at a point $z$ in a channel flow may be obtained from

$$\nu = \frac{\mu}{\rho} = \frac{\mu_f}{(1 - 2.5C)[\rho_f + (\rho_s - \rho_f)C]} \tag{7.91}$$

Although the prevailing opinion among researchers is that the von Karman constant has lower values in flows with suspended sediment than it does in clear water flows, this concept has not been uncontested. For example, Gust (1976) found no such results in experiments with suspensions of clay minerals in sea water. Itakura and Kishi (1980), in their studies of open-channel flow with suspended sediments, used a constant $\kappa$ in a model for velocity profile. The decreasing $\kappa$ value with sediment concentration was disputed by Coleman (1981). His argument is that the velocity profile with sediment suspension is not exactly given by Eq. 7.84 or by its defect law. Instead, the velocity-defect law should have the form

$$\frac{u_{\max} - u}{U_*} = \left\{ \left[ -\frac{2.303}{\kappa} \log \frac{z}{D} \right] + 2\frac{\Pi}{\kappa} \right\} - 2\frac{\Pi}{\kappa} \sin^2\left( \frac{\pi z}{2D} \right) \tag{7.92}$$

where $\Pi$ is the wake strength coefficient. The part of the above equation in square brackets is the original form of the von Karman–Prandtl velocity defect law. The augmentation terms have been added later by various workers in more recent development of the boundary-layer theory. These terms, which are included because of the existence of the wake flow region, are described in the literature on boundary-layer theory.

The part of Eq. 7.92 in curly brackets is the logarithmic part of the defect law in its later form including an additive intercept term $2\Pi/\kappa$. Since the last term in this equation disappears only at $z/D = 0$, the logarithmic part in curly brackets is only an asymptote that the velocity profile of Eq. 7.84 approaches as $z/D$ diminishes (see Fig. 7.20). This means that if $\kappa$ is to be defined from the slope of

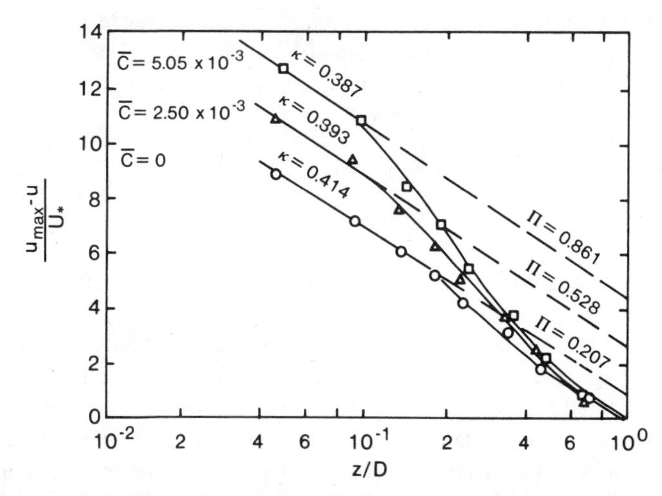


**Figure 7.20** Velocity-defect profile plots for different mean sediment concentrations for constant particle size of 0.105 mm (after Coleman, 1981).

the logarithmic part of Eq. 7.92, a straight-line fit to data points should be made at low values of $z/D$ where the last term in Eq. 7.92 is negligible in magnitude. This practice was not followed by previous investigators, who usually used the central part of the profile. At high values of $z/D$, the last term in Eq. 7.92 could be of an appreciable magnitude and the velocity profile thus deviates somewhat from being a straight line, as illustrated in Fig. 7.20. The value of $\Pi$ can be determined by extending the straight-line fit of the logarithmic part and calculating based on the intercept at $z/D = 1$, that is,

$$\Pi = \frac{\kappa}{2}\left(\frac{u_m - u}{U_*}\right)_{z/D=1} \tag{7.93}$$

Following this procedure, Coleman obtained $\kappa$ values from his experimental data and those of others to reach the conclusion that the $\kappa$ value is nearly constant at 0.4, independent of the amount of suspended sediment. In summary, Coleman argues that the changes in velocity profile with the sediment concentration is not caused by the variation in $\kappa$ but that it is associated with the wake strength coefficient. Using the data of Vanoni (1946) and Elata and Ippen (1961), Coleman developed a relation shown in Fig. 7.21, expressing $\Pi$ as a function of the depth-averaged concentration $\overline{C}$. A simple theoretical model, which was later developed by Parker and Coleman (1986), supports the Coleman analysis.

## 7.8 SEDIMENT TRANSPORT IN NONUNIFORM FLOW

Sediment transport described so far is for uniform flow that has a constant flow and transport rate along the channel. For nonuniform sediment transport, the spatial variation in sediment discharge is associated with changes in channel boundary which reflect sediment storage (and depletion). The rate of sediment storage is related to the spatial variation in sediment discharge by the continuity condition for bed sediment, which has the form

$$(1 - \lambda)\frac{\partial A_b}{\partial t} + \frac{\partial Q_s}{\partial x} - q_s = 0 \tag{7.94}$$

where $\lambda$ is the porosity of bed material, $A_b$ is the channel boundary within a reference frame, $Q_s$ is the bed-material load, and $q_s$ is the rate of lateral sediment inflow per unit channel length. The first term in Eq. 7.94 may be interpreted as the rate of change in channel boundary due to channel storage of sediment. This change is attributed to the spatial variation in sediment discharge represented by the second term, and to the lateral sediment inflow represented by the third term, of this equation. The factor $(1 - \lambda)$ converts the net volume of sediment used for $Q_s$ into bulk volume for $A_b$. A finite-difference solution technique for Eq. 7.94 is given in Sec. 13.5.

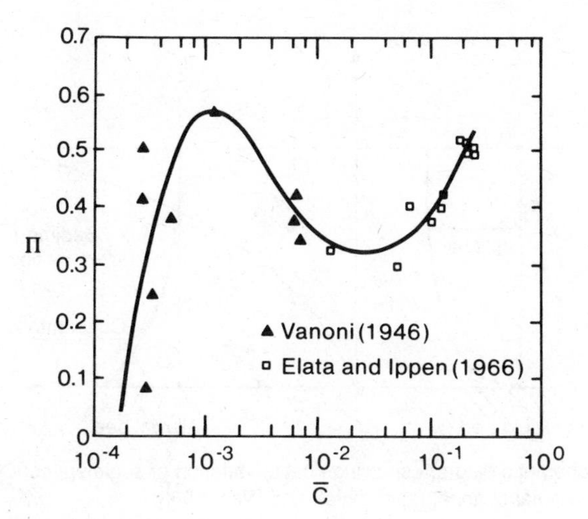


**Figure 7.21** Wake strength plotted against sediment concentration for different data sets (after Coleman, 1981).

For nonuniform sediment transport, the actual sediment discharge is also affected by diffusion. The diffusion process affects essentially the suspended load during deposition and entrainment because certain sediments, especially the fines, require considerable time or distance in settling or in attaining its transport capacity. Therefore, suspended load may not adjust immediately to the flow condition. The spatial development of suspended sediment distribution over an alluvial bed following the introduction of clear water is illustrated schematically in Fig. 7.22. When there is a rapid change in transport capacity of suspended load, the diffusion effect becomes significant. Diffusion is an important consideration in rivers carrying fine sediment and in reservoir sedimentation.

For nonuniform flow, a distinction has to be made between the actual transport rate $C_s'$ and the transport capacity $C_s$. Generally,

$$C_s' < C_s \quad \text{for } \partial U/\partial x > 0 \tag{7.95}$$

and

$$C_s' > C_s \quad \text{for } \partial U/\partial x < 0 \tag{7.96}$$

It requires a certain development distance to reach the equilibrium concentration. In mathematical modeling of sediment transport, sediment rate is usually computed at each discrete cross section; it needs to be corrected if the development length for sediment transport is longer than the grid size $\Delta x$.

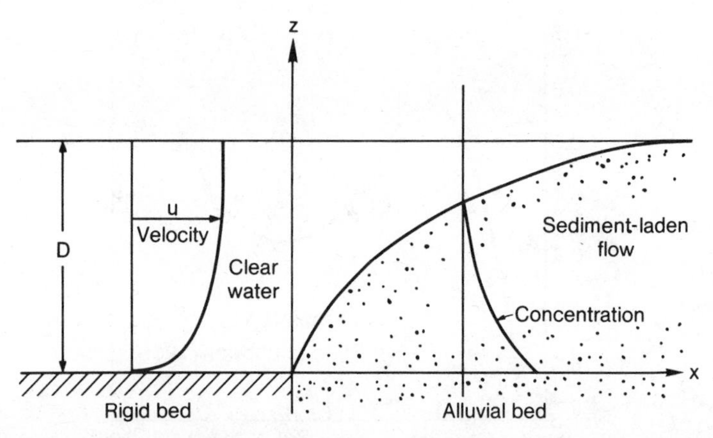


**Figure 7.22** Schematic diagram showing spatial variation of sediment concentration as clear water enters an alluvial channel (after Chien and Wan, 1983).

In the diffusion theory of sediment transport, the concentration of suspended load $C$ is described by the convection-diffusion equation (see Eq. 7.42):

$$u\frac{\partial C}{\partial x} + w\frac{\partial C}{\partial z} = \frac{\partial}{\partial z}\left(\varepsilon\frac{\partial C}{\partial z}\right) \tag{7.97}$$

The vertical velocity component $w$ is replaced by the fall velocity $-w_s$. The first term in Eq. 7.97 accounts for nonuniformity; it is zero for uniform flow for which the equation becomes the Rouse distribution of Eqs. 7.44 and 7.56.

In solving Eq. 7.97, appropriate boundary conditions and initial conditions are required. The first boundary condition states that there is zero net mass transfer at the water surface, that is,

$$\varepsilon\frac{\partial C}{\partial z} + w_s C = 0 \quad \text{at } z = D \tag{7.98}$$

The second boundary condition says that the net mass transfer at the bottom surface is due to the deposition and the entrainment rates, that is,

$$-\left(\varepsilon\frac{\partial C}{\partial z} + w_s C\right)_{z=0} = -\alpha w_s (C - C_*)_{z=0} \tag{7.99}$$

where $C_*$ is the equilibrium concentration of sediment or the potential carrying capacity of a specific flow, and $\alpha$ is a dimensionless coefficient that characterizes the rate of attaining the carrying capacity. The first term on the right-hand side of the foregoing equation, $-w_sC$, represents the deposition flux; the second term on the right-hand side, $w_sC_*$, is the entrainment flux. In other words, the rate of deposition (or entrainment) by the flow is proportional to the difference between the actual suspended load and the sediment transport capacity of the flow. Now, the upstream boundary condition is given by

$$C = C_0 \quad \text{at } x = x_0 \tag{7.100}$$

where the subscript 0 designates values at the upstream boundary.

Solutions of the convection-diffusion equation have been developed following one- and two-dimensional approaches. In a one-dimensional solution, the streamwise variation of depth-averaged or cross-sectionally averaged sediment concentration is considered. This approach has been employed by several investigators, for example, Mei (1969), Hjelmfelt and Lenau (1970), Zhang et al. (1983), and Cheng (1985). In a two-dimensional solution, for example, van Rijn (1986), vertical profiles of the sediment concentration and velocity $u$ under acceleration or deceleration are employed.

The one-dimensional approach is described in the following. Equation 7.97 can be written as

$$u\frac{\partial C}{\partial x} = \frac{\partial}{\partial z}\left(\varepsilon\frac{\partial C}{\partial z} + w_s C\right) \tag{7.101}$$

Integrating throughout depth and taking the depth-averaged values yields

$$\frac{d\overline{C}}{dx} = -\frac{\alpha w_s}{q}(\overline{C} - \overline{C}_*) \tag{7.102}$$

where the overbar is used to designate depth-averaged or cross-sectionally averaged quantities, and $q$ is the discharge per unit channel width. Equation 7.102, in which $\overline{C}$ is the dependent variable and the coefficients are functions of the independent variable $x$ only, is a linear differential equation of the first order. The general solution, which can be found in a standard text, has the form

$$\overline{C} = \left[c + \int \frac{\alpha w_s C_*}{q} \exp\left(\int \frac{\alpha w_s}{q} dx\right) dx\right] \exp\left(-\int \frac{\alpha w_s}{q} dx\right) \tag{7.103}$$

where $c$ is a constant which is obtained from the upstream boundary condition and which is given by Eq. 7.100 to be the concentration at the upstream section.

In order to apply it to a computation grid of cross sections, Eq. 7.103 is written in finite-difference form and in terms of the size fractions for sediment:

$$(\overline{C}_i)_{j+1} = \left[(\overline{C}_i)_j + \frac{\alpha w_{si}(\overline{C}_{*i})_{j+1/2}}{q_{j+1/2}} \exp\left(\frac{\alpha w_{si}}{q_{j+1/2}}\Delta x\right) \Delta x\right] \exp\left(-\frac{\alpha w_{si}}{q_{j+1/2}}\Delta x\right) \tag{7.104}$$

where $i$ is size-fraction index, $j$ is cross-section index counted from upstream to downstream, and $\Delta x$ is the grid spacing. Note that the mechanisms of deposition and entrainment are included in different terms of this equation.

Zhang et al. (1983) presented a simplified solution of Eq. 7.104, which, for a certain particle size, is given by

$$\overline{C}_{j+1} = (\overline{C}_*)_{j+1} + \exp\left(-\frac{\alpha w_s}{q}\Delta x\right)[\overline{C}_j - (\overline{C}_*)_{j+1}] \tag{7.105}$$

The exponential term in this equation is referred to as the *nonuniform transport coefficient;* its value is between 0 and 1.

The value of $\alpha$ must be determined separately for the cases of deposition and entrainment because of the different processes. The deposition case is much simpler, for which Zhang et al. obtained the following expression:

$$\alpha = 1 + \frac{K_1}{2} \tag{7.106}$$

where $K_1$ is the Peclet number, defined as $6w_s/\kappa U_*$. In the case of entrainment, they obtained the following relationship for $\alpha$.

$$\alpha = \frac{\pi^2}{K_1} + \frac{K_1}{4} \tag{7.107}$$

The diffusion during sediment entrainment from the alluvial bed is subject to the availability of sediment, which is closely related to bed-material composition and to the complex interaction of sorting and erodibility of the material. Analytical approach to such processes is given in the next section on sediment sorting.

## 7.9 SEDIMENT SORTING

The nonequilibrium, time-dependent sediment-transport processes are also complicated by sediment sorting in addition to diffusion. *Sediment sorting* refers to selective transport of sediment of different sizes. Predictive methods that keep account of bed-sediment composition while tracking bed-profile changes have been attempted by Bennett and Nordin (1977), Borah et al. (1982), and Lee and Odgaard (1986), among others. The Borah et al. method is described herein in terms of its components of residual transport capacity, active layer and armor, bed-material accounting, volume of entrainment, and numerical procedure.

### Residual Transport Capacity

In the case of nonequilibrium transport of graded sediments, a sediment-transport formula may only be used to determine the potential carrying capacity of a specific flow, but it is not directly applicable to estimate the capacity of individual size fractions. It is necessary to account for the extent that the stream's potential capacity is filled up by materials of all fractions that are already in motion. The *residual transport capacity* is defined as the stream's ability to carry any additional load of a particular size fraction in the presence of all of the size fractions already present in the flow. The formulation of the residual transport capacity is as follows.

Let $T_1$ be the potential sediment capacity of flow for a certain uniform size $d_1$ and the actual transport rate be $c_1$. Then $T_{r1} = T_1 - c_1 = T_1 - T_1\,(c_1/T_1)$ is the residual capacity of the flow for that size material. The residual capacity for size $d_1$ is also affected by the actual transport rates of other sizes. For a size $d_2$, the residual capacity for the $d_1$-size material in the presence of the load $c_2$ would be $T_{r1} = T_1 - T_1\,(c_2/T_2)$, where $c_2/T_2$ may be envisioned as that fraction of $T_1$ already consumed by the load $c_2$. If both sizes are simultaneously present in the flow, then $T_{r1} = T_1 - T_1\,(c_1/T_1) - T_1\,(c_2/T_2)$. This equation can be generalized to any size fraction $d_i$ and for $n$ size fractions as follows:

$$T_{ri} = T_i\left[1 - \sum_{j=1}^{n}\left(\frac{c_j}{T_j}\right)\right] = \Omega T_i, \qquad i = 1, 2, \ldots, n \tag{7.108}$$

The quantity within the bracket, or $\Omega$, is the remaining portion of $T_i$ for transporting additional material of size $d_i$. At any instant, $\Omega$ is assumed to be the same for all size fractions. $T_i$ is related to the flow properties and the $d_i$ size based on the sediment formula. A positive value indicates that the transport capacity has not been satisfied, that is, we have an eroding bed, provided that $T_i > 0$. Similarly a negative $\Omega$ means that the flow carries a load in excess of its potential capacity and will hence deposit the excess load on the bed, that is, we have an aggrading bed. When $\Omega = 0$, there is no net load change and the transport process remains in a pseudo-equilibrium condition.

## Active Layer and Armor

In the case of erosion, sediment transport is constrained by availability. In noncohesive sediments, the materials available for entrainment are essentially those exposed at the bed surface. *Active layer* refers to the surface layer from which the materials can be entrained by the flow. This layer is considered to be related to the height of bed forms, since they continuously mix all the sediment contained. It may also be regarded as a mixing zone below which the bed material remains undisturbed. For graded sediment, the slowly moving coarse sediment tends to collect at the base of the mixing zone, forming a band of large grains. During erosion, finer grains tend to be scoured first, leaving the coarse grains behind. For continued erosion, the finer particles are washed out of the mixing zone, several active layers may be scoured away, and the coarse area in the bed may become immobile. It thus forms an *armor coat* that protects the underlying material.

In the Borah et al. (1982) model, the mixing zone is pictured as a band of constant thickness divided into several layers, as shown in Fig. 7.23. The layer in contact with the flow is always referred to as the active layer. The thickness,

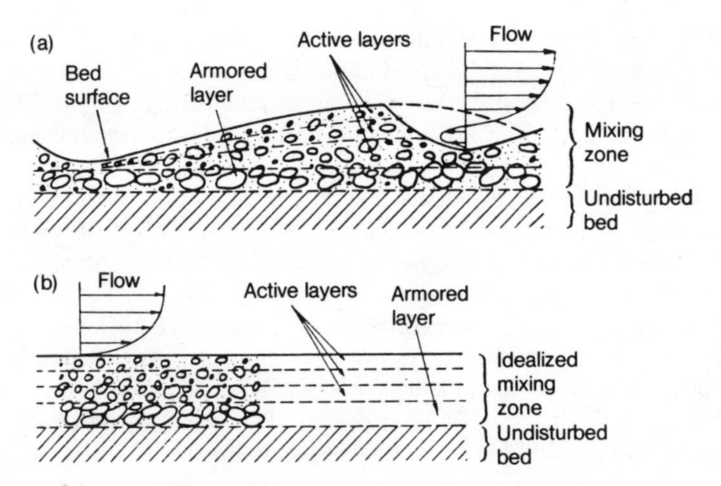


**Figure 7.23** Schematic representation of bed processes (after Borah et al., 1982).

porosity, and particle size distribution of this layer vary with the flow and bed evolution, but the layer is assumed to be homogeneous within itself at any given time. The active layer thickness is defined from volumetric consideration as

$$t = \frac{1}{\sum_{i=L}^{n} p_i} \frac{d_L}{1 - \lambda_L} \tag{7.109}$$

where $p_i$ is the size fraction, $i$ is size-fraction index, and $d_L$ and $\lambda_L$ are the size and porosity of fraction $L$, respectively. Fraction $L$ is the smallest size ($d_L$) of material the flow cannot transport, that is, $T_i = 0$ for $i \geq L$.

Equation 7.109 is a measure of the active-layer thickness when some of the fractions in the mixing zone cannot be moved by the flow. These fractions will contribute to the formation of an armor coat. This equation predicts a thinner active layer at low discharges for which $d_L$ is small and $\Sigma p_i$ is large. It predicts a thicker layer at high flows, reflecting the fact that a greater depth of bed can be sorted through by a higher flow during the same time period. The limit of $L$ is introduced as $n$, or the upper bound of active layer thickness when the flow is capable of transporting all fractions within the mixing zone. For example, in a uniform bed material ($p_n = 1$) with $\lambda_n = 0.5$, Eq. 7.109 gives (in the limit $L = n$), $t = 2d_n$, which is in agreement with the bed-layer thickness used by Einstein (see Sec. 7.3). The sediment contained in the active layer is the only material available for erosion during a simulation time step. When the bed is armored, no erosion can occur until the flow develops the necessary $T_r$ for moving sediment again.

### Bed-Material Accounting

The size composition and thickness of active layer are continuously tracked in a simulation. The entrainment frequency varies with the particle size. The smaller ones are more easily removed, but they also hide beneath larger particles. The size and hiding jointly affect entrainment frequency. In accounting for the change in size composition, consider a well-mixed active layer with three size fractions, $d_1 < d_2 < d_3$, being eroded by a flow with sufficient transport capacity to scour all three fractions shown in Fig. 7.24a. Scouring is imagined to begin with the entrainment of the $d_1$ particles exposed at the bed surface. Next, the $d_2$ particles at the bed surface are removed, followed by $d_1$ grains hidden underneath the $d_2$ grains. Finally, the $d_3$ particles are entrained, followed by the $d_1$ and $d_2$ particles underneath as well as by $d_1$ grains uncovered by the washing out of the $d_2$ particles. Thus one part of $d_1$ grains is removed from the surface, one part of $d_1$ particles is removed for every part of $d_2$ particles entrained by the flow, and one part of $d_2$ and two parts of $d_1$ grains are associated with the removal of every part of the $d_3$ grains. This ordering can be easily extended to $n$ size fractions, summarized in the following entrainment frequency matrix:

$$\mathbf{F} = [F_{ij}] = \begin{bmatrix} 1 & 0 & 0 & 0 & & 0 \\ 1 & 1 & 0 & 0 & & 0 \\ 2 & 1 & 1 & 0 & & 0 \\ 4 & 2 & 1 & 1 & & 0 \\ 8 & 4 & 2 & 1 & & 0 \\ 16 & 8 & 4 & 2 & & 0 \\ \vdots & \vdots & \vdots & \vdots & & \vdots \\ 2^{i-2} & 2^{i-3} & 2^{i-4} & 2^{i-5} & 2^{i-j-1} & 1 \end{bmatrix} \tag{7.110}$$

where $i, j = 1, 2, 3, \ldots, n$. In this matrix, the diagonal elements indicate that each size fraction at the bed surface is scoured once by itself. Other numbers in each row indicate the number of times each fraction $d_j$, $1 < j < i$, becomes available for entrainment, once $d_i$ is removed from the bed surface. The **F** matrix is time invariant because the graded materials are assumed to be uniformly mixed in the active layer.

### Volume of Entrainment

The amount of eroded materials can be determined from the frequencies $F_{ij}$ and the corresponding volumes of the fraction contained in the active layer. If $V_j$ designates the total volume of the $d_j$ size present in the active layer per unit channel

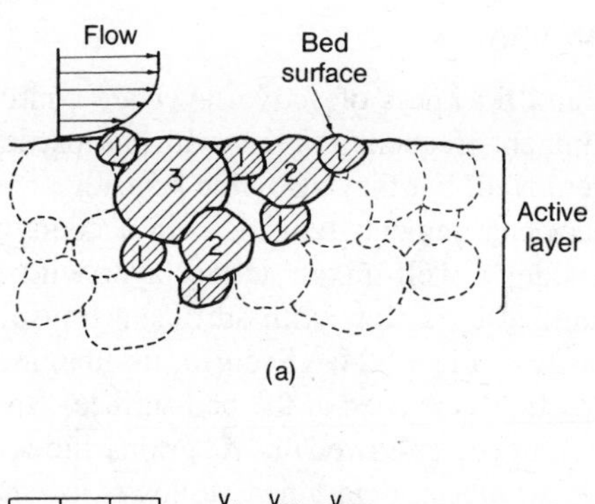


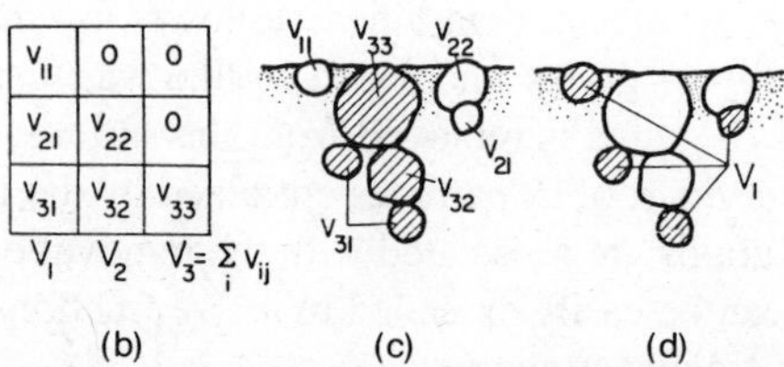


**Figure 7.24** Schematic representation of active-layer composition (after Borah et al., 1982).

length, $v_{ij}$ is the portion of $V_j$ that becomes available for entrainment when the exposed $d_i$ fraction is removed from the surface. Thus

$$V_j = \Delta A_b p_j (1 - \lambda_j) = \sum_{i=1}^{n} v_{ij} \tag{7.111}$$

where $\Delta A_b$ is the bed area of the active layer, equal to active-layer thickness times its share of channel width, and $v_{ij}$ is an element of volume entrainment matrix. The last term in the equation is the sum of all elements in column $j$, representing the total volume of each size fraction present in the active layer. The value of $v_{ij}$ is computed from the entrainment frequency and the total volume of each size fraction, or

$$v_{ij} = \frac{F_{ij} V_i}{\sum_{k=j}^{i} F_{kj} V_k} V_j, \qquad i = j, j+1, \ldots, n \quad j = 1, 2, \ldots, n \tag{7.112}$$

The entrained frequency is converted into volume by this equation. The volume entrainment matrix **V** $[V_{ij}]$ has the same size as the frequency matrix **F**. Adding up elements for each row $i$ gives the volume of potential erosion $\Sigma_{j=1}^{i} v_{ij}$ of sizes $d_1, d_2, \ldots,$ and $d_i$ with the removal of the largest size $d_i$ in that row. On the other hand, summation of all the elements in each column yields the total volume of each size class in the active layer. These concepts are schematically demonstrated in Fig. 7.24b, which shows that the **V** matrix associated with a small cluster of bed particles in Fig. 7.24c represents those that can be removed along with the third size fraction; that is, $v_{31}$ and $v_{32}$ are removed with the entrainment of $v_{33}$. The total volume of the first size fraction in the group, $V_1 = \Sigma_{i=1}^{n} v_{i1}$, is represented by the shaded grains of Fig. 7.24d.

The extent to which the volumes $v_{ij}$ will be actually entrained depends on the degree of detachability of the sediment particles and the time steps used in the simulation. Hence, an erodibility parameter $\varepsilon$ is introduced in the erosion computation, that is,

$$e_{ij} = \varepsilon v_{ij}, \qquad j = 1, 2, \ldots, i, \quad i = 1, 2, \ldots, n \tag{7.113}$$

where $e_{ij}$ are the elements of a so-called *volume erosion index,* and the matrix $\mathbf{e}$ represents the volumes of actual erosion from the corresponding elements of the $\mathbf{v}$ matrix. The erodibility parameter governs the amount of scour and is calibrated by matching the simulated results with the measurement.

### Numerical Procedure

Because of the complicated techniques and accounting procedure, the method for sediment sorting described in the foregoing text may only be employed following a numerical procedure. The changes of the quantities with time must usually be evaluated using small time increments. Basically, this method may be used in a mathematical model for routing graded sediment in alluvial channels described in Chapter 13. For each time increment, this method consists of the following major steps.

1. Computation of the residual transport capacities $T_i$ for different size fractions, $i = 1, 2, \ldots, n$, using a suitable sediment formula.
2. Determination of the change in sediment discharge (or concentration) with distance due to erosion and deposition. The continuity equation for sediment (Eq. 7.94) is employed in this step.
3. Computation for the separate cases of erosion, armoring, and deposition.
4. Updating sediment concentration.
5. Updating channel-bed profile.

## 7.10 SAMPLING FLUVIAL SEDIMENT

In days past, collection of sediments was accomplished with implements such as a bucket and a stopwatch. As time progressed, collection devices and methods were developed according to individual concerns. The result was a wide variety of non-comparable data and records whose accuracy was difficult to verify. The need for standardization of data collection procedures and instrumentation was recognized, and the U.S. Water Resources Council established the federally funded Federal Inter-Agency Sedimentation Project (FIASP) in 1939 to pursue this course. The FIASP's concerns are ongoing in the development and testing of new equipment and improvement of existing instruments used in the collection of sediment. A brief introduction to suspended-sediment samplers and bed-load samplers is given in this section. More details on fluvial sediment sampling are given in the ASCE Sedimentation Manual (1975).

## Suspended-Sediment Samplers

Suspended-sediment samplers are designed to collect water-sediment mixtures at flow velocity while causing minimal disturbance to stream conditions. A suspended-sediment sampler, such as that shown in Fig. 7.25, has a removable sample container that is suitable for transportation to the laboratory. It is usually constructed with fins or vanes to aid in orienting the nozzle into the flow and is designed to fill water and sediment at the local stream velocity smoothly without sudden inrush or gulping. To sample the greatest range over the stream vertical, the nozzle must approach the stream bed as close as possible. The depth that cannot be reached is about 3 1/2–6 in. for available samplers.

The sediment transported between the nozzle and the water surface is termed the *measured load*, and the part from the nozzle to the stream bed is called the *unmeasured load* (see Fig. 7.26). When bed load is not measured, the unmeasured load is estimated in order to obtain the total sediment load. A common method

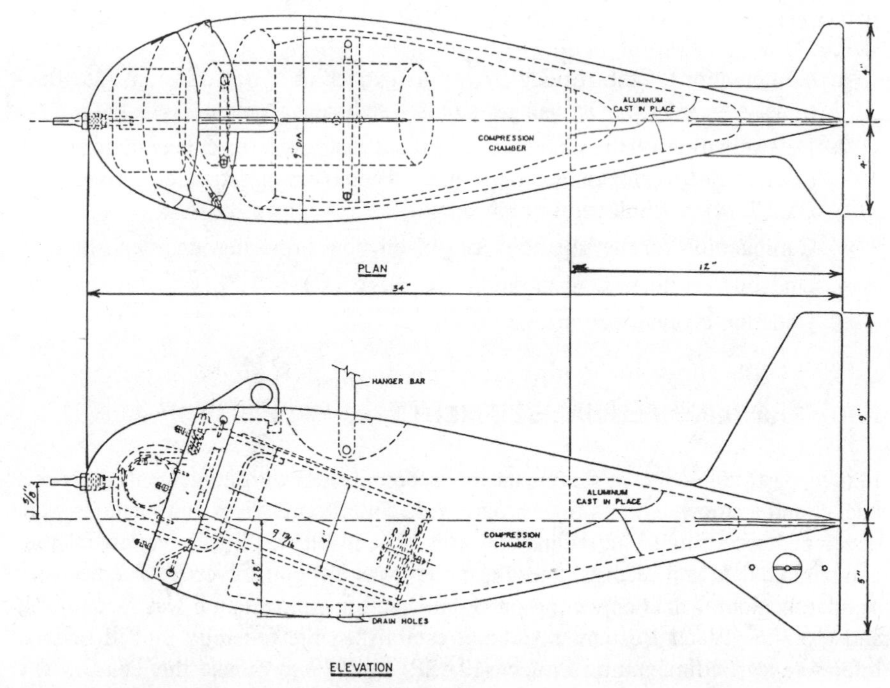


**Figure 7.25** Point-integrating suspended-sediment sampler, US P-63.

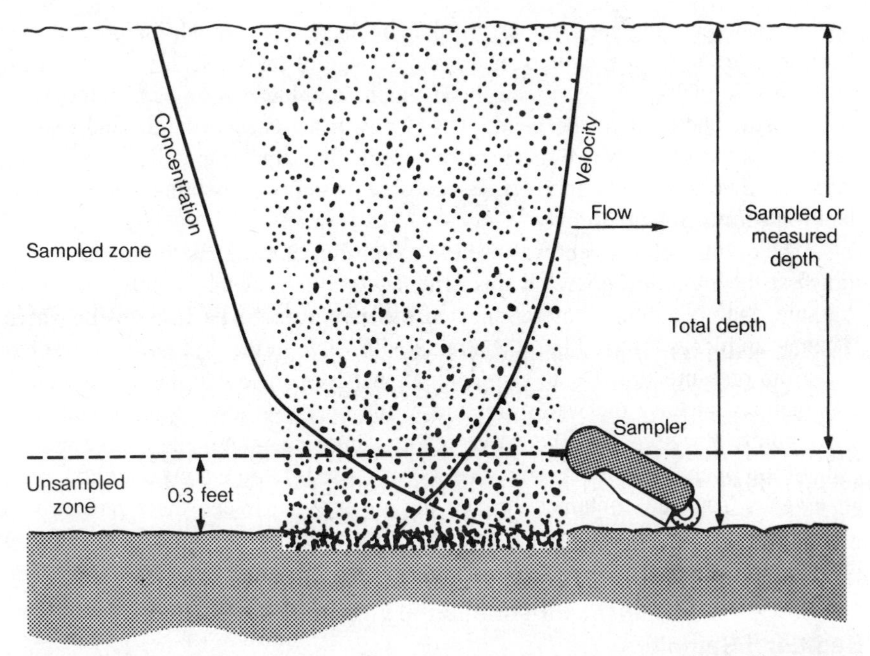


**Figure 7.26** Measured and unmeasured zones in a stream sampling vertical with respect to velocity of flow and sediment concentration by Culbertson from report by Guy and Norman (1976).

for estimating the unmeasured load is the *Modified Einstein Procedure* developed by Colby and Hembree (1955). The unmeasured load may also be taken as a percentage of the measured load, generally around 10%. Two types of suspended-sediment samplers — *depth-integrating samplers* and *point-integrating samplers* — are described in the following paragraphs.

Depth-integrating samplers are designed to collect a sample as they are lowered to the stream bed and raised to the surface at a uniform rate. Such samplers have changeable nozzle and sample-container sizes that are selected based on the stream velocity, stream depth, and the transit rate. The nozzle size must be so limited that it will admit a sample amount without causing the container to overfill for the chosen transit rate.

In operating a sampler, a clean bottle (sample container) is securely seated in the body of the sampler, the nozzle is attached, and the instrument is lowered at a uniform rate from the water surface to the stream bed. The flow enters the nozzle into the container isokinetically as air escapes through an exhaust tube. On contacting the stream bed, the sampler's direction is instantly reversed and the sampler is raised to the surface at the same, or some other, uniform rate. The sampler fills continually during the period submerged and must be removed from the stream prior to being completely filled. If the sample in the container reaches the level at the end of the nozzle, then inflow rate is not maintained and contamination occurs from flow entering the nozzle and exiting through the exhaust tube.

Point-integrating samplers are similar in shape to the depth-integrating samplers; they are used to sample over the same range of the vertical (from surface to within a few inches of the stream bed). Each sampler is equipped with a valve mechanism, enclosed in the head, that is power-actuated to begin and end the sampling process. The valve mechanism permits the sampler to collect flow at any point in the vertical, although it is also capable of sampling continuously as with a depth-integrating sampler.

The diving bell concept was used in the design of samplers to prevent sudden inrush at the sampling point. In this scheme, the body of the sampler contains a pressure chamber that is interconnected by tubing and a passage to the valve. When a sampler is lowered into the stream with valve in the first position, the passage from pressure chamber to sample container is open so that the air pressure in the chamber matches the water pressure at the nozzle at any depth. At the sampling point, a switch sets the valve to the sampling mode, the intake and air exhaust are open, and the pressure chamber is closed. During the sampling period, air escapes from the container through the air exhaust. When the sampling time terminates, the switch is released, the valve closes, and the sampler is raised from the stream.

## Bed-Load Samplers

A prominent bed-load sampler in wide use is the Helley-Smith sampler developed by the U.S. Geological Survey (Emmett, 1979), as shown in Fig. 7.27. This version has been altered several times since its introduction in order to adapt to various field uses. This model has a 3-in. × 3-in. entrance nozzle with a flared section at the rear to which a 0.2-mm mesh sample bag is attached. Later models use a 0.25-mm mesh bag. Sediment carried into the sampler is filtered out and

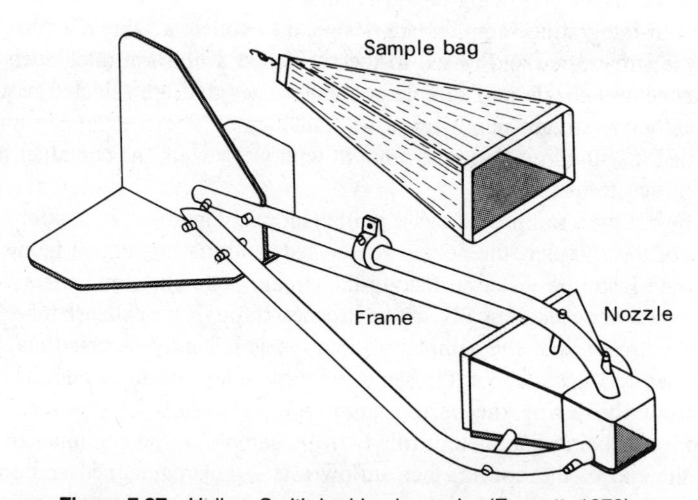


**Figure 7.27** Helley–Smith bed-load sampler (Emmett, 1979).

retained by the mesh bag. The frame of the sampler is equipped with a sliding bracket on the top member for cable suspension. There is a version adapted to a wading rod with aluminum frame and tubular rod. A scaled-up model with 6-in. × 6-in. entrance nozzle for larger sediment sizes and heavier samplers for deeper and swifter streams are also made.

In operating a cable-suspended model, the bed-load sampler is lowered to the stream bed, timed for a 30-sec duration, and retrieved. The total time for taking each sample is about 2–3 min for lowering, sampling, raising, emptying, and moving to a new location. Because temporal and spatial factors can result in significant variations in bed-load transport data, two traverses of the stream at sufficient locations should be conducted in order to obtain a reliable mean transport rate.

Because of the dynamic processes involved in sediment transport, different samplers, techniques, and methods of analysis will yield different rates at the same location. The local rate is also affected by bed forms. Also, because of the transient nature of streams, different rates could result in using the same instrumentation and analysis on samples from the same location taken at different times. In conclusion, sampling bed load is a difficult operation that requires experienced operators to obtain reliable results.

## REFERENCES

Ackers, P. and White, W. R., "Sediment Transport: A New Approach and Analysis," *J. Hydraul. Div. ASCE.* **99**(HY11), pp. 2041-2060, November 1973.

Alonso, C. V., "Selecting a Formula to Estimate Sediment Transport Capacity in Nonvegetated Channels," in *CREAMS (A Field Scale Model for Chemicals, Runoff, and Erosion from Agricultural Management System)*, W. G. Knisel, ed., U.S. Department of Agriculture, Conservation Research Report, No. 26, Chapter 5, pp. 426-439, May 1980.

ASCE, *Sedimentation Engineering*, Manuals and Reports on Engineering Practice, No. 54, Vito A. Vanoni, ed., 1975.

Bagnold, R. A., "An Approach to the Sediment Transport Problem from General Physics," *USGS Professional Paper 422-J*, 1966.

Bennett, J. S. and Nordin, C. F., "Simulation of Sediment Transport and Armouring, " *Hydrol. Sci. Bull.*, **XXII**(4), pp. 555-569, December 1977.

Bishop, A. A., Simons, D. B., and Richardson, E. V., "Total Bed-Material Transport," *J. Hydraul. Div. ASCE*, **91**(HY2), pp. 175-191, March 1965.

Bogardi, J., *Sediment Transport in Alluvial Streams*, Academiai Kiodo, Budapest, 1974.

Borah, D. K., Alonso, C. V., and Prasad, S. N., "Routing Graded Sediments in Streams: Formations," *J. Hydraul. Div. ASCE*, **108**(HY12), pp. 1486-1503, December 1982.

Brooks, N. H., "Laboratory Studies of the Mechanics of Streams Flowing Over a Movable Bed of Fine Sand," Ph.D. Thesis, California Institute of Technology, 1954.

Brooks, N. H., "Mechanics of Streams with Moveable Beds of Fine Sand," *Trans. ASCE*, **123**, pp. 526-594, 1958.

Brown, C. B., "Sediment Transportation," in *Engineering Hydraulics*, H. Rouse, ed., John Wiley & Sons, New York, 1950, Chapter XII.

Brownlie, W. R., "Prediction of Flow Depth and Sediment Discharge in Open Channels," Rept. No. KH-R-43A, W.M. Keck Laboratory of Hydraulics and Water Resources, California Institute of Technology, Pasadena, California, November 1981.

Brush, L. M., Ho, H. W., and Singamsetti, S. R., "A Study of Sediment in Suspension," International Association for the Science of Hydraulics, Commission on Land Erosion (Bari), Publication No. 59, 1962.

Chang, F-M., Simons, D. B., and Richardson, E. V., "Total Bed-Material Discharge in Alluvial Channels," Proceedings of the Twelfth Congress, IAHR, Fort Collins, Colorado, pp. 132-140, 1967.

Cheng, K. J., "An Integrated Suspended Load Equation for Non-equilibrium Transport of Non-uniform Sediment," *J. Hydrol.*, **79**, pp. 359-364, 1985.

Chien, N. and Wan, Z-H., *Mechanics of Sediment Transport*, Science Publication Co., 1983 (in Chinese).

Colby, B. R., "Discharge of Sands and Mean Velocity Relationships in Sand-Bed Streams," *USGS Professional Paper 462-A*, 1964.

Colby, B. R. and Hembree, C. H., "Computations of Total Sediment Discharge, Niobrara River near Cody, Nebraska," *USGS Water Supply Paper 1357*, 1955.

Coleman, N. L., "Velocity Profiles with Suspended Sediment," *J. Hydraul. Research*, **19**(3), pp. 211-229, 1981.

DuBoys, P., "Le Rhone et les Rivieres a Lit Affouillable," *Annales des Ponts et Chaussees*, **18**, Series 5, pp. 141-195, 1879.

Einstein, H. A., "Formulas for the Transportation of Bed Load," *Trans. ASCE*, **107**, pp. 561-573, 1942.

Einstein, H. A., "The Bed Load Function for Sediment Transportation in Open Channels," Technical Bulletin 1026, U.S. Department of Agriculture, 1950.

Einstein, H. A. and Chien, N. "Second Approximation to the Solution of the Suspended Load Theory," Research Report No. 3, University of California, Berkeley, California 1954.

Elata, C. and Ippen, A. T., "The Dynamics of Open Channel Flow with Suspensions of Neutrally Buoyant Particles," Technical Report. No. 45, Department of Civil and Sanitary Engineering, Massachusetts Institute of Technology, 1961, 69 pp.

Emmett, W. W., "A Field Calibration of the Sediment-Trapping Characteristics of the Helley–Smith Bedload Sampler," Open File Report 79-411, U.S. Geological Survey, 1979.

Engelund, F. and Hansen, E., *A Monograph on Sediment Transport in Alluvial Streams*, Teknisk Vorlag, Copenhagen, Denmark, 1967.

Fowler, L., "Notes on the Turbulence Function $\kappa$," Missouri River Division, U.S. Army Corps of Engineers, November 1953.

Garde, R. J. and Ranga Raju, K. G., *Mechanics of Sediment Transportation and Alluvial Stream Problems*, Wiley Eastern Ltd. New Delhi, 1977.

Gill, M. A., "Rationalization of Lacey's Regime Flow Equations," *J. Hydraul. Div.* ASCE, **94**(HY4), pp. 983-995, July 1968.

Graf, W. H., *Hydraulics of Sediment Transport*, McGraw-Hill, New York, 1971.

Gust, G., "Observations on Turbulent Drag Reduction in a Dilute Suspension of Clay in Sea-Water," *J. Fluid Mech.*, **75**(1), pp. 29-47, 1976.

Guy, H. P. and Norman, V. W., "Field Methods for Measurement of Fluvial Sediment," Techniques of Water-Resources Investigations of the U.S. Geological Survey, 3rd printing, U.S. Geological Survey, 1976, Chapter 2.

Guy, H. P., Simons, D. B., and Richardson, E. V., "Summary of Alluvial Channel Data from Flume Experiments, 1956–61," *USGS Professional Paper 462-I*, 1966.

Hjelmfelt, A. T. and Lenau, C. W., "Nonequilibrium Transport of Suspended Sediment," *J. Hydraul. Div. ASCE*, **96**(HY7), pp. 1567-1586, July 1970.

Hong, R.-J., Karim, M. F., and Kennedy, J. F., "Low Temperature Effects on Flow in Sand-Bed Streams," *J. Hydraul. Eng. ASCE*, **110**(2), pp. 109-125, February 1984.

Itakura, T. and Kishi, T., "Open Channel Flow with Suspended Sediments," *J. Hydraul. Div. ASCE*, **106**(HY8), pp. 1325-1343, August 1980.

Jobson, H. E. and Sayre, W. W., "Vertical Transfer in Open Channel Flow," *J. Hydraul. Div. ASCE*, **96**(HY3), 1970.

Lane, E. W., Carlson, E. J., and Hanson, O. S., "Low Temperature Increases Sediment Transportation in Colorado River," *Civ. Eng. ASCE*, **19**(9), pp. 45-46, September 1949.

Laursen, E. M., "The Total Sediment Load of Streams," *J. Hydraul. Div. ASCE*, **54**(HY1), pp. 1-36, February 1958.

Lee, H-Y. and Odgaard, A. J., "Simulation of Bed Armoring in Alluvial Channels," *J. Hydraul. Eng. ASCE*, **112**(9), pp. 794-801, September 1986.

Majumdar, H. and Carstens, M. R., "Diffusion of Particles by Turbulence: Effect on Particle Size," Water Resources Center, WRC-0967, Georgia Institute of Technology, Atlanta, 1967.

Mei, C. C., "Non-uniform Diffusion of Suspended Sediment," *J. Hydraul. Div. ASCE*, **95**(HY1), pp. 581-584, January 1969.

Meyer-Peter, E. and Muller, R., "Formulas for Bed-Load Transport," Paper No. 2, Proceedings of the Second Meeting, IAHR, 1948, pp. 39-64.

Nakato, T., "Numerical Integration of Eintein's Integrals," *J. Hydraul. Eng. ASCE*, **110**(12), pp. 1863-1868, December 1984.

Parker, G. and Coleman, N. L., "Simple Model of Sediment-Laden Flows," *J. Hydraul. Eng. ASCE*, **112**(5), pp. 356-375, May 1986.

Parker, G., Klingeman, P. C., and McLean, D. G., Bed Load and Size Distribution in Paved Gravel-Bed Streams," *J. Hydraul. Div. ASCE*, **108**(HY4), pp. 544-571, April 1982.

Ranga Raju, K. G., Garde, R. J. and Bhardwaj, R., "Total Load Transport in Alluvial Channels," *J. Hydraul. Div. ASCE*, **107**(HY2), pp. 179-191, February 1981.

Raudkivi, A. J., *Loose Boundary Hydraulics*, 2nd edition, Pergamon Press, Oxford, England, 1976.

Rottner, J., "A Formula for Bed-Load Transportation," *Houille Blanche*, **14**, pp. 285-300, 1959.

Rouse, H., "Modern Conceptions of the Mechanics of Fluid Turbulence," *Trans. ASCE*, **102**, pp. 463-505, 1937.

Rubey, W. W., "Equilibrium Conditions in Debri-Laden Streams," *Trans. AGU*, pp. 497-505, 1933.

Shen, H. W. and Hung, C. S., "An Engineering Approach to Total Bed Material Load by Regression analysis," in *Sedimentation (Einstein)*, H. W. Shen, ed., P.O. Box 606, Fort Collins, Colorado, 1972, Chapter 14.

Shen, H. W. and Hung, C. S., "Remodified Einstein Procedure for Sediment Load," *J. Hydraul. Eng.* ASCE, **109**(4), pp. 565-578, April 1983.

Shields, A., "Anwendung Aenlich Keitsmechanik und der Trubulenzfor-schung auf Die Geschiebebewegung," Mitteilungen de Preussischen Versuchsanstalt fur Wasserbau und Schiffbau, Berlin, Germany, 1936.

Simons, D. B. and Senturk, F., *Sediment Transport Technology*, Water Resources Publications, P.O. Box 2841, Littleton, Colorado, 1977.

Straub, L. G., "Missouri River Report," House Document 238, Appendix XV, Corps of Engineers, United States Department of the Army to 73rd United States Congress, 2nd Session, 1935, p. 1156.

Straub, L. G., "Effect of Water Temperature on Suspended Sediment Load in an Alluvial River," Paper D-25, Proceedings of the Sixth Meeting, IAHR, The Hague, The Netherlands, 1955.

Talapatra, S. C. and Ghosh, S. N., "Incipient Motion Criteria for Flow over a Mobile Bed Sill," *Proceedings of the Second International Symposium on River Sedimentation*, Nanjing, China, pp. 459-471, October 1983.

Toffaleti, F. B., "Definitive Computations of Sand Discharge in River," *J. Hydraul. Div. ASCE*, **95**(HY1), pp. 225-246, January 1969.

von Karman, T., "Turbulence and Skin Friction, "*J. of the Aeronautical Sciences*, **1**(1), pp. 1-20, January 1934.

van Rijn, L. C., "Sediment Transport: Bed Load Transport," *J. Hydraul. Eng. ASCE*, **110**(10), pp. 1431-1456, October 1984a.

van Rijn, L. C., "Sediment Transport II: Suspended Load Transport," *J. Hydraul. Eng.* ASCE, **110**(11), pp.1613-1641, November 1984b.

van Rijn, L. C., "Mathematical Modeling of Suspended Sediment in Nonuniform Flows," *J. Hydraul. Eng. ASCE*, **112**(6), pp. 433-455, June 1986.

Vanoni, V. A., "Some Experiments on the Transportation of Suspended Loads," *Trans. Am. Geophys. Union*, **20**(3), pp. 608-621, 1941.

Vanoni, V. A., "Transportation of Suspended Sediment by Water," *Trans. ASCE*, **111**, pp. 67-133, 1946.

Vanoni, V. A., "Some Effects of Suspended Sediment on Flow Characteristics," *Proceedings*, *Fifth Hydraulic Conference*, Bulletin 34, State University of Iowa, Studies in Engineering, Iowa City, Iowa, 1953, pp. 137-158.

White, W. R., Milli, H., and Crabbe, A. D., "Sediment Transport: An Appraisal of Available Methods," Vol. 2, Performance of Theoretical Method When Applied to Flume and Field Data, Report INT 119, Hydraulic Research Station, Wallingford, England, 1973.

Yalin, M. S., *Mechanics of Sediment Transport*, Pergamon Press, 1977.

Yang, C. T., "Unit Stream Power and Sediment Transport," *J. Hydraul. Div. ASCE*, **18**(HY10), pp. 1805-1826, October 1972.

Yang, C. T., "Incipient Motion and Sediment Transport," *J. Hydraul. Div. ASCE*, **99**(10), pp. 1679-1704, October 1973.

Yang, C. T. and Molinas, A. "Sediment Transport and Unit Stream Power Function," *J. Hydraul. Div. ASCE*, **108**(HY6), pp. 776-793, June 1982.

Yang, C. T., "Unit Stream Power Equation for Gravel," *J. Hydraul. Eng. ASCE*, **110**(HY12), pp. 1783-1798, December 1984.

Zhang, Q, Zhang, Z., Yue, J., Duan, Z., and Dai, M., "A Mathematical Model for the Prediction of the Sedimentation Process in Rivers," *Proceedings of the 2nd International Symposium on River Sedimentation*, Nanjing, China, 1983.

# 8

# FLOW IN CURVED RIVER CHANNELS

Any casual observer of a river would soon notice the sinuous channel pattern. In fact, nonbraided rivers are hardly straight over a length longer than a few channel widths. Because of the close interrelationship between river flow and river channel formation, many river channel features and processes, such as meander planform, bed topography, bank erosion, and lateral migration, are very much related to the dynamics of flow in curved channels, which, in turn, provides the basis of analysis and modeling. The flow in curved channels is under the influence of centrifugal acceleration, which induces (1) spiral motion in flow and (2) superelevation in water surface. Spiral motion, also known as helical motion, secondary currents, or transverse circulation, is in the direction normal to that of the primary (longitudinal) flow. Its occurrence is due to the difference in centrifugal acceleration $u^2/r$ ($u$ is the local longitudinal velocity, and $r$ is the radius of curvature) along a vertical line in the flow because of the vertical profile of $u$ in viscous fluid (see Fig. 8.1). For inviscid fluid without the velocity profile, spiral motion does not develop.

Spiral motion grows upon entering a bend. In a prismatic channel bend of sufficient length, the flow will eventually reach an equilibrium condition under which flow characteristics do not change from cross section to cross section. Such a flow is said to be fully developed. Because of the changing curvature of river channels, the spiral motion undergoes constant growth and decay. Therefore, an important area of investigation, included in this chapter, is the streamwise variation of the strength of spiral motion to which other flow and channel features are closely associated.

Basic equations governing the dynamics of flow in curved channels are derived in this chapter. Important features of the flow and sediment-transport processes, including transverse velocity profile, shear stress on channel bed, transverse bed slope, sediment size distribution, and energy expenditure, are described. In addition, streamwise variations of the spiral motion and other flow characteristics are also introduced.

## 8.1 BASIC EQUATIONS

Derivations of the equations of motion and continuity for curved channels are given by Rozovskii (1957), Rouse (1959), and Schlichting (1968), among others.

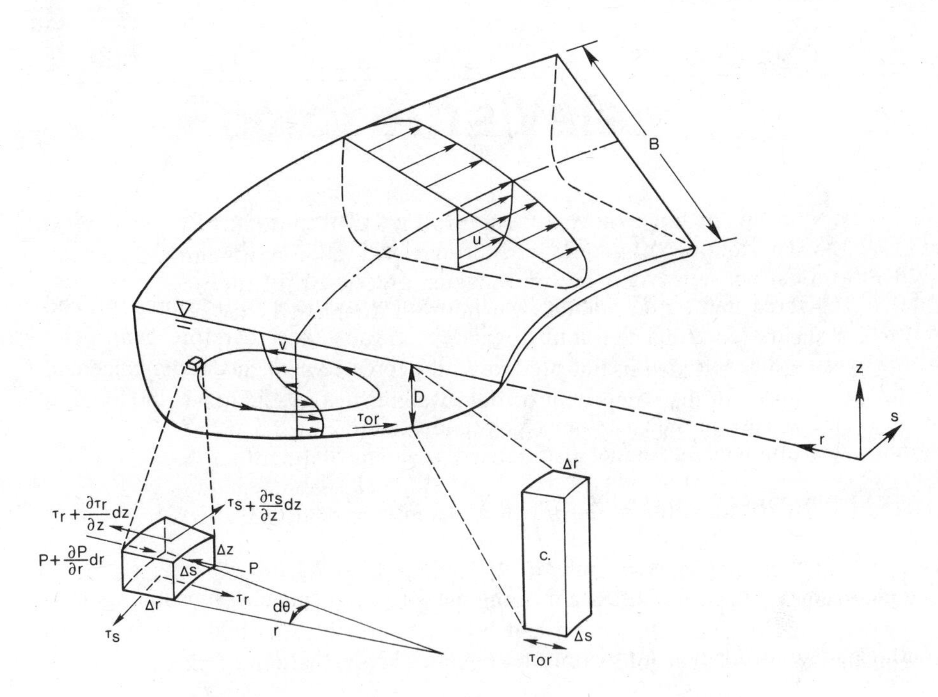


**Figure 8.1** Definition sketch for flow in a curved channel.

The analysis of flow in curved channels as presented herein is restricted to subcritical flow with hydrostatic pressure distribution. In river channels, the depth is generally much less than the width and radius of curvature. By assuming wide channels, bank effects are neglected in the analysis. In deriving the equation of motion, a differential element of fluid with dimensions $\Delta s$, $\Delta r$, and $\Delta z$, as shown in Fig. 8.1, is considered. The fundamental statement of Newton's law for an inertial reference is given in terms of momentum as follows:

$$d\mathbf{F} = dm\frac{D\mathbf{U}}{Dt} \tag{8.1}$$

where $\mathbf{F}$ is the force vector, $\mathbf{U}$ is the velocity vector, $t$ is time, and $dm$ is the mass of the differential element $\Delta s \Delta r \Delta z$. This mass is a part of the velocity field $\mathbf{U}$ that can be written in terms of the tangential, radial, and vertical components as follows:

$$\mathbf{U} = u\mathbf{i}_s + v\mathbf{i}_r + w\mathbf{i}_z \tag{8.2}$$

where $u$, $v$, and $w$ are velocity components, and $\mathbf{i}_s$, $\mathbf{i}_r$ and $\mathbf{i}_z$ are unit vectors in the respective directions of $s$, $r$, and $z$.

The right-hand quantity in Eq. 8.1 is the substantial derivative, that is, the derivative following the motion of the fluid element, whose components in polar–

cylindrical coordinates are

$$a_s = \frac{Du}{Dt} = \frac{\partial u}{\partial t} + u\frac{\partial u}{\partial s} + v\frac{\partial u}{\partial r} + w\frac{\partial u}{\partial z} + \frac{uv}{r} \tag{8.3}$$

$$a_r = \frac{Dv}{Dt} = \frac{\partial v}{\partial t} + u\frac{\partial v}{\partial s} + v\frac{\partial v}{\partial r} + w\frac{\partial v}{\partial z} - \frac{u^2}{r} \tag{8.4}$$

The force vector, $d\mathbf{F}$, acting on the element includes surface pressure and surface shear. Note that the centrifugal acceleration, and therefore centripetal force, is already included in Eq. 8.4. Now, the force components in the tangential direction consist of the component of fluid weight, $\rho g S \Delta s \Delta r \Delta z$ ($S$ is the longitudinal slope), and the shear force, which is given by

$$\left[\left(\tau_s + \frac{\partial \tau_s}{\partial z}dz\right) - \tau_s\right] dsdr = \frac{\partial \tau_s}{\partial z} dsdrdz \tag{8.5}$$

Other tangential surface forces are negligible for wide channels. Substituting these forces in the tangential direction and Eq. 8.3 into Eq. 8.1 yields the following equation for the $s$ direction:

$$\frac{\partial u}{\partial t} + u\frac{\partial u}{\partial s} + u\frac{\partial u}{\partial r} + w\frac{\partial u}{\partial z} = -\frac{uv}{r} + gS + \frac{1}{\rho}\frac{\partial \tau_s}{\partial z} \tag{8.6}$$

The net pressure force in the radial direction is attributed to the superelevation, which causes a difference in hydrostatic pressure on the sides of the prism, that is,

$$\left[P - \left(P + \frac{\partial P}{\partial r}dr\right)\right] dsdz = -\rho g S_r dsdrdz \tag{8.7}$$

where $S_r$ is the transverse water-surface slope. The radial component of the shear force is

$$\left[\left(\tau_r + \frac{\partial \tau_r}{\partial z}dz\right) - \tau_r\right] ds\,dr = \frac{\partial \tau_r}{\partial z} dsdrdz \tag{8.8}$$

Other radial surface forces are negligible under the assumption of wide channel. Substituting the radial quantities of Eqs. 8.4, 8.7, and 8.8 into Eq. 8.1 yields the following equation for the $r$ direction

$$\frac{\partial v}{\partial t} + u\frac{\partial v}{\partial s} + v\frac{\partial v}{\partial r} + w\frac{\partial v}{\partial z} = \frac{v^2}{r} - gS_r + \frac{1}{\rho}\frac{\partial \tau_r}{\partial z} \tag{8.9}$$

The continuity equation has the form

$$\frac{\partial v}{\partial r} + \frac{v}{r} + \frac{\partial v}{\partial s} + \frac{\partial w}{\partial z} = 0 \tag{8.10}$$

Transverse inclination of the water surface, or superelevation, in a channel bend can be obtained from the balance of radial forces acting on the column of fluid with depth $D$. If the transverse force contributed by the bed is neglected, then pressure force associated with transverse surface inclination is balanced by the centripetal force, that is,

$$\int_0^D \frac{v^2}{r} \rho \, dsdrdz - \rho g S_r \, dsdrdz = 0 \tag{8.11}$$

Thus,

$$S_r = \frac{\int_0^D u^2 \, dz}{gr} = \frac{C_r U^2}{gr} \tag{8.12}$$

where $C_r$ is a correction factor and $U$ is the depth-averaged velocity used to replace the local velocity $u$. By assuming $C_r = 1$, the superelevation $\Delta Z$ between the outside bank and inside bank can be approximated as

$$\Delta Z = \int_{r_1}^{r_2} S_r \, dr = \int_{r_1}^{r_2} \frac{U^2}{gr} dr \simeq \frac{\overline{U}^2 B}{g r_c} \tag{8.13}$$

where $r_1$ is the inner radius, $r_2$ is the outer radius, $r_c$ is the center radius, $B$ is the surface width and $\overline{U}$ is the cross-sectionally averaged velocity.

## 8.2 TRANSVERSE VELOCITY PROFILES FOR FULLY DEVELOPED FLOW

A number of transverse velocity profiles for steady, fully developed flow in curved channels have been developed (see, e.g., Rozovskii, 1957; Yen, 1972; Kikkawa et al., 1976; and Falcon-Ascanio and Kennedy, 1983). Most relationships are obtained based on the semiempirical theories of turbulence, using the equations of motion (Eqs. 8.6 and 8.9) as the analytical basis. For steady flow, the time derivatives in Eqs. 8.6 and 8.9 drop out. For fully developed flow, $\partial u/\partial s = 0$ and $\partial v/\partial s = 0$. In addition, velocity components $v$ and $w$ are small in comparison to $u$ in a wide channel. After dropping second order terms the equations of motion become

$$gS + \frac{1}{\rho}\frac{\partial \tau_s}{\partial z} = 0 \tag{8.14}$$

and

$$\frac{v^2}{r} - gS_r + \frac{1}{\rho}\frac{\partial \tau_r}{\partial z} = 0 \tag{8.15}$$

Equation 8.14, in essence, gives the shear stress $\tau_s$ as $\rho g(D - z)$ whereas Eq. 8.15 provides the radial force distribution along the vertical. By assuming isotropic turbulence, the shear stresses $\tau_s$ and $\tau_r$ in turbulent flow can be expressed in terms of the eddy viscosity and the respective velocity gradients, that is,

$$\tau_s = \varepsilon\frac{\partial u}{\partial z} \quad \text{and} \quad \tau_r = \varepsilon\frac{\partial v}{\partial z} \tag{8.16}$$

The eddy viscosity $\varepsilon$ is determined from Eqs. 8.14 and 8.16 by a prescribed profile of the longitudinal velocity along the vertical, $u = Fn(z)$. Then this value of $\varepsilon$ is substituted into Eq. 8.16 for $\tau_r$, with which the transverse velocity profile $v = Fn(z)$ is obtained by integration. The integration constant is evaluated at the water surface.

Rozovskii (1957) assumed the following logarithmic longitudinal velocity distribution:

$$\frac{u}{U} = 1 + \frac{g^{1/2}}{\kappa C}(1 + \ln \eta) \tag{8.17}$$

where $U$ is the depth-averaged longitudinal velocity, $\kappa$ is the von Karman constant (which has the approximate value of 0.4 in clear fluid), $C$ is the Chezy coefficient, and $\eta = z/D$. Using this velocity distribution, Rozovskii derived the following formula for transverse (radial) velocity profile in the case of a smooth bottom:

$$\frac{v}{U} = \frac{1}{\kappa^2}\frac{D}{r}\left[F_1(\eta) - \frac{g^{1/2}}{\kappa C}F_2(\eta)\right] \tag{8.18}$$

where $F_1(\eta)$ and $F_2(\eta)$ as functions of the relative depth $\eta$ are given respectively as

$$F_1(\eta) = \int \frac{2 \ln \eta}{\eta - 1}\, d\eta \tag{8.19}$$

$$F_2(\eta) = \int \frac{\ln^2 \eta}{\eta - 1} d\eta \tag{8.20}$$

These functions are shown graphically in Fig. 8.2.

In the case of rough channel bottom, Rozovskii obtained the following equation:

$$\frac{v}{U} = \frac{1}{\kappa^2} \frac{D}{r} \left\{ F_1(\eta) - \frac{g^{1/2}}{\kappa C} [F_2(\eta) + 0.8(1 + \ln \eta)] \right\} \tag{8.21}$$

Velocity profiles based on this equation for $C = 60$ and $C = 30$ in metric units are shown in Fig. 8.3. With changes in roughness, the velocity profiles vary mainly in the bottom region and remain almost constant throughout the flow depth, agreeable with Rozovskii's experimental findings.

The transverse velocity profile by Kikkawa et al. (1976) was derived from the equation of motion. The eddy viscosity is assumed to be the same as that of the straight, two-dimensional channel. They obtained the formula

$$\frac{v}{U} = F^2 \frac{1}{\kappa} \frac{D}{r} \left[ F_A(\eta) - \frac{1}{\kappa} \frac{U_*}{U} F_B(\eta) \right] \tag{8.22}$$

where $U$ is the cross-sectionally averaged velocity, $F$ is the radial distribution of depth-averaged velocity normalized by $U$, $U_*$ is the average shear velocity of the section,

$$F_A(\eta) = -15\left(\eta^2 \ln \eta - \frac{1}{2}\eta^2 + \frac{15}{54}\right) \tag{8.23}$$

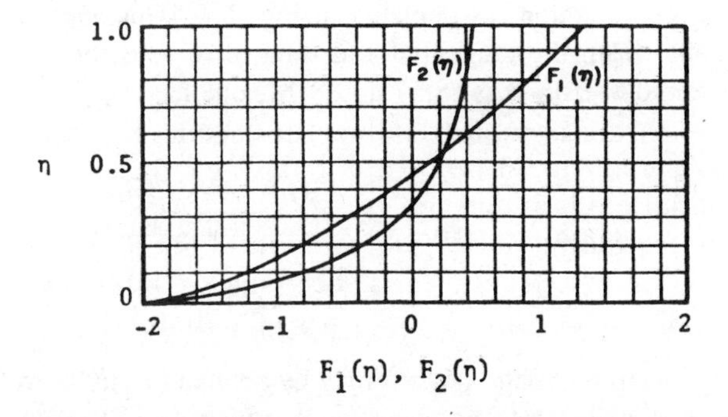


**Figure 8.2** Graphs of functions $F_1(\eta)$ and $F_2(\eta)$ by Rozovskii.

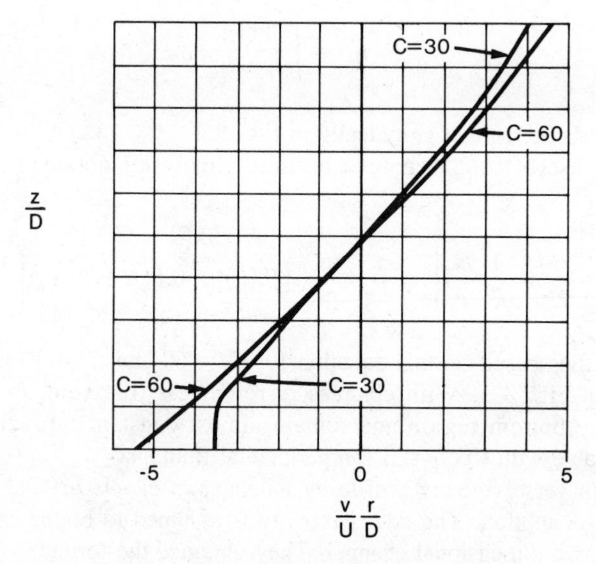


**Figure 8.3** Vertical distributions of transverse velocity from Rozovskii's equation for rough boundary.

and

$$F_B(\eta) = \frac{15}{2}\left(\eta^2 \ln^2\eta - \eta^2 \ln \eta + \frac{1}{2}\eta^2 - \frac{19}{54}\right) \tag{8.24}$$

Equation 8.22, similar to Eq. 5.18 by Rozovskii, indicates that the transverse velocity is directly related to $D/r$ and the longitudinal velocity. Experimental verification of this formula was provided by Kikkawa (1976)

Because of the transverse velocity component, the velocity vector near the channel bed is not in the tangential direction, but rather assumes an angle of deviation $\delta$ from this direction, as depicted in Fig. 8.4. This angle at any depth is readily obtainable from the tangential and transverse velocity profiles along the vertical. For example, from Eqs. 8.17 and 8.18, one has

$$\tan \delta = \frac{v}{u} = \frac{1}{\kappa^2}\frac{D}{r}\frac{F_1(\eta) - \dfrac{g^{1/2}}{\kappa C}F_2(\eta)}{1 + \dfrac{g^{1/2}}{\kappa C}(1 + \ln \eta)} \tag{8.25}$$

A similar relationship for rough bottom may be obtained from the velocity profiles given by Eqs. 8.17 and 8.21. The value of tan $\delta$ near the channel bed is evaluated by setting $\eta$ to zero. For two values of the Chezy coefficient ($C = 60$ and 30)

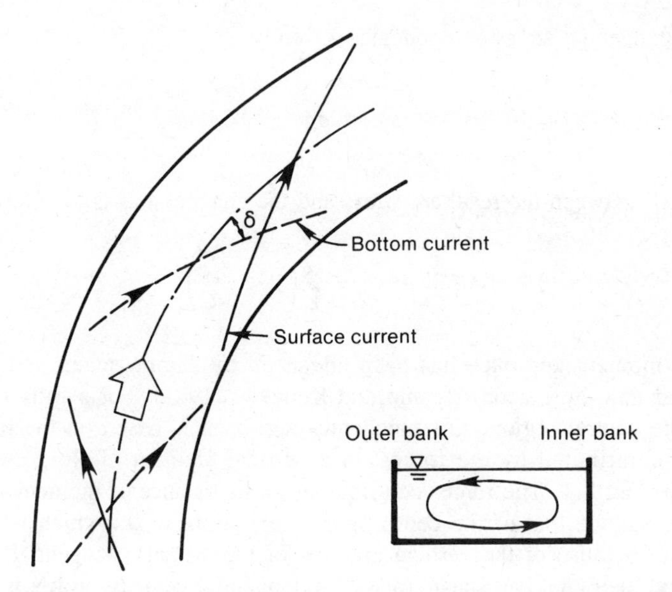


**Figure 8.4** Sketch of bottom and surface currents with angle of deviation from tangential direction of channel.

Rozovskii obtained the corresponding values

$$\tan \delta \simeq 11\frac{D}{r} \tag{8.26}$$

which led him to the conclusion that channel roughness only has minor effect on this angle. If the transverse velocity profile for rough bottom is used, a similar relationship is also obtained. This result has been proved satisfactory when compared with both laboratory and field data (Kondrat'ev et al., 1959).

## 8.3 BOUNDARY SHEAR STRESS

Boundary shear stress in curved channels is usually analyzed in terms of its tangential (longitudinal) and radial (transverse) components. The radial shear stress is closely related to the transverse sediment movement, bed topography, and grain size distribution, and it heavily influences the radial velocity profile $v(r,\eta)$. The radial shear stress in fully developed transverse flow has been determined based on the radial velocity gradient (see Eq. 8.16) or the moment balance described later. From a radial velocity profile similar to Rozovskii's (Eq. 8.21), Jansen et al. (1979) obtained the following equation for the radial component of boundary shear stress:

$$\tau_{0r} = -\rho D\frac{U^2}{r}\left[2\left(\frac{g^{1/2}}{\kappa C}\right)^2 - 2\left(\frac{g^{1/2}}{\kappa C}\right)^3\right] \tag{8.27}$$

Since the tangential stress component is given by

$$\tau_{0s} = \rho g \frac{U^2}{C^2} \tag{8.28}$$

the angle $\delta'$ between the resultant stress and the channel axis is therefore

$$\tan \delta' = -\frac{2}{\kappa^2}\frac{D}{r}\left(1 - \frac{g^{1/2}}{\kappa C}\right) \tag{8.29}$$

The moment approach has been advanced by Zimmermann and Kennedy (1978) and then by Falcon-Ascanio and Kennedy (1983). For a fully developed flow in the central region, this mechanics approach is based on the balance of moments contributed by the forces on a vertical element of fluid $D\Delta s\Delta r$, as depicted in Fig. 8.1. The forces contributing to the balance of moments about the centroid $c$ include that of the centrifugal acceleration, $u^2/r$, which is vertically nonuniform because of the vertical gradient of the primary (tangential) velocity, $u(r,\eta)$, and the radial bed shear stress. The tangential velocity profile is based on the power-law equation

$$\frac{u}{U} = \frac{1+m}{m}\left(\frac{z}{D}\right)^{1/m} \tag{8.30}$$

where $m$ is the reciprocal of the exponent and is related to the friction factor $f$ and von Karman's $\kappa$ by

$$m = \kappa\left(\frac{8}{f}\right)^{1/2} \tag{8.31}$$

The force due to centrifugal acceleration contributes a moment given by the expression

$$\rho \int_0^D \frac{u^2}{r}\left(z - \frac{D}{2}\right) dsdrdz \tag{8.32}$$

The radial pressure-gradient force passes through the centroid of the element and therefore contributes no moment. The vertical shear stress is small because of the small radial velocity gradient and its moment contribution has been shown to be of a higher order (Falcon-Ascanio and Kennedy, 1983). Therefore, the moment in Eq. 8.32 is balanced primarily by the radial bed shear stress, that is,

$$\frac{1}{2} D \tau_{0r} dsdr = \rho \int_0^D \frac{u^2}{r}\left(z - \frac{D}{2}\right) rd\theta\, drdz \tag{8.33}$$

After simplification, it becomes

$$\tau_{0r} = \frac{2\rho}{rD} \int_0^D u^2 \left( z - \frac{D}{2} \right) dz \tag{8.34}$$

Substituting the velocity profile for $u$ (Eq. 8.30) into Eq. 8.34 and integrating yields

$$\tau_{0r} = \frac{1 + m}{(2 + m)m} \rho \frac{D}{r} U^2 \tag{8.35}$$

The application of this radial bed stress on the transverse bed slope is given in the following section.

## 8.4 TRANSVERSE BED SLOPE AND GRAIN-SIZE DISTRIBUTION

The radial bed shear stress gives rise to transverse sediment movement that will establish a transverse bed slope, see Figs. 2.7 and 8.5. In the complex interaction

**Figure 8.5** Transverse bed slope in curved channel.

between the flow and alluvial bed, the grain size also becomes nonuniformly distributed with coarser particles usually found in the thalweg. Analyses of the flow-bed interaction in alluvial channel bends have been made by NEDECO (1959), Yen (1970), El-Khudairy (1970), Engelund (1974), Kikkawa, et al. (1976), Bridge (1977), Zimmermann and Kennedy (1978), Odgaard (1981, 1982, 1984), Falcon-Ascanio and Kennedy (1983), and Parker and Andrews (1985), among others. Considerable success has been made in the analysis of the transverse bed slope in fully developed flow. Progress has also been made on the steady-state grain-size distribution in river bends. Certain studies pertaining to the transverse bed slope and grain-size distribution are described in the following subsections.

## Falcon-Ascanio–Kennedy Approach

In the approach by Falcon-Ascanio and Kennedy (1983), the local bed inclination $\beta(r)$ (see Fig. 8.6) is determined based on the assumption that the radial component of fluid force balances the submerged weight component down the transverse bed slope, that is,

$$\tau_{0r} = z_b(1 - \lambda)(\rho_s - \rho)g \sin \beta \tag{8.36}$$

where $z_b$ is the bed-layer thickness and $\lambda$ is the bed-layer porosity. This thickness was developed by Karim (1981) from inferential evidence; it is expressed as

$$z_b = d\frac{U_*}{U_{*c}} \tag{8.37}$$

where $d = d_{50}$ is the median bed-material size, $U_*(r) = U/(f/8)^{1/2}$ is the local shear velocity, and $U_{*c}$ is the critical shear velocity for incipient particle motion.

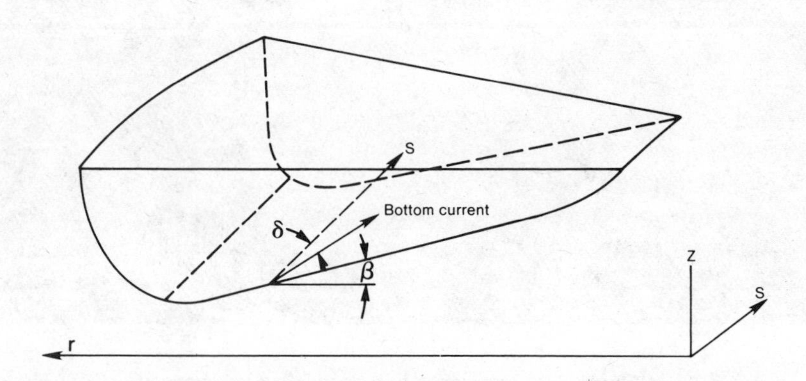


**Figure 8.6** Sketch showing direction of bottom currents on transverse bed slope.

In terms of the critical Shields stress $\tau_{*c} = \tau_c/(\rho_s - \rho)\, gd$, $U_{*c}$ may be expressed as

$$U_{*c} = \left(\frac{\tau_c}{\rho}\right)^{1/2} = \left(\frac{\rho_s - \rho}{\rho} gd\tau_{*c}\right)^{1/2} \tag{8.38}$$

Substitution of Eqs. 8.35, 8.37, and 8.38 into Eq. 8.36 and incorporation of the simplified Nunner's relation $m = 1/f^{1/2}$ yields

$$S_t = \sin\beta \simeq \frac{dD}{dr} = \frac{D}{r}\mathbf{F}_d \frac{(8\tau_{*c})^{1/2}}{1-\lambda}\frac{1+f^{1/2}}{1+2f^{1/2}} \tag{8.39}$$

where $S_t$ is the transverse bed slope and $\mathbf{F}_d$ is the particle densimetric Froude number defined by

$$\mathbf{F}_d = \frac{U}{\{[(\rho_s - \rho)/\rho]gd\}^{1/2}} \tag{8.40}$$

Equation 8.39 is substantiated with measured transverse bed slopes in flume experiments with sand.

The transverse bed profile may be obtained from Eq. 8.39 by integration, provided the transverse variations in velocity and grain size are given or determined. For this purpose, the local depth-averaged velocity $U(r)$ is related to the local longitudinal slope by the Darcy–Weisbach equation $U(r) = (8\tau_0/\rho f)^{1/2}$. In order for the transverse slopes of the bed and water surface to be constant along the channel, the local longitudinal slope $S$ of both must follow the relationship (Yen and Yen, 1971)

$$S = S_c \frac{r_c}{r} \tag{8.41}$$

where the subscript $c$ designates centerline values. Therefore

$$U(r) = \left[\frac{8}{f} gSD(r)\right]^{1/2} = \left[8S_c \frac{r_c}{r}\frac{gD(r)}{f}\right]^{1/2} \tag{8.42}$$

Note that this equation neglects the bed-shear-stress reduction factor, which takes into account the transport of primary flow momentum out of the central region to the vicinity of the outer bank, where it is balanced by the bank stress. Substituting Eqs. 8.42 into Eq. 8.39 and integrating the resulting expression for $dD/dr$ yields

$$\frac{1}{D^{1/2}} - \frac{1}{(D_c)^{1/2}} = \left[\frac{1}{r^{1/2}} - \frac{1}{(r_c)^{1/2}}\right]\frac{(8\tau_{*c})^{1/2}}{1-\lambda}\frac{1+f^{1/2}}{1+2f^{1/2}}\left\{\frac{8S_c r_c g}{fg[(\rho_s-\rho)/\rho]d}\right\}^{1/2} \tag{8.43}$$

This equation gives a slightly convex profile, as exemplified in Fig. 8.7 for the Missouri River data with $(8\tau_{*c})^{1/2}/(1 - \lambda) = 1.3$ and the given grain-size distribution.

## Engelund–Bridge Approach

The approach by Engelund (1974) was later modified and extended by Bridge (1977). This approach for steady, fully developed flow is based on the argument that the net rate of transverse sediment transport is zero under the dynamic equilibrium. Although the channel migrates laterally, the migration rate is small compared with the movement of bed-load particles and thus the assumption of a stationary bed is justified.

The transverse sediment movement on an inclined channel bed is under the influence of spiral motion and gravity. As shown in Fig. 8.6, the bottom currents are at an angle of deviation $\delta$ from the longitudinal direction. But under the dynamic equilibrium, the particles move in the longitudinal direction. Therefore, the force on the particle due to the bottom currents is balanced by the submerged weight directed down the transverse bed slope, that is,

$$F_D \sin \delta = (W - F_L) \sin \beta \tag{8.44}$$

where $F_D$ is the drag on a single particle in the direction of shear, $W$ is the submerged weight of particle, and $F_L$ is the lift force. In the presence of bed forms,

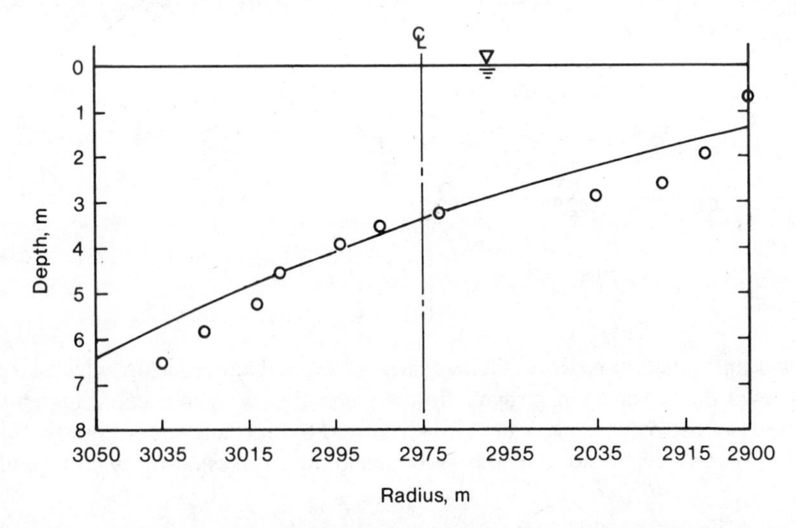


**Figure 8.7** Comparison of measured transverse river bed profile with analytical prediction by Falcon-Ascanio and Kennedy (1983). (Copyrighted by and reprinted with the permission of Cambridge University Press).

the angle $\beta$ would assume an average value. The bed slope in the longitudinal direction is assumed negligible. Under such an equilibrium, the balance of forces acting on a particle in the downstream (longitudinal) direction is

$$F_D \cos \delta = (W - F_L) \cos \beta \tan \phi \tag{8.45}$$

where $\phi$ is the angle of repose of sediment and tan $\phi$ is the dynamic friction coefficient. Combining Eq. 4.44 and 4.45 gives

$$\tan \delta = \frac{\tan \beta}{\tan \phi} \tag{8.46}$$

This equation, derived by Engelund (1974), is valid as long as $\beta$ is small in comparison to $\phi$. Its accuracy becomes questionable when an appreciable amount of suspended sediment is present.

Bridge's approach on the transverse bed slope is based on Eq. 8.46 and on the angle of deviation $\delta$ obtained by Rozovskii (1957), as given in Eq. 8.26. Substituting Eq. 8.26 and $\tan \beta = dD/dr$ into Eq. 8.46 yields

$$\frac{dD}{dr} = 11 \frac{D}{r} \tan \phi \tag{8.47}$$

After integration, this equation becomes

$$D = cr^{11 \tan \phi} \tag{8.48}$$

The integration constant $c$ is evaluated at the channel centerline, thus

$$\frac{D}{D_c} = \left(\frac{r}{r_c}\right)^{11 \tan \phi} \tag{8.49}$$

This equation has been shown to fit laboratory and natural point-bar profiles very well when tan $\phi$ is about 0.4–0.5. Although $\phi$ varies with particle size, the effect of particle size on transverse bed profile is not explicitly expressed in Eq. 8.49.

The transverse distribution of grain size, in Bridge's approach, is determined based on the assumption that particles at different flow depths along the bed surface all travel in the longitudinal direction under the lateral equilibrium. Therefore, the grain sizes must be distributed in such a way that this equilibrium, or zero lateral transport, is maintained. Near the concave bank, where the depth is greater, grain particles are under larger longitudinal and transverse shear. This larger transverse force is balanced by heavier particles such that they maintain the longitudinal direction of movement. The mathematical derivation is given as follows.

For any given particle on the transverse bed surface, the balance of transverse drag and weight component down the slope is given as

$$\pi\left(\frac{d}{2}\right)^2 \tau_0 \tan\delta = \frac{4}{3}\pi\left(\frac{d}{2}\right)^3 (\rho_s - \rho)g \sin\beta \tag{8.50}$$

where $\tau_0$ is the longitudinal bed shear stress. Because $\beta$ is small, $\sin\beta \simeq \tan\beta = dD/dr$, and Eq. 8.50 reduces to

$$\tan\beta = \frac{dD}{dr} = \frac{3\tau_0 \tan\delta}{2dg(\rho_s - \rho)} \tag{8.51}$$

From Eqs. 8.46 and 8.51, one has

$$d = \frac{3\tau_0}{2(\rho_s - \rho)g \tan\phi} \tag{8.52}$$

From Eq. 8.42 for fully developed flow, a relationship for $\tau_0$ may be written

$$\tau_0 = \rho g D S_c \frac{r_c}{r} \tag{8.53}$$

From Eqs. 8.52 and 8.53, an approximate relationship for grain-size distribution is obtained:

$$d = \frac{3\rho D S_c r_c}{2(\rho_s - \rho)r \tan\phi} \tag{8.54}$$

The value of $d$ for any local depth $D$ given by Eq. 8.49 can thus be determined. Note that this equation requires no reference particle size, such as $d_c$.

## Odgaard Approach

Odgaard (1981, 1982, 1984) made studies of the transverse bed profile, velocity, and grain-size distribution. He formulated an alternative approach based on a theory for the critical condition of sediment motion. As a basic assumption, the size of the bed-forming particles is such that the particles are just about to move in the longitudinal direction by rolling about their points of support. The reasoning of this assumption is that when the grain shear stress $\tau_0'$ exceeds the critical shear stress $\tau_c$, the immobile bed will continue to be acted upon by $\tau_c$. The excess shear $\tau_0' - \tau_c$ is carried as drag on the moving bed particles and only indirectly is transferred to the bed by occasional encounters.

The Odgaard approach for the transverse bed slope may be considered as a modification of the Falcon-Ascanio–Kennedy approach. Although the derivations are basic similar, the major difference is in Odgaard's use of the critical condition based on the grain roughness. The transverse bed slope so derived has the following form:

$$\sin \beta = \frac{3\alpha}{2} \frac{D}{r} \frac{U^2}{[(s - 1)gd_{cr}]^{1/2}} \frac{1 + m'}{m'(2 + m')} \tag{8.55}$$

where $\alpha$ is the projected area–volume ratio for a particle, normalized by that for a sphere, $s = \rho_s/\rho$ is the specific gravity of a particle; $d_{cr}$ is the diameter of a particle whose motion is impending; and $m'$ is the reciprocal of the velocity exponent for grain roughness. The value of $m'$ is related to the critical Shields stress $\tau_{*c} = \tau_c/[(\rho_s - \rho)gd_{cr}]^{1/2}$ as follows:

$$m' = \kappa \frac{U}{[(s - 1)gd_{cr}\tau_{*c}]^{1/2}} \tag{8.56}$$

Since the angle $\beta$ is related to the velocity $U$ and particle diameter $d_{cr}$ in Eq. 8.55, the transverse variations of $U$, $d_{cr}$ and $m'$ are required in using this equation to obtain the transverse bed profile. Such variations are described in the following text. From Eq. 8.42, the velocity distribution can be expressed with reference to the channel centerline quantities as follow:

$$\frac{U}{U_c} = \frac{m'}{m'_c}\left(\frac{D}{D_c}\right)^{1/2}\left(\frac{r_c}{r}\right)^{1/2} \tag{8.57}$$

Manning's $n$ varies as the 1/6 power of the grain size in Strickler's formula for grain roughness (see Eq. 3.27). Therefore, $m'$ is assumed to vary with the relative depth $d/D$ to the 1/6 power; and then Eq. 8.57 becomes

$$\frac{U}{U_c} = \left(\frac{d_c}{d}\right)^{1/6}\left(\frac{D}{D_c}\right)^{2/3}\left(\frac{r_c}{r}\right)^{1/2} \tag{8.58}$$

This relationship for the radial variation of $U$ is supported by the results from Odgaard's Sacramento River bend study.

The grain-size distribution in a bend, with graded bed material, is not uniform and is characterized by coarser sediment near the concave bank. In a semiempirical approach, Odgaard (1982) assumed straight profile for the transverse bed based on physical evidence; he also assumed that the critical Shields stress $\tau_{*c}$ is proportional to $d^{-2/3}$ from previous studies. From these assumptions, the distribution of grain size was obtained as follows:

$$\frac{d}{d_c} = \left(\frac{D}{D_c}\right)^{5/3}\left(\frac{r_c}{r}\right)^{3/2} \tag{8.59}$$

Substituting Eq. 8.59 into Eq. 5.58 then yields

$$\frac{U}{U_c} = \left(\frac{D}{D_c}\right)^{7/18} \left(\frac{r_c}{r}\right)^{1/4} \tag{8.60}$$

In the application of Odgaard's method, average values of depth, velocity, and grain size for the cross section are required. In the procedure of application, the cross-sectional average value of $m'$ is first computed according to Eq. 8.56, and then the transverse bed slope is obtained from Eq. 8.55. For a reliable estimation, Eq. 8.55 should be calibrated against river data in advance.

## 8.5 LATERAL BED-LOAD TRANSPORT

The role of secondary currents in curved alluvial channels consists of moving particles away from the concave bank and toward the convex bank. This lateral movement has effects on point-bar formation, sediment sorting, lateral migration, bank erosion, and width variation. At the time of transverse equilibrium, the lateral bed-load transport is counterbalanced by the transverse bed slope. Methods for computing lateral bed-load discharge have been developed by Engelund (1974), Kikkawa et al. (1976), Ikeda (1982), and Parker (1984), among others. Such methods are essential for the understanding of the processes and mathematical modeling.

Lateral bed-load transport in a curved alluvial channel is caused by the secondary currents and is countered by the transverse bed slope. In a study by Ikeda (1982), the lateral bed-load transport on the transverse slope in a straight channel is given in functional form as

$$\frac{q'_*}{\tan \beta} = F\left(\frac{\tau_*}{\tau_{*c}}\right) \tag{8.61}$$

and

$$q'_* = \frac{q'_b}{[s - 1)gd^3]^{1/2}} \tag{8.62}$$

where $q'_*$ is the dimensionless lateral bed-load discharge per unit width, $q'_b$ is the lateral bed-load discharge per unit width, $\tau_*$ is the dimensionless shear stress or the Shields stress, and $\tau_{*c}$ is the critical Shields stress.

In order to establish the equation for the functional relationship, wind tunnel tests were performed by Ikeda using 0.26-mm and 0.42-mm sand. Based on these experimental results and other data from a water flume, the dimensionless bed-load discharge was obtained as shown in Fig. 8.8; it is given by the following

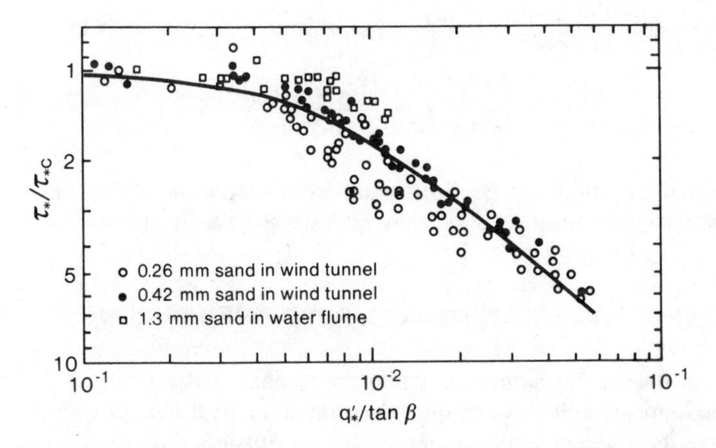


**Figure 8.8** Universal relation between $q_*'/\tan\beta$ and $\tau_*/\tau_{*c}$ by Ikeda (1982).

DuBoys-type equation:

$$\frac{q_*'}{\tan\beta} = 0.0085\left[\frac{\tau_*}{\tau_{*c}}\left(\frac{\tau_*}{\tau_{*c}} - 1\right)\right]^{0.5} \tag{8.63}$$

In curved alluvial channels, prediction of lateral bed-load transport must also include the effects of secondary currents in addition to the transverse bed slope. Following this concept, Parker (1984) suggested a general equation, on the basis of the approach by Kikkawa et al. (1976), for the lateral bed-load discharge as follows:

$$\overset{\text{I}}{\frac{q_b'}{q_b} = \frac{q_*'}{q_*}} = \overset{\text{II}}{\tan\delta} - \overset{\text{III}}{\frac{1 + (C_L/C_D)\tan\phi}{\tan\phi}\left(\frac{\tau_{*c}}{\tau_*}\right)^{1/2}\tan\beta} \tag{8.64}$$

where $q_b$ and $q_*$ are the longitudinal bed-load discharges, $C_L$ is the lift coefficient, and $C_D$ is the drag coefficient. In this equation, the lateral–longitudinal rate ratio (term I) is related to the deviation of bottom velocity from the longitudinal velocity due to secondary currents (term II) and the deflection of particle path due to the transverse bed slope (term III). The former contributes to the particle movement up the lateral slope, whereas the latter is responsible for the transport down the slope. In the case of lateral equilibrium, $q_s'$ is 0 and Eq. 8.64 is reduced to the form

$$\tan\beta = \frac{\tan\phi}{1 + (C_L/C_D)\tan\phi}\left(\frac{\tau_*}{\tau_{*c}}\right)^{1/2}\tan\delta \tag{8.65}$$

The transverse bed slope given in this equation bears resemblance to Eqs. 8.39 and 8.55.

In the absence of secondary currents, $\delta$ is zero and Eq. 8.64 becomes

$$\frac{q_b'}{q_b \tan \beta} = \frac{q_*'}{q_* \tan \beta} = \frac{1 + (C_L/C_D) \tan \phi}{\tan \phi} \left( \frac{\tau_{*c}}{\tau_*} \right)^{1/2} \tag{8.66}$$

Parker tested the validity of Eq. 8.66 using Ikeda's wind-tunnel data and then concluded that there is general consistency between Eqs. 8.63 and 8.66.

## 8.6 ENERGY EXPENDITURE IN CURVED OPEN CHANNELS

The rate of energy expenditure in curved open channels is a subject of scientific investigation because it plays an important role in alluvial river processes, and it is required in the energy (or power) approach. The effect of a curve, as explained by Chow (1959), is similar to that produced in a straight channel by increased roughness. Therefore, the power dissipated in a curved channel is greater than that in a straight channel of the same depth, velocity, and surface roughness. At subcritical conditions, such flow resistance will cause decreased flow velocities and increased depths upstream, as illustrated in Fig. 8.9. The backwater effect represents an upstream storage of potential energy which supplies the excess energy dissipation through the curve. The increase in power expenditure in a curved channel may be attributed to the following causes:

1. Internal fluid friction due to secondary currents.
2. Boundary resistance associated with transverse shear.

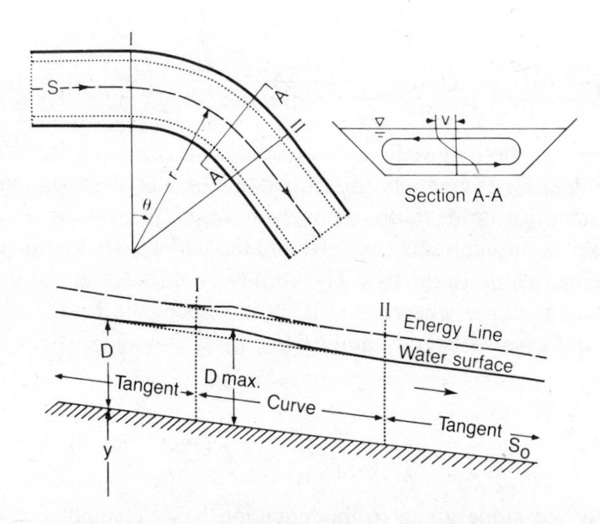


**Figure 8.9** Schematic of flow around channel bend.

3. Eddy loss resulting from flow separation in sharp bends.
4. Eddy loss due to sudden jumps occurring at high Froude numbers.

Different aspects of energy loss in curved channels have been studied by Bagnold (1960), Leopold et al. (1960), Rozovskii (1957), Ippen and Drinker (1962), Yen (1965), Hayat (1965), Soliman and Tinney (1968), Chang (1983), and Pang (1983), among others. Certain approaches are described herein.

With other conditions being equal, a meandering channel has been found to be steeper than a straight channel because of the necessity of overcoming the additional resistance due to channel curvature, based on a field study by Leopold et al. (1964) and laboratory studies by Bagnold (1960), Leopold et al. (1960), and Toebes and Sooky (1966). Certain such studies have related the energy expenditure in meandering channels to channel curvature, flow depth, Froude number, and other variables. The data indicate that channel curvature alone can produce losses of the same order as those due to longitudinal resistance.

## Rozovskii Approach

Analyses of energy expenditure due to internal turbulent shear and boundary resistance was presented by Rozovskii (1957), who derived the following equation for the fully developed flow in wide rectangular channel bends:

$$S'' = \left(12\frac{g^{1/2}}{C} + 30\frac{g}{C^2}\right)\left(\frac{D}{r_c}\right)^2 \mathbf{F}^2 \tag{8.67}$$

where $S''$ is the energy gradient due to secondary currents, $g$ is the gravitational acceleration, $C$ is the Chezy coefficient, $D$ is the depth of flow, $r_c$ is the center radius of curvature, $\mathbf{F} = U/(gD)^{1/2}$ is the Froude number, and $U$ is the mean flow velocity. The first term on the right-hand side of Eq. 8.67 represents the energy loss due to internal turbulent friction associated with transverse circulation; the second term is the loss resulting from the transverse boundary shear. This equation shows that the energy loss due to channel curvature is a direct function of the depth–radius ratio and the Froude number and that it increases with channel roughness. Because of the complexity of the derivation and the lack of experimental verification, Rozovskii concluded that the problem of energy loss in bends could not be considered as solved and was in need of further study. The major limitation is that the turbulent shear may not be adequately evaluated because of experimental difficulties.

## Chang Approach

Analytical and experimental studies of velocity distributions in curved open channels have been quite extensive (see Sec. 8.2). Similar relationships have been established for the respective longitudinal and transverse velocity distributions in the fully developed region. The Chang (1983) approach is based on the established velocity distributions with which the rate of work done by the fluid is obtained. The rate of energy expenditure, synonymous to the rate of work done, is equal to the combined losses due to internal fluid friction and boundary resistance for steady flows.

The flow to be analyzed is shown schematically in Fig. 8.1. The conditions of steady, subcritical, and fully developed flow are considered. In a curved channel, the rate of energy expenditure per unit channel length, $P$, can be considered as consisting of two components: that associated with longitudinal resistance, $P'$, and that due to transverse circulation, $P''$, or

$$P = P' + P'' \tag{8.68}$$

Now, since

$$P = \gamma QS, \qquad P' = \gamma QS', \quad \text{and} \quad P'' = \gamma QS'' \tag{8.69}$$

one has

$$S = S' + S'' \tag{8.70}$$

where $\gamma$ is the specific weight of fluid, $Q$ is the flow discharge, $S$ is the total energy gradient, $S'$ is the longitudinal energy gradient, and $S''$ is the transverse energy gradient. Since energy is a scalar quantity, its components so defined must not be misinterpreted as vector quantities.

Energy expenditure at a cross section is analyzed based on the work done by the fluid, which is equal to the losses through internal friction and boundary resistance. The fact that the work done represents the combined internal and boundary losses may be illustrated first by the flow in a straight channel. Now, the rate of work done is obtained from

$$P = \int \mathbf{U} \cdot d\mathbf{F} \tag{8.71}$$

where $\mathbf{U}$ is the velocity vector and $\mathbf{F}$ is the force vector. If the velocity vector is replaced by the local longitudinal velocity $u$, and the force is replaced by the tractive force, the rate of work done becomes the rate of energy expenditure, $\gamma QS$, that is,

$$P = \int_B \int_D \gamma u S \, dz dy = \gamma QS \tag{8.72}$$

where $B$ is the channel width, $D$ is the channel depth, $y$ is the horizontal coordinate, and $z$ is the vertical coordinate. The rate of energy expenditure, $\gamma QS$, is equal to the combined internal and boundary losses.

In a curved channel, the energy dissipated through transverse circulation is obtained from its transverse and vertical velocity components and the forces in their respective directions, namely, the centripetal force and the gravitational force. For an angular increment $d\theta$, the rate of energy loss is

$$P'' = \int \mathbf{U} \cdot d\mathbf{F} = \int_{r_1}^{r_2} \int_0^D \rho\left(v\frac{u^2}{r} + wg\right) rd\theta dzdr \tag{8.73}$$

where $r_1$ is the innermost channel radius, $r_2$ is the outermost channel radius, and the radial velocity $v$ is positive in the outward direction. At first glance, it appears as if this equation does not include the turbulent shear and channel roughness to which the energy loss is related. In reality, the effects of the turbulent shear and channel roughness are implicitly reflected in the velocity distributions. In Eq. 8.73, the vertical velocity is very small in a wide channel, and the work done by upward and downward vertical velocities is largely mutually canceled, hence its contribution is neglected. For a unit channel length or $r_c d\theta = 1$, one has

$$P'' = \int_{r_1}^{r_2} \int_0^D \rho v \frac{u^2}{r_c} dzdr \tag{8.74}$$

where the local depth $D$ is a function of $r$ as it varies with the cross-sectional geometry and transverse water-surface profile associated with the superelevation.

The distributions of velocity components $u$ and $v$ are required to determine the energy expenditure using Eq. 8.74. The longitudinal velocity varies in both the vertical and transverse directions, of which the vertical distribution given by Eq. 8.30 is used. From Eq. 8.57, the transverse distribution of $U$ normalized by cross-sectional mean velocity $\overline{U}$ is

$$F = \frac{U}{\overline{U}} = \left[\left(\frac{D}{D_c}\right)\left(\frac{r_c}{r}\right)\right]^{1/2} \tag{8.75}$$

The vertical variation of transverse velocity $v$ is based on that given by Eq. 8.22.

Equation 8.74 together with the velocity distributions form the general analytical model for curved channels with a given cross-sectional shape. Because of the nonlinearity in velocity distributions and the general cross-sectional shape, this equation may not be integrated in a closed form; it is therefore solved by numerical integration. However, a simplified model for the energy expenditure due to channel curvature can be obtained in a closed form if the following simplifications are made: (1) The cross section is wide and rectangular in shape so that the wall effect is negligible, (2) superelevation and transverse variations in $u$ and $v$ are not considered, and (3) the distribution of transverse velocity component is linearized. Vertical distribution of the transverse velocity may be approximated by a straight line. Such a linear approximation for Eq. 8.22 is

$$v = \frac{2}{\kappa}\left(3.75 - \frac{1.875}{\kappa}\frac{U_*}{U}\right)\left(z - \frac{D}{2}\right)\frac{U}{r_c} \tag{8.76}$$

Substituting this equation into Eq. 8.74 and integrating yields

$$P'' = \gamma Q\left(\frac{2.07f + 4.68f^{1/2} - 1.83f^{3/2}}{0.565 + f^{1/2}}\right)\left(\frac{D}{r_c}\right)^2 \mathbf{F}^2 \tag{8.77}$$

where $\mathbf{F}$ is the Froude number. This simplified analytical model predicts transverse energy loss similar to that of Rozovskii's formula (Eq. 8.67).

Now, the rate of total energy expenditure including longitudinal and transverse losses is

$$P = \gamma QS = \gamma Q\frac{f}{8}\mathbf{F}^2 \tag{8.78}$$

The ratio of transverse loss to total loss may be obtained from Eqs. 8.77 and 8.78.

The simplified analytical model given in Eq. 8.77 identifies those variables to which the energy expenditure is related. The transverse loss is directly related to the depth–radius ratio, the Froude number, and channel roughness. The ratio of transverse to total loss is a direct function of the depth–radius ratio but is an inverse function of channel roughness. In other words, transverse loss becomes more important in curved channels with a sharp curvature, greater depth, and lower roughness. Of course, the effects of channel width and cross-sectional shape on energy loss are not reflected in the simplified model but rather in the general analytical model. The computing procedure for flow through curved channels is given in the following section.

## 8.7 STREAMWISE VARIATION OF SPIRAL MOTION

River channels are characterized by the streamwise changing curvature, to which variations of flow dynamics, bed topography, and pattern of lateral migration are closely related. Determination of these changing features is based on the fluid dynamics governing the development of spiral motion along the channel. Spiral motion may exit in straight, as well as curved, channels. As flow enters a curved channel, however, spiral motion induced by the centripetal force becomes more pronounced. After leaving a curved reach, the flow will gradually return to the parallel pattern. The growth and decay of transverse flow was investigated by

Rozovskii (1957). As explained by him, the growth of transverse flow, upon entering a bend, is due to the centripetal force but is counteracted by the internal viscous force. After leaving a bend exit, the centripetal force essentially disappears and the transverse velocity decays under the viscous effect. The turbulent shear is the internal viscous resistance that the flow has to overcome in transforming from parallel flow into the spiral pattern and vice versa.

The original analysis of transverse flow development by Rozovskii was limited to cases of simple channel bend preceded or followed by tangents. Extension of Rozovskii's study to river channels of streamwise changing curvature was made by Chang (1984a, 1984b). Chang's approach is based upon the equation of motion for the transverse velocity $v$ (Eq. 8.9). Since velocity components $v$ and $w$ are small in comparison to $u$, the terms $v\,(\partial v/\partial r)$ and $w\,(\partial v/\partial z)$ are of a higher order and are hence neglected. The remaining equation is given as follows:

$$\overset{\text{I}}{u\frac{dv}{ds}} = \overset{\text{II}}{\frac{u^2}{r}} - \overset{\text{III}}{gS_r} + \overset{\text{IV}}{\frac{\partial}{\partial z}\left(\varepsilon\frac{\partial v}{\partial z}\right)} \tag{8.79}$$

This equation, with its four terms labeled, provides the basis of analysis for the growth and decay of the transverse velocity. The mechanism of flow development may be described by the physical significance of each term. The spatial variation of transverse velocity (term I) is related to the interactive forces associated with centrifugal acceleration (term II), transverse water-surface slope (term III), and turbulent shear (term IV). While centrifugal acceleration, being a positive term, causes $v$ to grow, transverse water-surface slope and turbulent shear, being negative terms, are responsible for its decay. The centrifugal acceleration is imposed upon the flow by the channel curvature. The transverse circulation grows when the summation of terms II, III, and IV is positive—that is, when the centripetal force overcomes the combined resistance due to transverse water-surface slope and fluid friction. The transverse circulation decays under the opposite condition. An equilibrium is established when the summation of these terms is zero. Such a situation occurs in a fully developed flow for which $v$ is a constant, or when $v$ is at a maximum near an apex, or at a minimum near a crossover. The growth and decay may be generalized to include circulations in opposite directions. If the initial circulation is in the reverse direction to the channel curvature, that is, the value of $v$ is negative, then the effects on the circulation from these factors are exactly the opposite.

The transverse velocity varies along the vertical and radial directions; therefore, Eq. 8.79 is evaluated at the water surface along the channel centerline. Since wall effects are not considered in this equation, it is limited to wide channels with a width–depth ratio greater than about 5.

Term IV in Eq. 8.79, the turbulent shear, may be related to the friction factor following a procedure suggested by Odgaard (1986). In his approach, the vertical variation of the transverse velocity component is approximated by the linear profile

$$\frac{v}{\tilde{v}} = 2\left(\frac{z}{D} - \frac{1}{2}\right) \tag{8.80}$$

where $\tilde{v}$ is the transverse velocity at the water surface. This profile is a good approximation to those shown in Figs. 8.3 and 8.10. From Eq. 8.80 and the parabolic ε-profile, $\varepsilon = \kappa U_* z(1 - z/D)$ given in Eq. 7.52, one has

$$\frac{\partial}{\partial z}\left(\varepsilon \frac{\partial v}{\partial z}\right)_{z=D} = -\kappa\left(\frac{f}{2}\right)^{1/2} \frac{\tilde{u}\tilde{v}}{D} \frac{m}{1+m} \tag{8.81}$$

where $m = \kappa(8/f)^{1/2}$, and $\tilde{u}$ and $\tilde{v}$ are surface velocities along channel centerline. The longitudinal surface velocity $\tilde{u}$ in this equation can be related to the mean velocity $U$ (see Eq. 8.30) as follows:

$$\frac{\tilde{u}}{U} = \phi \tag{8.82}$$

where $\phi$ is a function of the friction factor $f$. From Eq. 8.13, the transverse water-surface slope may be approximated as

$$S_r = \frac{U^2}{g r_c} \tag{8.83}$$

After substituting Eqs. 8.82 and 8.83 into Eq. 8.79 and rearranging terms, the following equation is obtained:

$$\frac{d\tilde{v}}{ds} + \frac{\kappa}{D}\left(\frac{f}{2}\right)^{1/2} \frac{m}{1+m} \tilde{v} = \left(\phi - \frac{1}{\phi}\right)\frac{U}{r_c} \tag{8.84}$$

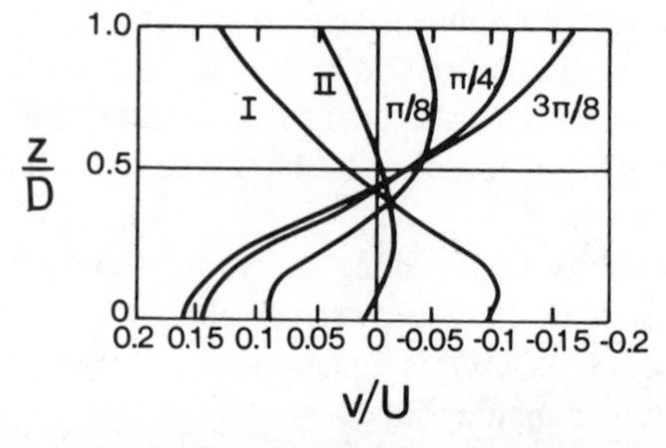


**Figure 8.10** Profiles of transverse velocity measured by Yen (1965) at different locations along centerline of curved flume in Fig. 8.11.

The parameter $\phi - 1/\phi$ is evaluated using the boundary condition of developed transverse flow for which relationships for the velocity profile are available. From the velocity profile given in Eq. 8.22, the following equation for transverse surface velocity is obtained:

$$\frac{\tilde{v}}{U} = \frac{1}{\kappa}\frac{D}{r_c}\left[\frac{10}{3} - \frac{1}{\kappa}\frac{5}{9}\left(\frac{f}{2}\right)^{1/2}\right] \tag{8.85}$$

From this equation and Eq. 8.84 with $d\tilde{v}/ds = 0$ for developed transverse flow, the parameter $\phi - 1/\phi$ is obtained. Substituting $\phi - 1/\phi$ into Eq. 8.84 gives

$$\frac{d\tilde{v}}{ds} + \frac{\kappa}{D}\left(\frac{f}{2}\right)^{1/2}\frac{m}{1+m}\tilde{v} = \left(\frac{f}{2}\right)^{1/2}\left[\frac{10}{3} - \frac{1}{\kappa}\frac{5}{9}\left(\frac{f}{2}\right)^{1/2}\right]\frac{U}{r_c} \tag{8.86}$$

This equation, in which $v$ is the dependent variable and the coefficients are functions of the independent variable $s$ alone, is a linear differential equation of the first order. The general solution, which can be found in a standard text, has the form

$$\tilde{v} = \left[c + \int F_1(f)\frac{U}{r_c}\exp\left(\int F_2(f)\,ds\right)ds\right]\exp\left(-\int F_2(f)\,ds\right) \tag{8.87}$$

where

$$F_1(f) = \left(\frac{f}{2}\right)^{1/2}\left[\frac{10}{3} - \frac{1}{\kappa}\frac{5}{9}\left(\frac{f}{2}\right)^{1/2}\right], \qquad F_2(f) = \frac{\kappa}{D}\left(\frac{f}{2}\right)^{1/2}\frac{m}{1+m} \tag{8.88}$$

The constant $c$ is evaluated from the boundary condition at the initial upstream point, where $\tilde{v}$ equals the initial transverse velocity $\tilde{v}_i$. The value of $c$ is obtained to be $\tilde{v}_i$.

Equation 8.87 provides the spatial variation of transverse velocity along the channel centerline. Growth of this velocity is attributed to channel curvature. The transverse velocity reaches an equilibrium, that is, it becomes fully developed when the effects from channel curvature are in balance with transverse turbulent shear. In the absence of channel curvature, Eq. 8.87 becomes

$$\tilde{v} = \tilde{v}_i \exp\left(-\int F_2(f)\,ds\right) \tag{8.89}$$

This equation provides the decay of transverse velocity in the downstream tangent of a bend. The rate of decay is directly related to channel roughness and inversely related to flow depth.

## 8.8 COMPUTATION OF FLOW THROUGH CURVED CHANNELS

The flow through curved channels is characterized by the streamwise variations in water surface, longitudinal and transverse velocity, flow resistance, strength of circulation, and so on. Such variations in flow characteristics may be computed using equations for longitudinal and transverse flows that have so far been presented. In order to obtain solutions at discrete cross sections along a channel, the governing equations are written in finite-difference forms. The energy equation for the longitudinal direction (Eq. 3.58) has the form

$$Z_j + \alpha \frac{U_j^2}{2g} = Z_{j+1} + \alpha \frac{U_{j+1}^2}{2g} + (\bar{S}' + \bar{S}'')\,\Delta s \tag{8.90}$$

where $\alpha$ is the energy coefficient defined in Eq. 3.56, $j$ is the $s$-coordinate index, and the overbar denotes the average value for those at $j$ and $j + 1$. The finite-difference equation for the transverse velocity (Eq. 8.87) is

$$\tilde{v}_{j+1} = \left[ \tilde{v}_j + \bar{F}_1(f) \frac{\bar{U}}{\bar{r}_c} \exp\,[\bar{F}_2(f)\,\Delta s]\,\Delta s \right] \exp\,[-\bar{F}_2(f)\,\Delta s] \tag{8.91}$$

Equation 8.91 provides the streamwise variation in transverse surface velocity. In the process of growth and decay, this transverse velocity is less than the developed velocity. The transverse velocity profiles described in Sec. 8.2 are for fully developed transverse flow for which there is zero net transverse discharge across the depth; that is, the outward discharge is balanced by the inward discharge. It is assumed that transverse circulation retains its shape during growth and decay; then, the circulation may be represented by its surface velocity. Thus, the transverse energy gradient $S''$ in Eq. 8.90 can be evaluated using Eq. 8.74 or 8.77.

Because of the spatial variation in superelevation, there is a small lateral discharge across the channel centerline associated with the change in superelevation. The assumption of retaining the shape of circulation during growth and decay ignores the small lateral discharge. However, this discrepancy can be circumvented if Eq. 8.91 is evaluated along the discharge centerline across which there is zero net lateral discharge. The discharge centerline tends to move closer to the concave bank upon entering a bend.

### Method of Computation

For a given channel configuration, roughness and discharge, Eq. 8.90 is solved using the standard-step method for water-surface computation described in Sec. 3.8. For subcritical flow, which is under downstream control, the computation is executed from downstream to upstream. However, in evaluating Eq. 8.91 for transverse velocity, the spatial variation of $\tilde{v}$ needs to be evaluated from upstream toward downstream. Since the flow profile must satisfy both equations, the final solution is obtained by successive approximations. In the procedure, the transverse velocity is assumed to be zero initially, then an approximate flow profile is obtained using Eq. 8.90 from downstream to upstream. With the computed hydraulic parameters, transverse velocity $\tilde{v}$ is then evaluated using Eq. 8.91 from upstream to downstream. Then the energy gradient $S''$ in Eq. 8.90 is computed using Eq. 8.74 or 8.77. The results converge rapidly after a few round sweeps along the channel. For any finite-difference scheme, the consistency and accuracy of computation depend on the grid size, which in this case is $\Delta s$. From the experience of this analysis, computed results approach constancy if $\Delta s < 2D$.

## Sample Computations and Results

The mathematical method was applied to compute flow characteristics through curved channels used, respectively, by Yen (1965) and by Ippen and Drinker (1962). In the first case, detailed measurements on transverse velocities were made. In the second case, flow resistance in terms of boundary shear was measured. For both cases, the backwater caused by curved channels was small; the energy coefficient $\alpha$ was near unity.

The laboratory flume used by Yen had two identical 90° curves of reversed direction connected by a 14-ft tangent, as shown in Fig. 8.11. The trapezoidal channel had a bottom width of 6 ft and 1-on-1 side slopes. Channel roughness in terms of Manning's $n$ was measured to be 0.0103. Five runs were made in the flume for different subcritical flow conditions. Longitudinal and transverse velocity distributions were measured in the tangent and the second curve. Measured transverse velocity profiles at different sections along the channel centerline are shown in Fig. 8.10. The decay from I to $\pi/8$ is followed by growth after $\pi/8$ and the fully developed flow after $\pi/4$. For this experiment, the transverse velocity profile maintains a more or less similar pattern during growth and decay. Because of the mild curvature, flow separation was not observed. Transverse surface velocities along channel centerline computed using the mathematical method are in general agreement with measured values for all five runs. Results for Run 3 are presented in Fig. 8.11 as a sample. This run had a channel bed slope of 0.00072, an average centerline depth of 0.512 feet, a mean velocity of 2.27 ft/sec, and a Froude number of 0.58.

As shown in Fig. 8.11, spatial variation of the transverse-velocity–mean-velocity ratio, $\tilde{v}/U$, is significant, characterized by a multistage development along the measured reaches. After leaving the first curve, transverse velocity decays gradually in the straight reach from I to II. Upon entering the second curve of opposite curvature at II, the transverse flow rapidly reverses its direction and starts to grow in the new direction. It reaches an approximate constant, that is, developed condition, after the midpoint of this curve. For the developed flow, which extends to the curve exit at III, the ratio $\tilde{v}/U$ has a value of about 15%.

Simulated results pertaining to spatial variations of friction factor and energy gradient are also shown in Fig. 8.11. These values vary in direct relation to transverse velocity. Partitioning of total energy gradient $S$ into longitudinal energy gra-

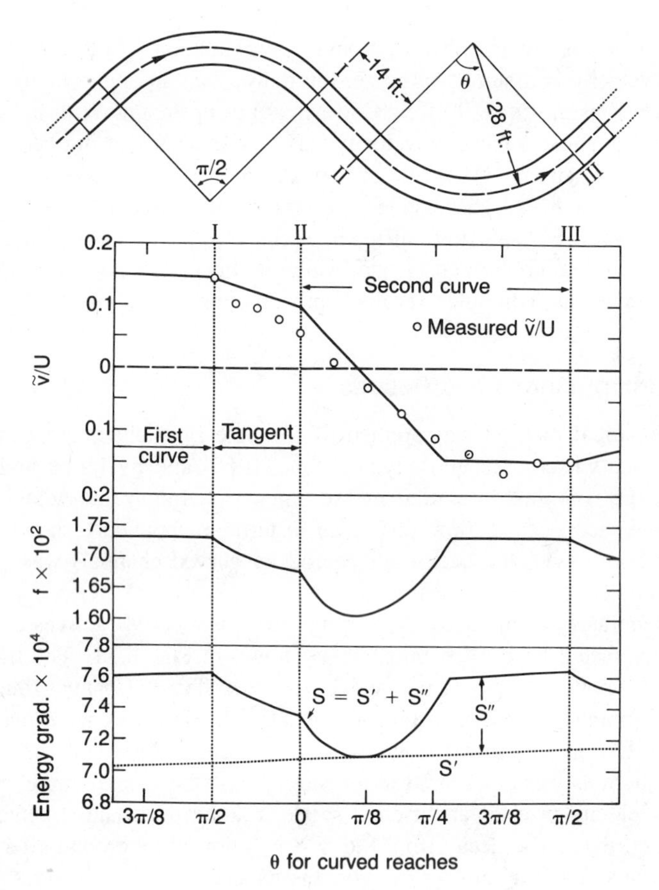


**Figure 8.11** Variations of transverse surface velocity, friction factor, and energy gradient along curved channel.

dient $S'$ and transverse energy gradient $S''$ indicates that $S'$ stays essentially constant and that the increased $f$ and $S$ are primarily caused by transverse energy loss. For the developed flow, transverse loss in terms of $S''/S'$ is computed to be 7.2% for the rather mild curves and shallow depths.

Experiments by Ippen and Drinker (1962) were conducted in two trapezoidal concrete flumes with 1-on-2 side slopes. Each flume had a 60° bend, preceded and followed by straight reaches, similar to that shown in Fig. 8.9. Channel roughness in terms of Manning's $n$ was between 0.009 and 0.01. Boundary shear stresses were measured by using a surface pitot tube. Mean shear stress at a section, $\tau_0$, was then determined by averaging local shear stresses over the wetted perimeter.

The mathematical model was applied to simulate flow characteristics for Runs 1 and 6 of their experiments as samples. A brief summary of channel geometries and test conditions, including bottom width, average depth, radius of curvature, discharge, average velocity, channel bed slope, and Froude number, is listed in Table 8.1. Computed results pertaining to spatial variations of transverse surface velocity, energy gradient, and boundary shear stress are presented in Fig. 8.12. The boundary shear is computed based on the energy gradient, and the equation $\tau_0 = \gamma RS$.

The transverse velocity starts to grow upon entering the curve, and it begins to decay after the curve exit. The transverse velocity does not become fully developed because of insufficient curve length. The occurrence of transverse circulation induces an increased total energy gradient $S$. This increase is primarily attributed to a greater transverse energy gradient $S''$ because the longitudinal energy gradient $S'$ stays approximately the same for these cases. The maximum transverse loss in terms of $S''/S'$ is computed to be 26% for Run 1 and 22% for Run 6. Locations of the shear maxima are at the curve exit, and relatively higher shears persist for a considerable distance in the downstream tangent. Computed shear distributions

**TABLE 8.1 Summary of Channel Geometries and Sample Test Conditions by Ippen and Drinker (1962)**

| Run Number (1) | Bottom Width (in.) (2) | Average Depth (in.) (3) | $r_c$ (in.) (4) | $Q$ (cfs) (5) | Average Velocity (ft/sec) (6) | $S_0 \times 10^5$ (7) | F (8) |
|---|---|---|---|---|---|---|---|
| 1 | 24 | 2.98 | 60 | 0.85 | 1.36 | 64 | 0.53 |
| 6 | 12 | 3.02 | 70 | 0.45 | 1.19 | 55 | 0.48 |

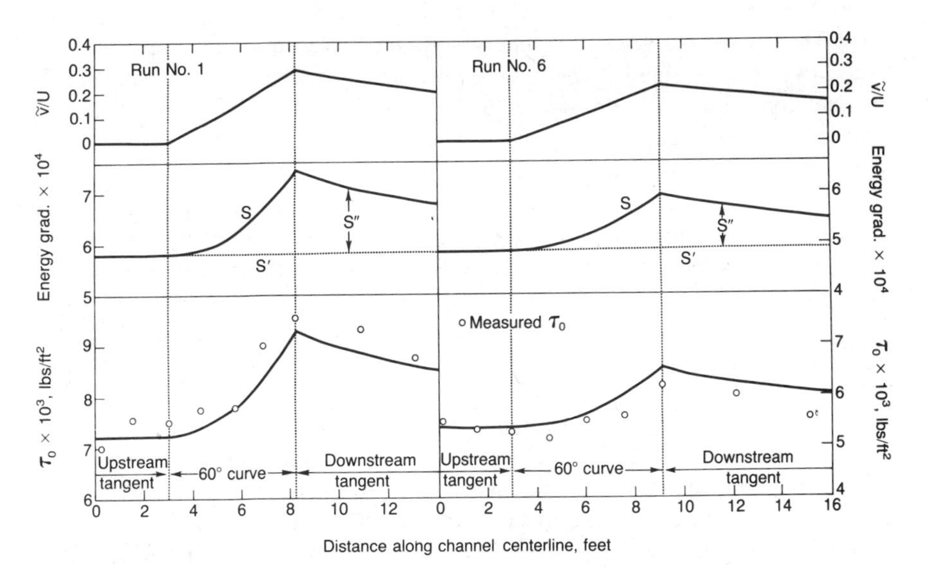


**Figure 8.12** Spatial variations of transverse surface velocity, energy gradient, and boundary shear stress along two channel bends.

compare favorably with experimental data. The greater excess shear predicted for Run 1 is associated with a greater value of $(D/r_c)^2$ and width-depth ratio. Minor flow separation downstream of the curve was observed in Run 1. Along a similar vein, Rozovskii (1957) carried out an experiment in a single 90° bend with movable bed and fixed wall. His results show that intensive scour occurs near the bend exit, consistent with the location of the maximum shear.

## 8.9 TRANSVERSE FLOW AND CROSS-STREAM FLOW IN RIVER CHANNELS

The equations for transverse velocity profiles have been compared well with laboratory and certain field measurements, as reported by Rozovskii (1957), Yen (1965), and Kikkawa et al. (1976), among others. However, for more irregular sections, the results are not as good. Thorne et al. (1983) made measurements of bend flow hydraulics on the Fall River at low stage. They used a two-component electromagnetic current meter, which measures mutually perpendicular velocity components. Sample results at a bend apex and a crossover are shown in Fig. 8.13. These results exemplify the usually pronounced transverse circulation in a bend and the coexistence of small secondary flow cells induced by local irregularities. Near a crossover, the secondary flow is generally weak, characterized by multiple cells.

The secondary flow, or transverse flow, must be distinguished from the cross-stream flow. The former refers to velocity component normal to the primary flow. The latter is the velocity component perpendicular to the tangential direction of the river channel, and it contributes a net lateral discharge with respect to the channel centerline. The cross-stream flow results from the angular difference between the discharge centerline and channel centerline; it occurs as the thalweg swings from one bank to the other, crossing the channel centerline.

Much of the current bend theory has been called into question by Dietrich et al. (1979) and by Thorne and Rais (1983), because they have found significant net cross-stream discharge in natural meanders. At the upstream part of a bend, such as section A–A in Fig. 8.14, the cross-stream discharge is typically 10% of the primary flow. This observation reveals another phenomenon that should be related to the development of helical motion. The cross-stream flow should also affect the pattern of meander bend migration.

In this discussion, the flow path is distinguished from the channel centerline, as shown schematically in Fig. 8.14. The flow path, defined as the discharge centerline, has zero net transverse discharge; that is, the inward and outward transverse flow at different depths across this line are in balance. Under this condition, the transverse flow may be considered as a circulation as shown in Fig. 8.14a. The circulation, by definition, has no net lateral discharge. The flow path is generally in line with the locus of the high velocity.

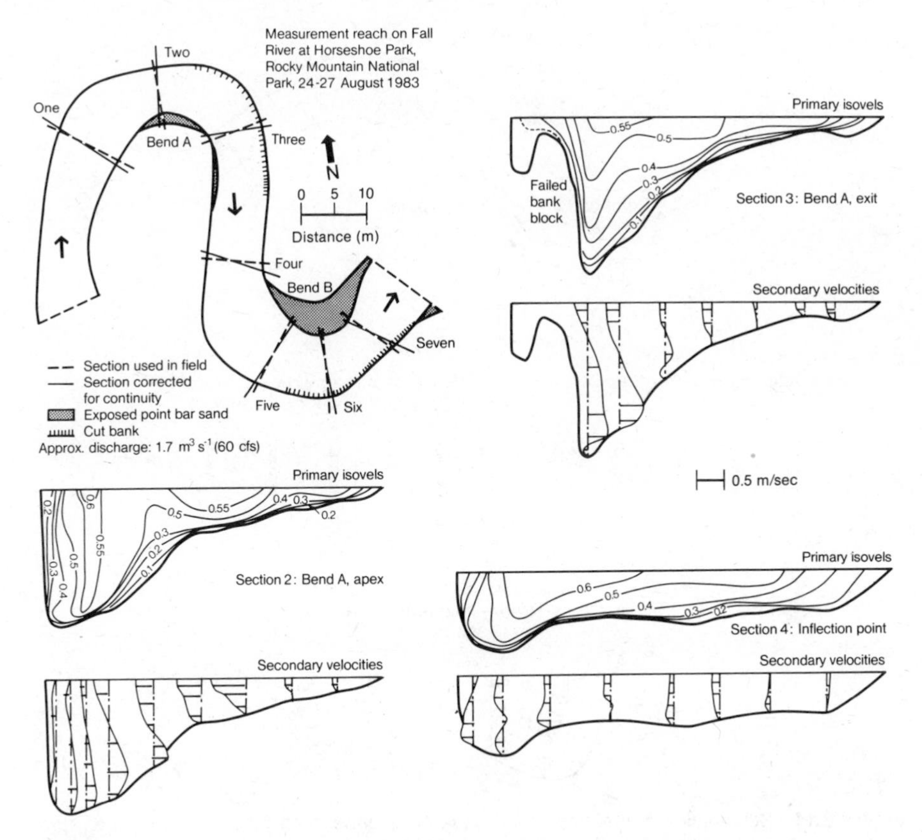


**Figure 8.13** Measurement reach of Fall River and primary and secondary velocities at three cross sections (after Thorne et al. 1983).

The channel path is the channel centerline. The angular difference between flow path and channel path contributes to cross-stream discharge. The cross-stream flow is responsible for the change in superelevation and it contributes to the redistribution of the longitudinal shear. To depict the schematic flow pattern, two cross sections are selected: Section A–A is normal to the channel path and section A′–A′ is normal to the flow path. The pattern of secondary flow for section A′–A′ is shown in Fig. 8.14a, the inward and outward flows are in balance to form a circulation if one ignores the local flow irregularities. Because of the angular difference, the primary flow has a cross-stream component on section A–A, as depicted in Fig. 8.14b. The cross-stream flow pattern on section A–A due to the primary flow and secondary circulation is shown in Fig. 8.14c. The secondary circulation exerts a downward action on the concave bank and moves sediment from the concave bank to the convex bank. Depending on the relative transport rate due to cross-stream component of the primary flow and the secondary circulation, the sediment may move either away or toward a bank.

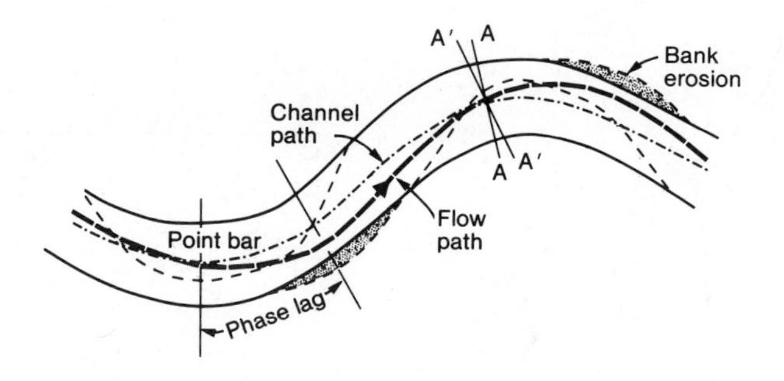


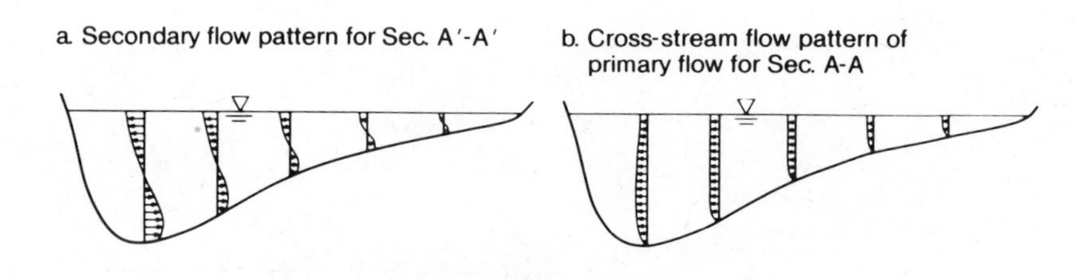


c. Cross-stream flow pattern for Sec. A-A

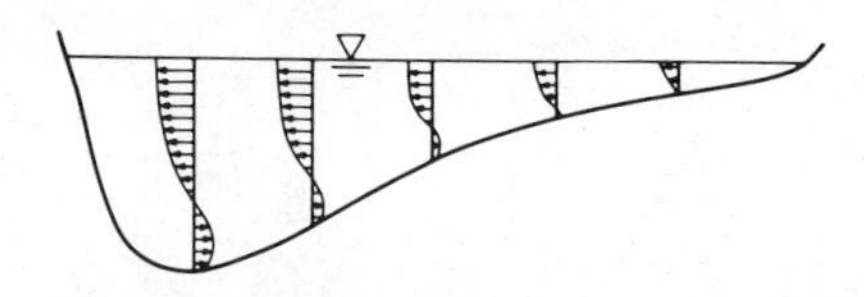

**Figure 8.14** Schematics of meander bend showing channel path (channel centerline), flow path (discharge centerline), and flow patterns.

## REFERENCES

Bagnold, R. A., "Some Aspects of the Shape of River Meanders," *USGS Professional Paper 282-E,* 1960.

Bakhmeteff, B. A. and Allen, W., "The Mechanism of Energy Loss in Fluid Friction," *Proc. ASCE*, **71**(2), pp. 129–166, February 1945.

Bridge, J. S., "Flow, Bed Topography, Grain Size and Sedimentary Structure in Open Channel Bends: A Three-Dimensional Model," *Earth Surf. Proc.*, **2**, pp. 401–416, 1977.

Chang, H. H., "Energy Expenditure in Curved Open Channels," *J. Hydraul. Eng. ASCE*, **109**(7), pp. 1012–1022, July 1983.

Chang, H. H., "Regular Meander Path Model," *J. Hydraul. Eng. ASCE* **110**(10), pp. 1398–1411, October 1984a.

Chang, H. H., "Variation of Flow Resistance through Curved Channels," *J. Hydraul. Eng. ASCE*, **110**(12), pp. 1772–1782, December 1984b.

Chow, V. T., *Open Channel Hydraulics*, McGraw-Hill, New York, 1959.

Dietrich W. E., Smith, J. D., and Dunne, T., "Flow and Sediment Transport in a Sand Bedded Meander," *J. Geol.*, **87**, pp. 305–315, 1979.

El-Khudairy, M., "Stable Bed Profiles in Continuous Bend," Ph.D. Thesis, University of California at Berkeley, 1970.

Engelund, F., "Flow and Bed Topography in Channel Bends," *J. Hydraul. Div. ASCE*, **100**(HY11), pp. 1631–48, November 1974.

Falcon-Ascanio, M. and Kennedy, J. F., "Flow in Alluvial-River Curves," *J. Fluid Mech.*, **133**, pp. 1–16, 1983.

Hayat, S., "The Variation of Loss Coefficient with Froude Number in an Open Channel Bend," Master's Thesis, University of Iowa, 1965.

Ikeda, S., "Lateral Bed Load Transport on Side Slopes," *J. Hydraul. Eng. ASCE*, **108**(HY11), pp. 1369–1373, November 1982.

Ippen, A.T. and Drinker, P. A., "Boundary Shear Stresses in Curved Trapezoidal Channels," *J. Hydraul. Div. ASCE*, **88**(HY5), pp. 143–179, September 1962.

Jansen, P. Ph., van Bendegom, L., van den Berg, J., de Vries, M., and Zanen, A., *Principles of River Engineering*, Pitman, London, 1979.

Karim, M. F., "Computer-Based Predictors for Sediment Discharge and Friction Factor of Alluvial Streams," Ph.D. Thesis, Civil Engineering Department, University of Iowa, 1981.

Kikkawa, H., Ikeda, S., and Kitagawa, A., "Flow and Bed Topography in Curved Open Channels," *J. Hydraul. Div. ASCE*, **102**(HY9), pp.1327–1342, September 1976.

Kondrat'ev, N. E., ed., *River Flow and River Channel Formation*, 1959, translated from Russian and published by the Israel Program for Scientific Translations, Jerusalem, 1962.

Leopold, L. B., Bagnold, R. A., Wolman, M. G., and Brush, L. M. Jr., "Flow Resistance in Sinuous or Irregular Channels," *USGS Professional Paper 282-D,* 1960.

Leopold, L. B., Wolman, M. G., and Miller, J. P., Fluvial Professes in Geomorphology, W. H. Freeman, San Francisco, California, 1964.

NEDECO (Netherlands Engineering Consultants), *River Studies and Recommendations on Improvement of Niger and Benue*, North Holland Publ. Co., Amsterdam, 1959.

Odgaard, A. J., "Transverse Bed Slope in Alluvial Channel Bends," *J. Hydraul. Div. ASCE*, **107**(HY12), pp. 1677–1694, December 1981.

Odgaard, A. J., "Bed Characteristics in Alluvial Channel Bends," *J. Hydraul. Div. ASCE*, **108**(HY11), pp. 1268–1281, November 1982.

Odgaard, A. J., "Grain-Size Distribution of River-Bed Armor Layers," *J. Hydraul. Eng. ASCE*, **110**(9), pp. 1267–1272, September 1984.

Odgaard, A. J., "Meander-Flow Model I: Development," *J. Hydraul. Eng. ASCE*, **112**(12), pp. 1117–1136, December 1986.

Pang, B-D., "Experimental Study on the Overbank Flow and Its Energy Loss," Selected Research Papers, Academia Sinica, XI, *Sedimentation*, 1983, pp. 192–213, (in Chinese).

Parker, G., discussion of "Lateral Bed Load Transport on Side Slopes" by *S*. Ikeda, *J. Hydraul. Eng. ASCE*, **110**(2), pp. 197–199, February 1984.

Parker, G. and Andrews, E. D., "Sorting of Bed load Sediment by Flow in Meander Bends," *Water Resour. Res.*, **21**(9), pp. 1361–1373, September 1985.

Rouse, H., ed., *Advanced Mechanics of Fluids*, John Wiley & Sons, New York, 1959.

Rozovskii, I. L., "Flow of Water in Bends of Open Channels," the Academy of Sciences of the Ukrainian SSR, 1957, translated from Russian by the Israel Program for Scientific Translations, Jerusalem, Israel, 1961 (available from Office of Technical Services, U.S. Department of Commerce, Washington, D.C., PST Catalog No. 363, OTS 60-51133).

Schlichting, H., *Boundary-Layer Theory*, 6th edition, McGraw-Hill, New York, 1968.

Soliman, M. M. and Tinney, E. R., "Flow Around 180° Bends in Open Rectangular Channels, " *J. Hydraul. Div. ASCE*, **94**(HY4), pp. 893–908, July 1968.

Thorne, C. R. and Rais, S., "Secondary Current Measurements in a Meandering River," *River Meandering*, Proceedings of the Conference Rivers '83. New Orleans, Louisiana, October 1983, pp. 675–686.

Thorne, C. R., Rais, S., Zevenbergen, L. W., Bradley, J. B., and Julien, P. Y., "Measurement of Bend Flow Hydraulics on the Fall River at Low Stage," WRFSL Project Report No. 83-9p, Colorado State University, November 1983.

Toebes, G. H. and Sooky. A. A., "Hydraulics of Meandering Rivers with Flood Plains," Conference Reprint 351, ASCE Water Resources Engineering Conference, Denver, Colorado, May 16–20 1966, 37 pp.

Yen, B. C., "Characteristics of Subcritical Flow in a Meandering Channel," Institute of Hydraulic Research, the University of Iowa, Iowa City, 1965, 155 pp.

Yen, B. C., "Spiral Motion of Developed Flow in Wide Curved Open Channels," *Sedimentation (Einstein)*, H. W. Shen, ed., P.O. Box 606, Fort Collins, Colorado, 1972, Chapter 22.

Yen, C. L., "Bed Topography Effect on Flow in a Meander," *J. Hydraul. Div. ASCE*, **96**(HY1), pp. 57–73, January 1970.

Yen, C. L. and Yen, B. C., "Water Surface Configuration in Channel Bends," *J. Hydraul. Div. ASCE,* **97**(HY2), pp. 303–321, February 1971.

Zimmermann, C. and Kennedy, J. F., "Transverse Bed Slope in Curved Alluvial Streams," *J. Hydraul. Div. ASCE*, **104**(HY1), pp. 33–48, January 1978.

# PART III

# REGIME RIVERS AND RESPONSES

# 9

# ANALYTICAL BASIS FOR HYDRAULIC GEOMETRY

Analytical determination of the hydraulic geometry for river channels may be accomplished if the applicable physical relations are sufficient to describe the unknowns or degrees of freedom in channel morphology. The degrees of freedom of a river, in the broad sense, consist of the width, depth, channel slope, and bank slope. Applicable physical relations for determining the hydraulic geometry are presented in this chapter. Whereas the channel depth can be determined from the flow-resistance relation, that is, the stage–discharge predictor, for a given discharge, a physical relation that governs the stable width formation is not nearly as obvious. An important missing link that hindered rational effort by earlier researchers was the physical condition governing the width formation of alluvial channels. An alluvial river is unconstrained, in the long term, in developing its stable width to which the depth, slope, velocity, and flow resistance are closely related. Indeed, all boundaries of an alluvial river are free surfaces. The question of what mechanism underlies the width formation must be answered before a rational determination of channel geometry can be accomplished. Otherwise, river mechanics could only be at the unsatisfactory state of development.

## 9.1 APPLICABLE PHYSICAL RELATIONSHIPS

Physical relations useful for the analysis of hydraulic geometry are grouped into four categories pertaining to fluid dynamics, sediment transport, bank stability, and dynamic equilibrium, respectively. Although the role of each category of physical relations can be separately identified, they are nevertheless closely interrelated in the fluvial processes. Under the dynamic equilibrium, the dynamics of two-phase (liquid–solid) flow and the channel form have adapted to a state of mutual compatibility. The physical relationships for steady uniform flow are described as follows:

1. *Relationships Related to Fluid Dynamics.* The steady forms of the flow continuity equation and the flow-resistance relationship (see Secs. 3.2, 3.5,

3.6, and 6.5) are included in this category. Because it is a stage–discharge predicator, the flow-resistance relationship is useful for depth determination.

2. *Sediment-Transport Relationships*. The mobile channel bed, as molded by the interaction of flow and bed load, must be compatible with sediment transport mechanics. Therefore, appropriate bed-load equations are useful in determining the hydraulic geometry. For a prescribed cross section, discharge, and load, the channel slope can be computed using the sediment-transport equation.
3. *Criteria for Bank Stability*. The submerged bank slope is related to the bank stability, which varies with the bank material. For noncohesive (sandy or gravel) banks, the bank slope is limited by the angle of repose of the material. For cohesive (silt and clay) banks, the bank slope can be steeper, generally in direct relation to the silt–clay content. Vegetation growth also enhances bank stability. Although bank stability will certainly affect the hydraulic geometry, it does not constrain the development of the stable channel configuration in the long run. In other words, bank stability is a necessary factor for width formation, but it is not the sufficient factor. It is also interesting to see, in Fig. 11.10, that the bank material is related to the flow conditions of rivers.
4. *Relationships for the State of Dynamic Equilibrium*. An alluvial channel in dynamic equilibrium must comply with certain physical relationships governing the equilibrium state. A relationship of this type may, in turn, be employed to provide the stable width. Details of this development are given in Sec. 9.2.

The foregoing physical relations are equal in number to the four degrees of freedom (width, depth, channel slope, and bank slope) for stable alluvial channels. Therefore, they are sufficient for the determination of the hydraulic geometry. Their applications are described in Chapters 10 and 11.

## 9.2 PHYSICAL RELATIONSHIPS PERTAINING TO STABLE WIDTH

Different forms of physical relationships and analyses for the stable width of alluvial channels have evolved; all of them are based on the consideration of dynamic equilibrium. For example, the stable width of straight alluvial channels was analyzed by Parker (1978) on the premise that there is a balance in the lateral exchange of sediment in the channel flow between the bank zone and the central region. Lateral sediment exchange, in his analysis, was related to the mechanics of internal river flow.

The energy, or power, approach was also employed to determine the self-formed channel width for alluvial rivers in dynamic equilibrium. Along a regime river reach, the steady flows of water and sediment are in balance with the rates supplied, with no water or sediment storage in the reach. If the reach is short such that it has streamwise constancy in discharge, the state of dynamic equilibrium must satisfy the following three conditions:

1. Equal sediment discharge along the channel.
2. Minimum stream power per unit channel length subject to the constraints.
3. Streamwise uniformity in power expenditure or energy gradient subject to the constraints.

The first condition is apparent because sediment storage is absent in a regime channel. The second and third conditions are energy- or power-based hypotheses that are described in the following as the first and second hypothesis in river channel formation.

### The First Hypothesis: Minimum Stream Power per Unit Channel Length

The definition of this hypothesis by Chang (1979a) is as follows:

> For an alluvial channel, the necessary and sufficient condition of equilibrium occurs when the stream power per unit channel length $\gamma QS$ is a minimum subject to given constraints. Hence, an alluvial channel with water discharge $Q$ and sediment load $Q_s$ as independent variables tends to establish its width, depth, and slope such that $\gamma QS$ is a minimum. Since $Q$ is a given parameter, minimum $\gamma QS$ also means minimum channel slope $S$.

Note that the power associated with sediment load is usually negligible. From this definition, this concept is applicable to streams with uniform flow. Its application for determining the stable width of alluvial channels is shown schematically in Fig. 9.1 and illustrated numerically by Example 10.2. For gradually varied flow, it is assumed that the stream power at a cross section is the same as for a uniform flow having the same discharge and energy gradient of the section. With this assumption, this concept may be used at any given cross section in a gradually varied flow. Under such a situation, $S$ represents the energy gradient.

The hypothesis of minimum stream power for a hydraulic system may be derived from the principle of virtual work for a mechanical system. The principle of virtual work, as defined by Langhaar (1962), is given as follows:

> For an unchecked mechanical system, the necessary and sufficient condition for equilibrium is $\delta W \equiv 0$, provided that $\delta W$ exists.

In this statement, $\delta W$ is the virtual work done by forces (external and internal) that act on the system caused by a virtual displacement consistent with the constraints. The sign $\equiv$ signifies that the equality holds for all admissible virtual displacements. A system is said to be unchecked if the forces change continuously when a virtual displacement is executed.

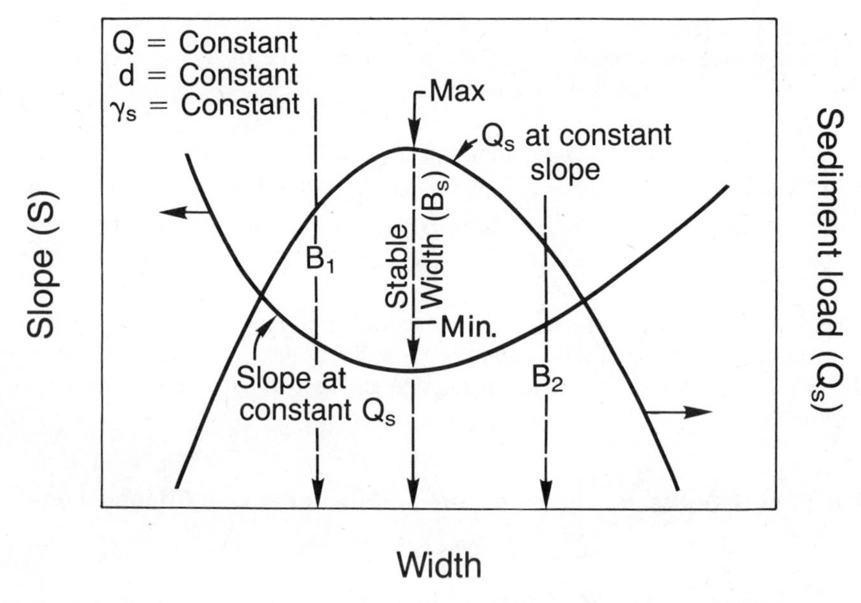


**Figure 9.1** Comparison of minimum stream power and maximum sediment efficiency for stable width determination.

This principle may be generalized immediately to apply to systems that translate at constant velocity, since uniform translation such as channel flow in a moving frame is equivalent to a static system. It may also be applied to a fixed control volume in a flow field as long as the mass inside remains constant. Also, the work or energy may be replaced by power when the rate of energy is of interest. Now, consider a control volume of steady uniform flow in a channel with a unit length. With the introduction of a virtual displacement per unit time (virtual velocity) to the control volume, the change in forces is continuous for every admissible path subject to the constraints. Therefore, under the equilibrium condition, the virtual stream power $\delta P$ caused by a virtual velocity must satisfy the condition $\delta P \equiv 0$, which represents an extremum. Since a maximum has no meaning, it therefore represents a minimum.

## The Second Hypothesis: Minimum Stream Power for a Channel Reach

This hypothesis is an extension of the first hypothesis and is given by Chang (1982) as follows:

> The equilibrium geometry of an alluvial channel reach of equal discharge is so adjusted that the power expenditure is a minimum subject to the given constraints. Minimum power expenditure for the channel reach is equivalent to equal power expenditure per unit channel length or uniform energy gradient along the channel.

The stream power—that is, the rate of energy expenditure—of a channel reach is given by

$$P = \int_L \gamma QS \, ds \tag{9.1}$$

where $P$ is the total stream power of a reach in energy per unit time, $L$ is reach length, and $s$ is the coordinate along the flow direction. Concepts similar to this hypothesis can be found in earlier literature, for example, Gilbert (1880), Langbein (1964), Leopold et al. (1964), and Inglis (1974).

The physical condition of equal power expenditure is satisfied by any uniform flow because it has a constant channel slope equal to the energy gradient in addition to having a constant geometry and roughness. The energy gradient $S$ may also be a constant in a gradually varied flow if the channel geometry and roughness are properly adjusted. If the energy gradient is approximated by the water-surface slope, then uniform energy gradient is equivalent to the straight-line water-surface profile along the channel. River channel adjustment toward dynamic equilibrium is usually more rapid in sand-bed rivers than in gravel-bed rivers. Water-surface profiles measured in sand-bed rivers without physical constraints are more or less straight lines. The water-surface profile for gravel-bed rivers is also nearly straight over riffles and pools at the bankfull discharge, as reported by, for example, Leopold and Wolman (1957) and Keller (1971), but at a low flow, the profile is usually not straight when channel formation in inhibited or constrained by the coarse material.

The linear water-surface profile may not be established in alluvial rivers under the influence of such physical constraints as bridge constrictions, grade controls, bedrock outcroppings, and so on. But the sediment equilibrium, that is, equal sediment discharge along the channel, may still be established. It therefore follows that equal sediment discharge and equal power expenditure are independent conditions.

The second hypothesis can be verified by introducing a small perturbation into the alluvial channel flow with initially equal $\gamma QS$ along the reach. If the perturbation, such as a small constriction, results in a local rise in power expenditure, it must also cause a backwater effect, which represents upstream power storage to supply the increase in power expenditure. Thus the total power loss increases because of the additional conversion—contraction and expansion—power loss.

### Other Related Concepts

In recognition of the lack of sufficient conditions for determining the hydraulic geometry, earlier investigators speculated that river morphology is related to the principle of least work or minimum energy. For example, Mackin (1948) stated that the river channel, as an agent of transportation, must be so adjusted that somehow it contributes to the transport of debris. Other extremal hypotheses have also been proposed, such as the hypothesis of minimum unit stream power by

Yang (1976), the hypothesis of maximum sediment efficiency by Kirkby (1977) and later by White, et al. (1982), and the hypothesis of maximum friction factor by Davies and Sutherland (1983). Rammett (1980), Kondap and Garde (1980), and Hancu and Batuca (1980) applied the variational principle to determine the hydraulic geometry of stable alluvial channels on the basis of minimum power or maximum sediment efficiency.

The general theory of minimum energy and energy dissipation rate are explained by Yang and Song (1979, 1986). They also used examples of application to demonstrate the flexibility of the theory.

For maximum sediment efficiency, it may be stated that, with a prescribed discharge, channel slope, and sediment size, the stable geometry is so adjusted that the sediment discharge is maximized, compatible with the physical relations for flow resistance and sediment transport. This hypothesis is useful if the channel slope is given as an independent variable, such as in the case of a laboratory flume, or if the channel slope may be assumed to have a constant value.

The hypotheses of minimum stream power and maximum sediment efficiency are now compared. For the former, a given sediment load is transported by the stable channel at the minimum stream power or slope; for the latter, the stable channel on a fixed slope moves the maximum load. Both imply maximum efficiency for the stable channel as an agent of transportation. Actually, applications of these hypotheses yield identical results. The equivalence of these two concepts is illustrated schematically in Fig. 9.1; its proof was made by White et al. (1982) using the variational principle.

# 10

# DESIGN OF STABLE ALLUVIAL CHANNELS

The design of stable sand-bed canals and gravel-bed channels that are subject to both scour and silt are described in this chapter. Such alluvial channels are characterized by a mobile bed and are thus different from those immobile bed channels that are subject only to scour, which were discussed in Chapter 5. Alluvial channels with a certain bed load, including most of the sand-bed channels and many gravel-bed channels, are subject to both scour and silt. Typically, a channel of this type has a noncohesive bed because cohesive materials of silt and clay in the mobile bed will become a part of the wash load and will thus be removed by the flow. Sand and other coarser materials, on the other hand, may stay in the bed. It is through this sorting process that the mobile bed has predominantly noncohesive materials. Bank materials of sand-bed channels, however, are often cohesive. In fact, silt–clay content of the bank material has close correlation with river morphology and is in some way related to the hydraulics of river flow, which will be described in Sec. 11.3.

The research on hydraulic geometry of stable alluvial canals has relied heavily on canal data, both field and laboratory. Compilations of such data have been made by Blench and Simons (1974) and by Brownlie (1981), among others. These data have contributed significantly to the advancement of related research.

In a typical design problem, the design discharge and the slope of the terrain are prescribed; and the width, depth, and slope of the channel need to be determined. Geometric parameters of typical canal sections and their trapezoidal approximations are shown in Fig. 10.1. The slope of a canal is limited by the slope of the terrain. Within this limit, the channel slope can be reduced at will with the use of grade-control, or drop, structures. The energy head at a drop structure may be utilized for the generation of hydroelectric power, as illustrated in Fig. 10.2.

Methods for the design of stable alluvial canals were developed following the regime approach and the rational approach. In the regime approach, relationships for the channel width, depth, and slope were established based on measurements from stable alluvial canals and rivers. In the rational approach, on the other hand, the hydraulic geometry of stable channels is deduced from theoretical consider-

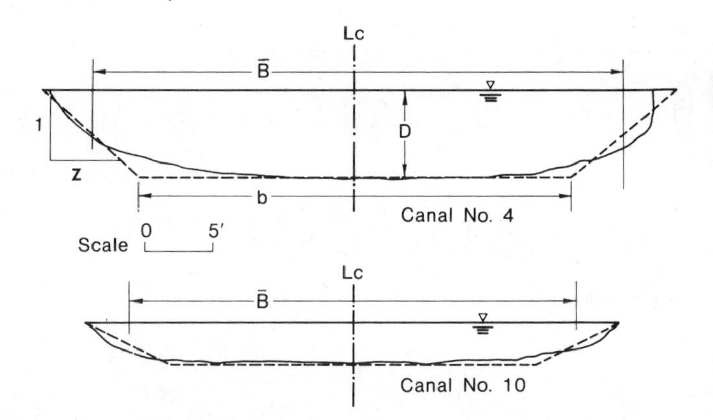


**Figure 10.1** Sample canals and their trapezoidal approximations based on Simons and Bender data (Blench and Simons, 1974).

ations of the hydraulics of flow and sediment transport under the dynamic equilibrium. Regime methods and a rational method for the design of individual sand-bed canals are first described, followed by the design of canals connected in a system, the maturing of canals, and the stable geometry of active gravel-bed streams.

## 10.1 REGIME METHODS FOR STABLE ALLUVIAL CANAL DESIGN

The regime approach, also called the *regime theory*, was originated by engineers working on the irrigation canals in India and Pakistan as they gradually observed certain correlations among the variables for stable sand-bed canals. Since then, different methods reflecting increasing knowledge and greater sophistication on canal design have evolved. The methods by Lacey (1930, 1935, 1958), Blench (1952, 1970), and Simons and Albertson (1960) are described herein. Each method usually consists of three formulas for the stable width, depth, and slope. The limit of application of each method is generally not specified. But experiences from the India–Pakistan canals have shown that the straight alignment of these canals can be maintained with minor maintenance if the Froude number is less than about 0.3. At higher values, the canals have the tendency to meander, demonstrated by bank cutting and formation of a sinuous thalweg. For this reason, the canals are designed to have Froude numbers less than about 0.3.

### Lacey's Regime Method

The regime method by Lacey was published from 1930 to 1958 in a series of papers; this method may be briefly summarized by three regime equations. The

**Figure 10.2** All American Canal and drop structure and power station (courtesy of Imperial Irrigation District).

first one, which may be interpreted as the flow-resistance equation, has the two following versions:

$$U = 1.15(f\overline{D})^{1/2} \tag{10.1}$$

$$U = \frac{1.346}{N_a}\overline{D}^{1/4}R^{1/2}S^{1/2} \tag{10.2}$$

where $U$ is the mean velocity in feet per second, $\overline{D}$ is the mean depth in feet, defined as cross-sectional area of flow divided by the surface width $B$, $R$ is the hydraulic radius, and $S$ is channel slope. The variable $f$ is Lacey's silt factor and $N_a$ is his absolute rugosity, defined respectively as

$$f = 1.6d^{1/2} \tag{10.3}$$

and

$$N_a = 0.0225f^{1/4} \tag{10.4}$$

where $d$ is median bed-material size in millimeters. Equation 10.2 is similar to Manning's equation for flow resistance. When these two equations are compared, it indicates that Manning's $n$ is a function of $N_a$, $R$, and $\overline{D}$, that is, the roughness of the alluvial bed is not a constant but is flow-related.

The second regime equation by Lacey has the form

$$B = 2.67Q^{1/2} \tag{10.5}$$

where $B$ is the surface width in feet, and $Q$ is water discharge in cubic feet per second. It is interesting to note that stable channel width based on this relationship is a function of the discharge alone. This is in sharp contrast to certain rivers for which the width is strongly dependent upon the slope in addition to the discharge. An explanation for this difference is given in Sec. 11.3, which is on the analysis of alluvial rivers.

The third equation by Lacey provides the channel slope:

$$S = \frac{f^{5/3}}{1830Q^{1/6}} \tag{10.6}$$

According to this equation, channel slope is inversely proportional to discharge raised to the power of 1/6 and is directly related to the sediment size.

For a given discharge and median sediment size, the three independent equations by Lacey provide the required stable width, depth, and slope. No specific relation is given for the bank slope of the channel. Most of the canals he studied have cohesive banks. Application of Lacey's method should be limited to the ranges of bed-material size between 0.15 and 0.40 mm and discharge between

5 and 5000 cfs. In estimating the bed-material size for canal design, it is also important to consider the maturing from a newly excavated channel to the quasi-equilibrium, to be discussed in Sec. 10.4.

### Blench's Regime Method

Since the introduction of Lacey's method, much more development work has been done, particularly by Blench (1952, 1970), who extended previous regime methods to include cases of different bank materials. His regime method consists of three independent conditions: the bed factor, the side factor, and the flow-resistance equation, from which the width, depth, and slope of the channel may be determined.

The bed factor $F_b$ is given as

$$F_b = \frac{U^2}{D} \tag{10.7}$$

where $U$ is the mean velocity (in feet per second) and $D$ is the depth of flow (in feet) . The side factor $F_s$ is

$$F_s = \frac{U^3}{\overline{B}} \tag{10.8}$$

where $\overline{B}$ is the average width equal to the cross-sectional area divided by $D$. The flow-resistance equation has the form

$$\frac{U^2}{gDS} = 3.63\left(1 + \frac{C}{2330}\right)\left(\frac{U\overline{B}}{\nu}\right)^{1/4} \tag{10.9}$$

where $\nu$ is kinematic viscosity and $C$ is suspended-sediment concentration in parts per million (ppm) by weight. This concentration in alluvial canals consists largely of wash load. Respective empirical values for the bed factor and side factor provided by Blench are

$$F_b = 1.9d^{1/2} \tag{10.10}$$

$$\begin{aligned} F_s &= 0.1 \quad \text{for slightly cohesive banks} \\ &= 0.2 \quad \text{for medium cohesive banks} \\ &= 0.3 \quad \text{for highly cohesive banks} \end{aligned} \tag{10.11}$$

where $d$ is median sediment size in millimeters.

From the three independent conditions represented by Eqs. 10.7, 10.8, and 10.9, channel width, depth, and slope are obtained as

$$\overline{B} = \left(\frac{F_b Q}{F_s}\right)^{1/2} \tag{10.12}$$

$$D = \left(\frac{F_s Q}{F_b^2}\right)^{1/3} \tag{10.13}$$

and

$$S = \frac{(F_b)^{5/6}(F_s)^{1/12}\nu^{1/4}}{3.63(1 + C/2330)gQ^{1/6}} \tag{10.14}$$

These three equations provide the dimensions of a stable canal under a given discharge, sediment concentration, sediment size, and bank cohesiveness. Through the use of the side factor, this method accounts for the effect of bank cohesiveness on the channel geometry. It can be seen from these equations that the stable width decreases while the depth increases with an increase in bank cohesiveness. Since bank slope of the canal is in direct relation to bank cohesiveness, canals with more cohesive banks have smaller widths and greater depths.

## Simons and Albertson's Regime Method

The foregoing design methods by Lacey and Blench are limited to canals with a sand bed and cohesive banks. Simons and Albertson (1960) extended this scope to canals of different characteristics, using data of the India–Pakistan canals and additional ones collected in Colorado, Wyoming, and Nebraska. Based on these data, they classified canals into the following five types: (1) sand bed and banks, (2) sand bed and cohesive banks, (3) cohesive bed and banks, (4) coarse noncohesive material, and (5) same as (2) but with heavy sediment loads, 2000–8000 ppm. The hydraulic geometry is distinguished according to the canal types so classified.

Regime equations by Simons and Albertson are given in English units and may be grouped into three categories. Equations in the first category are related to the regime width, including:

$$P = K_1 Q^{0.5} \tag{10.15}$$

$$\overline{B} = 0.9P = 0.9K_1 Q^{0.5} \tag{10.16}$$

and

$$\overline{B} = 0.92B - 2.0 \tag{10.17}$$

where $P$ is the wetted perimeter, $\overline{B}$ is the average width, $B$ is the surface width, and $K_1$ is a coefficient related to the canal type. Figure 10.3 shows the stable width as a function of discharge and canal type represented by Eq. 10.16, together with the ranges of such variables in the data used. For each canal type the stable width is proportional to $Q^{0.5}$, and for the same $Q$, the width has a decreasing trend with bank cohesiveness or bed material size reflected in the value of $K_1$. Equations of the second group give the channel depth:

$$R = K_2 Q^{0.36} \tag{10.18}$$

$$D = 1.21R \quad \text{for } R < 7 \text{ ft} \tag{10.19}$$

and

$$D = 2 + 0.93R \quad \text{for } R \geq 7 \text{ ft} \tag{10.20}$$

where $R$ is the hydraulic radius and $K_2$ is a coefficient related to the canal type. Equations of the third group are flow-resistance formulas:

$$U = K_3(R^2 S)^m \tag{10.21}$$

$$\frac{U^2}{gDS} = K_4\left(\frac{U\overline{B}}{\nu}\right)^{0.37} \tag{10.22}$$

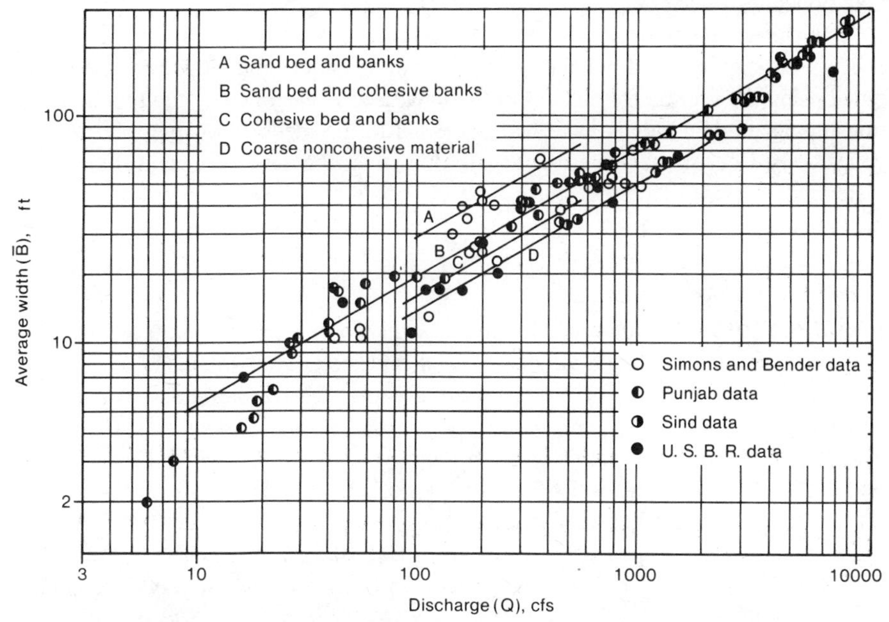


**Figure 10.3** Variation of average width with discharge and type of canal (after Simons and Albertson, 1960).

where $m$ is an exponent, and $K_3$ and $K_4$ are coefficients. Equation 10.21 is similar to Lacey's resistance formula (Eq. 10.1) in form. Equation 10.22 expresses the friction factor as a function of the Reynolds number and is only applicable to the first three types of canals. Coefficients and exponents for these equations are tabulated below:

| Coefficient | Channel Type | | | | |
|---|---|---|---|---|---|
| | 1 | 2 | 3 | 4 | 5 |
| $K_1$ | 3.5 | 2.6 | 2.2 | 1.75 | 1.7 |
| $K_2$ | 0.52 | 0.44 | 0.37 | 0.23 | 0.34 |
| $K_3$ | 13.9 | 16.0 | — | 17.9 | 16.0 |
| $K_4$ | 0.33 | 0.54 | 0.87 | — | — |
| $m$ | 0.33 | 0.33 | — | 0.29 | 0.29 |

This method reflects the effects of bed and bank materials on the stable channel geometry in addition to the discharge. For the same canal type, the width and depth are in direct relation to the discharge, and they are essentially independent of the slope. For the same discharge, canals with cohesive banks are smaller in width, deeper in depth, and flatter in slope than those with noncohesive banks. This comparison is attributed to the fact that cohesive banks are generally steeper than noncohesive ones.

In the application of this method, the design discharge and the canal type need to be specified beforehand; then the width, depth, and slope may be computed using the regime formulas. For the data used in establishing the regime relationships, sand-bed canals have bed materials in the medium to fine sand range, cohesive-bed canals have bed materials finer than the sand size, and the coarse materials, or gravel, for the fourth type of canals are between 20 and 82 mm in median size.

► ***EXAMPLE 10.1.*** Design a stable alluvial canal to convey the discharge of 2100 cfs for the median bed-material size of 0.34 mm, cohesive banks, and a sediment concentration of 200 ppm admitted into the canal. Obtain the width, depth, and slope using the respective methods of Lacey, Blench, and Simons and Albertson. The data of this problem are taken from Punjab Canal No. 13, which has a surface width of 125 ft, a depth of 7.45 ft, and a slope of 0.00018. Compare the computed values with the measured ones.

First, the Lacey method is used. From Eq. 10.5, $B = 2.67Q^{1/2} = 122.4$ ft; from Eq. 10.3, $f = 1.6d^{1/2} = 0.933$. From Eq. 10.1 and the value of $f$ we obtain

$$\frac{Q}{A} = 1.11\left(\frac{A}{B}\right)^{1/2} \tag{10.23}$$

where $A$ is the cross-sectional area of flow. Substituting $Q = 2100$ and $B = 122.4$ into Eq. 10.23 yields

$$A = 760 \text{ ft}^2$$

With the values of $A$ and $B$ now obtained, the depth of flow may be computed from the geometric relationship for a trapezoid:

$$A = (B - 2zD)D + zD^2$$

where $z$ is the bank slope in $z$ horizontal units to 1 vertical unit. Assume a $z$ value of 1.5 for the cohesive bank material; the value of $D$ is obtained to be 6.8 ft.

Second, for the Blench method, the side factor of 0.2 is selected and the bed factor is obtained from

$$F_b = 1.9d^{1/2} = 1.11$$

From Eq. 10.12, $\overline{B} = 108$ ft; from Eq. 10.13, $D = 7.0$ ft. Assume a $z$ value of 1.5 for the side slope; the surface width is obtained from $B = \overline{B} + zD = 118.5$ ft. From Eq. 10.14,

$$S = \frac{1.11^{5/6} \times 0.2^{1/12} \times 10^{-5/4}}{3.63 \times 1.0858 \times g \times 2100^{1/6}} = 0.000118$$

Third, in using the Simons and Albertson method, this canal is considered to be type 2. From Eq. 10.16 and the $K_1$ value of 2.6, $\overline{B} = 107.2$ ft; from Eq. 10.17, $B = 114.4$ ft. From Eq. 10.18 and the $K_2$ value of 0.44, $R = 6.9$ ft; from Eq. 10.19, $D = 8.4$ ft.

Now, the mean velocity is given by

$$U = \frac{Q}{\overline{B}D} = 2.34 \quad \text{ft/sec}$$

From Eq. 10.21 with $K_3 = 16.0$ and $m = 0.33$, $S = 0.000062$.

The widths and depths obtained using these methods compare favorably with the measured values. Such comparisons are not surprising because the Punjab Canal data were used in establishing these regime equations. Comparisons for the computed and measured slopes show considerable percentage variations. Such discrepancies should not be taken too seriously because a very flat slope is sensitive to variation.

## 10.2 RATIONAL METHOD FOR STABLE ALLUVIAL CANAL DESIGN

For stable sand-bed canals, a rational design method was developed by Chang (1980a, 1985b) based on the physical relations of sediment transport, flow resistance and dynamic equilibrium described in Sec. 9.2. In this method, the channel configuration is approximated as a trapezoid, as shown in Fig. 10.1, which is uniquely defined by its width (surface width $B$ or average width $\bar{B}$), depth $D$, channel slope $S$, and side slope $z$. A typical design chart (see Figs. 10.6 and 10.8) is provided, in which the width, depth, slope, water discharge, bed-material size, and bed load (or sediment concentration) are interrelated. To design a canal using this method, the water discharge, sand size, and sediment concentration admitted into the canal need to be specified and the side slope needs to be estimated on the basis of the bank material. Then stable width, depth, and slope of the channel can be obtained within the limit of application. In the following, the analytical development, a numerical example, variation of stream power with channel width, stable width in relation to bank material, and design chart of this method are described.

## Analytical Development

For stable alluvial canals the independent variables, including the water discharge $Q$, sediment inflow rate $Q_s$, and sediment characteristics represented by its median size $d$, are determined at the entrance of each canal and at the intake for the canal system. From the concept of minimum stream power described in Sec. 9.2, the velocity, width, and depth of the canal in dynamic equilibrium must be such that the inflow rates of water and sediment are transported at the minimum channel slope while compatible with the physical relations of flow resistance, sediment transport, and bank stability.

The three dependent variables of width, depth, and slope can be obtained using the three physical relations of flow resistance, bed-load transport, and minimum stream power for a certain bank stability. Note that the velocity is not an additional degree of freedom because it is determined from the discharge, width, and depth. The bed-load equation is used because bed load is responsible for molding the channel shape. In the analytical development, Lacey's resistance formula (Eq. 10.2) was used for sand-bed canals, but it can be replaced by any other valid relation such as Engelund's relation or Brownlie's equation. For bed-load computation, the DuBoys formula (Eq. 7.5) was used, but it can be replaced by any other valid formula. Of course, somewhat different results can be expected if a different resistance relation or bed-load equation is employed. To account for the bank effect, the hydraulic radius should be used in place of the local depth in the flow-resistance and bed-load formulas. In a stable canal, bed-load transport is essentially limited to the bottom width $b$ shown in Fig. 10.1.

For a set of independent variables $Q$, $Q_s$, and $d$ together with an assumed bank slope $z$ reflecting the bank stability, the dependent variables $B$, $D$, and $S$ are obtained following the computing steps given in Fig. 10.4. A brief description of these steps is given as follows:

1. A set of independent variables $Q$, $Q_s$, and $d$ are first selected as input variables. The side slope $z$ is selected on the basis of the bank material.

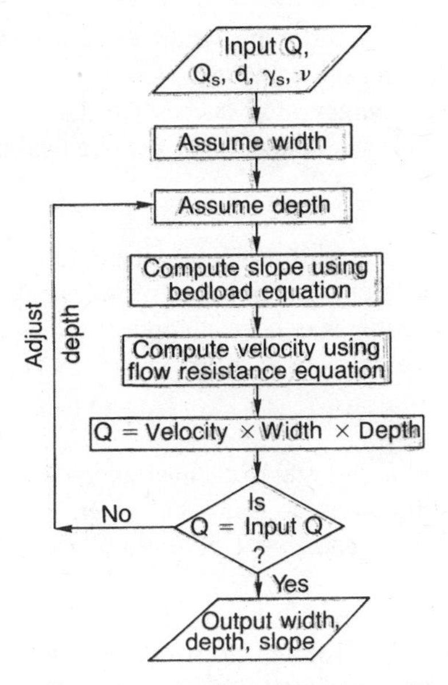


**Figure 10.4** Flow chart showing major steps of computation.

2. A series of incremental channel widths $B$ are assumed. For each width, the depth is computed using the flow-resistance relation following a trial-and-error procedure, and channel slope is calculated using the bed-load formula. In order to compute the slope using the DuBoys formula, it is desirable to express the slope in terms of the specified variables. For bed-load transport on the bottom width $b$, the DuBoys formula can be written as

$$Q_s = q_s b = \frac{0.173}{d^{4/3}} \gamma RS(\gamma RS - \tau_c)b \tag{10.24}$$

It can be rearranged as a quadratic equation for $S$:

$$c_1 S^2 - c_2 S - c_3 = 0 \tag{10.25}$$

where

$$c_1 = \gamma^2 R^2, \qquad C_2 = \gamma R \tau_c, \quad \text{and} \quad c_3 = \frac{Q_s d^{3/4}}{0.173b} \tag{10.26}$$

Then, $S$ can be obtained from the equation

$$S = \frac{c_2 + (c_2^2 + 4c_1c_3)^{1/2}}{2c_1} \tag{10.27}$$

3. The slopes computed at different widths are compared, and the minimum slope is selected among the computed slopes. The computation can be carried to the desired accuracy after successive iterations using smaller width increments. The stable width and depth are obtained from those at the minimum slope.

The foregoing procedure outlined lends itself to computer solution, especially if many cases for different combinations of independent variables need to be made. Although the procedure has been computerized, the following numerical example based on hand calculation is given in order to illustrate the computing steps.

▶ ***EXAMPLE 10.2.*** Determine the stable channel geometry and slope following the foregoing procedure outlined for the following given conditions: $Q$ = 1000 cfs, $Q_s$ = 0.05 cfs, $d$ = 0.3 mm, and $z$ = 1.5. Assume a specific gravity of 2.65 for the bed material.

For these given values, the absolute rugosity $N_a$ in Lacey's resistance equation is first computed from Eq. 10.4 as 0.0218; the critical shear $\tau_c$ in DuBoys formula is obtained from Eq. 7.9 as 0.0182 lb/ft$^2$. The values of variables are given in Table 10.1 following the order of computation and are explained as follows:

The discharge computed in col. 13 is compared with the given discharge of 1000 cfs. If these values are not the same, a new depth must be assumed in col. 2. The trial-and-error procedure is repeated until these two values are within the desired accuracy. The $D$ values listed in lines 1 and 2 reflect the trials, but starting from the third line, only the correct depths are listed.

The computed values of $S$, $D$, $B/D$, $U$, and $n$ as functions of the width are shown graphically in Fig. 10.5. For this sample case, the stable width which corresponds to the minimum slope is determined to be 87 ft using 1-ft width increments. The values of other variables at this width are given in Table 10.1.

## Variation of Power Expenditure with Channel Width

Results of the numerical example shown in Fig. 10.5 are used as the basis of discussion for the variation of power expenditure with channel width. For the specified values of $Q$, $Q_s$, $d$, and $z$, the variation of $S$ or $\gamma QS$ with $B$ has a minimum under certain counteractive factors, explained as follows. From a large width, say 150 ft, at which the bank effect is small, channel slope decreases with width reduction because the concentration of flow improves the sediment efficiency so that the given discharge and load are transported at a flatter slope or less power expenditure per unit channel length. However, this increase in transport rate with width decrease is countered by channel banks. Because of the lower shear stress, the surface areas of stable banks contribute little or nothing to bed-load transport. With a decrease in width, the weight of these ineffective portions of the channel increases while the effective width (bottom width) becomes smaller. Conse-

**TABLE 10.1 Computation for the Numerical Example**

| $B$ (ft) (1) | $D$ (ft) (2) | $b$ (ft) (3) | $A$ ($ft^2$) (4) | $P$ (ft) (5) | $R$ (ft) (6) | $c_1$ ($\times 10^{-4}$) (7) | $c_2$ (8) | $c_3$ ($\times 10^3$) (9) | $S$ ($\times 10^4$) (10) | $\overline{D}$ (11) | $U$ (12) | $Q$ (13) | $n$ (14) | **F** (15) |
|---|---|---|---|---|---|---|---|---|---|---|---|---|---|---|
| 150 | 3.00 | 141 | 437 | 152 | 2.88 | 3.23 | 3.27 | 0.831 | 2.188 | 2.91 | 2.02 | 885 | | |
| 150 | 3.50 | 140 | 506 | 153 | 3.32 | 4.28 | 3.77 | 0.837 | 1.907 | 3.37 | 2.10 | 1064 | | |
| 150 | 3.32 | 140 | 481 | 152 | 3.16 | 3.90 | 3.59 | 0.837 | 1.997 | 3.21 | 2.08 | 1000 | 0.0218 | 0.20 |
| 140 | 3.48 | 130 | 469 | 143 | 3.30 | 4.24 | 3.75 | 0.904 | 1.967 | 3.35 | 2.13 | 1000 | 0.0217 | 0.21 |
| 130 | 3.67 | 119 | 457 | 132 | 3.45 | 4.64 | 3.92 | 0.985 | 1.938 | 3.51 | 2.19 | 1000 | 0.0216 | 0.21 |
| 120 | 3.88 | 108 | 443 | 122 | 3.62 | 5.11 | 4.11 | 1.08 | 1.912 | 3.69 | 2.26 | 1000 | 0.0215 | 0.21 |
| 110 | 4.13 | 97.6 | 429 | 113 | 3.81 | 5.66 | 4.33 | 1.20 | 1.888 | 3.90 | 2.33 | 1000 | 0.0214 | 0.21 |
| 100 | 4.43 | 86.7 | 413 | 103 | 4.03 | 6.31 | 4.57 | 1.35 | 1.869 | 4.13 | 2.42 | 1000 | 0.0213 | 0.21 |
| 90 | 4.79 | 74.6 | 396 | 92.9 | 4.27 | 7.09 | 4.85 | 1.55 | 1.859 | 4.40 | 2.52 | 1000 | 0.0211 | 0.21 |
| 87 | 4.91 | 72.3 | 391 | 90.0 | 4.35 | 7.35 | 4.93 | 1.62 | 1.858 | 4.49 | 2.56 | 1000 | 0.0211 | 0.21 |
| 80 | 5.23 | 64.3 | 377 | 83.2 | 4.54 | 8.02 | 5.15 | 1.82 | 1.862 | 4.72 | 2.65 | 1000 | 0.0210 | 0.21 |
| 70 | 5.81 | 52.6 | 356 | 73.5 | 4.84 | 9.12 | 5.50 | 2.23 | 1.893 | 5.08 | 2.81 | 1000 | 0.0208 | 0.22 |
| 60 | 6.59 | 40.2 | 330 | 64.0 | 5.16 | 10.4 | 5.86 | 2.91 | 1.982 | 5.50 | 3.03 | 1000 | 0.0206 | 0.23 |
| 50 | 7.74 | 26.8 | 297 | 54.7 | 5.43 | 11.5 | 6.17 | 4.37 | 2.238 | 5.94 | 3.37 | 1000 | 0.0204 | 0.24 |

Col. 1. Assumed surface width of channel.
Col. 2. Assumed central depth of trapezoidal channel.
Col. 3. Bottom width of trapezoidal channel computed from $B - 2zD$.
Col. 4. Cross-sectional area of flow computed from $bD + zD^2$.
Col. 5. Wetted perimeter obtained from $b + 2(1 + z^2)^{1/2}D$.
Col. 6. Hydraulic radius obtained from $A/P$.
Cols. 7, 8, and 9. Coefficients $c_1$, $c_2$, and $c_3$ in the DuBoys formula (Eq. 10.26).
Col. 10. Channel slope computed from the DuBoys formula (Eq. 10.27).
Col. 11. Mean depth used in Lacey's formula (Eq. 10.2) computed from $A/B$.
Col. 12. Mean velocity computed from Lacey's formula (Eq. 10.2) for the computed values of $N_a$, $\overline{D}$ in Col. 11, $R$ in Col. 6, and $S$ in Col. 10.
Col. 13. Water discharge obtained from the values of $A$ in Col. 4 and $U$ in Col. 12.
Col. 14. Manning's coefficient computed from Manning's formula using the values of $R$ in Col. 6, $S$ in Col. 10, and $U$ in Col. 12.
Col. 15. Froude number computed according to $U/(gA/B)^{1/2}$.

quently, it will eventually require a steeper slope to transport the given discharge and load. Under these two counteractive factors, an alluvial channel develops a stable width that coincides with the minimum channel slope.

Although there exists a minimum slope in the slope–width relationship, it nevertheless is not a very distinct minimum because other slopes in the vicinity are only slightly greater. The shape of this mild curve suggests that channel width can be nearly stable within a certain range. This feature partly accounts for the variation in channel geometry between analytical prediction and field data. It should also be recognized that the accuracy of any extremal method is affected by the flow-resistance and bed-load relations used. In fact, somewhat different results can be generated by different relations. Because the bed-load formula is usually limited in accuracy, it undoubtedly affects the results. However, the stable geometry, as determined at the slope extremum among the comparative values, is depen-

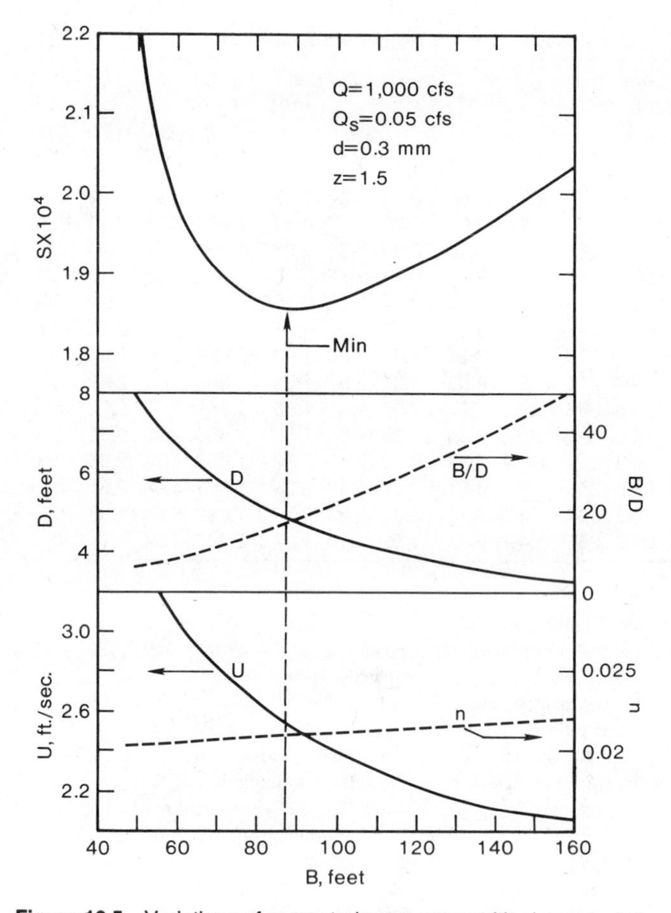


**Figure 10.5** Variations of computed parameters with channel width.

dent on the variation of bed load with the flow and geometric conditions. The pattern of variation is more reliable than the actual load predicted by the formula.

## Stable Width in Relation to Bank Material

The side slope in stable alluvial canals is directly related to the strength of the bank material in resisting erosion. Steep side slopes represent strong cohesive bank materials; and flatter side slopes denote weak bank materials that are less cohesive. The strong materials are associated with narrower channels and weak materials with wider channels, when other conditions are equal.

The hydraulic geometry obtained in Example 10.2 is for the side slope of 1.5 to 1. Similar results can be obtained for other side slopes. The effects of side slope on the hydraulic geometry can be discussed from the viewpoint of hydraulic efficiency or minimum stream power. From basic hydraulics, the best hydraulic section has a side slope angle of 60°, or it is represented by 1 horizontal unit to $3^{1/2}$ vertical units. This side slope is seldom exceeded in earth channels. Therefore, an increase in side slope is usually associated with better hydraulic and sediment efficiency and a corresponding decrease in slope, or power expenditure per unit channel length. The channel width decreases and the depth increases with an increase in side slope. These adjustments in hydraulic geometry are often induced by the formation of side berms resulting from the deposition of cohesive sediment in the maturing process.

## Design Chart for Stable Alluvial Canals

The foregoing computing procedure was employed to compute different sets of channel width $B$, depth $D$, and slope $S$ for different input sets of $Q$, $Q_s$, and $d$. The $z$ value of 1.5 for the side slope was used in this case. The resulting values of $B$ and $D$ are shown as functions of $Q$, $S$, and $d$ in Fig. 10.6, which is the design chart for stable alluvial canals. The values of $Q_s$ (bed load) and $U$ are shown as functions of $Q$, $S$, and $d$ in Fig. 10.7. For the information given in Fig. 10.6, concentrations of bed-material load computed using the Engelund–Hansen formula are shown in Fig. 10.8. The computed results indicate that, at the same $S$, the values of $B$, $D$, $Q_s$, and $U$ vary approximately in proportion to $d^{1/2}$; therefore, the two variables $S$ and $d$ can be combined into one variable $S/d^{1/2}$ in these figures.

Figures 10.6–10.8 are for alluvial canals with a mobile sand bed in the lower flow regime of ripples or dunes. The lower boundary of each relationship is at the

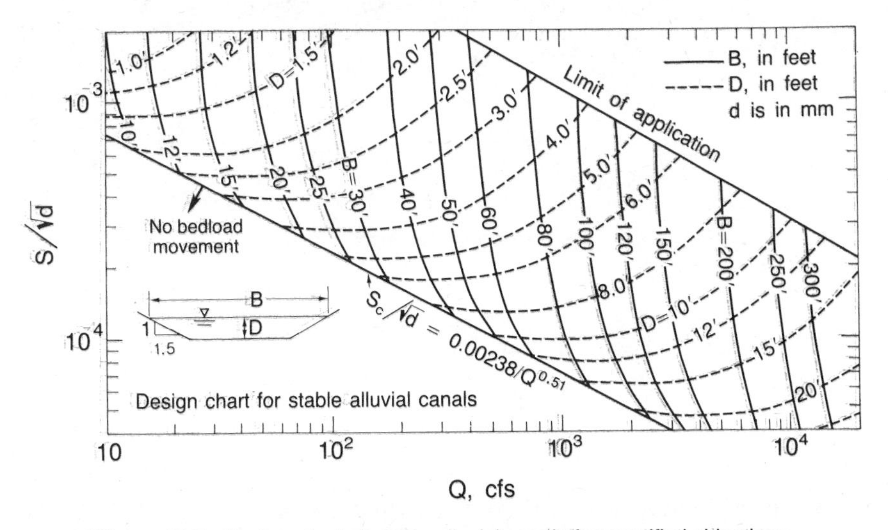


**Figure 10.6** Design chart of stable alluvial canals for specified side slope.

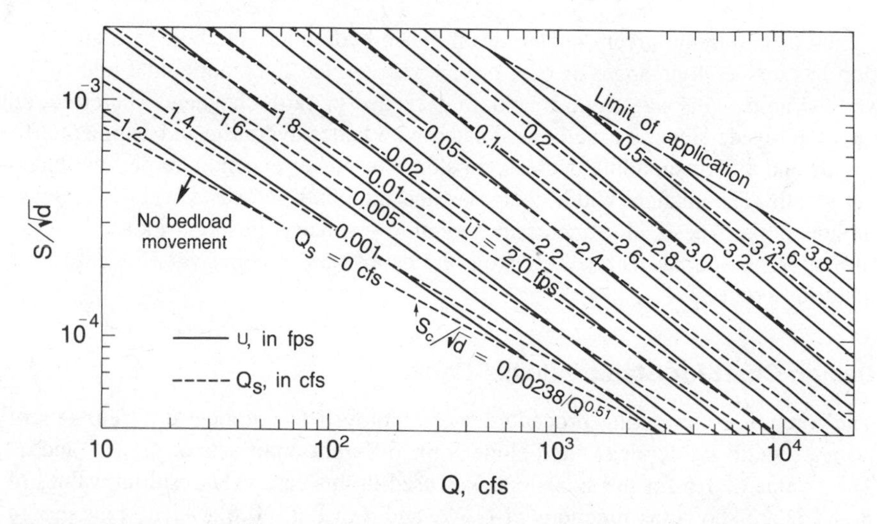


**Figure 10.7** Bed load and velocity as functions of water discharge, slope, and sediment size.

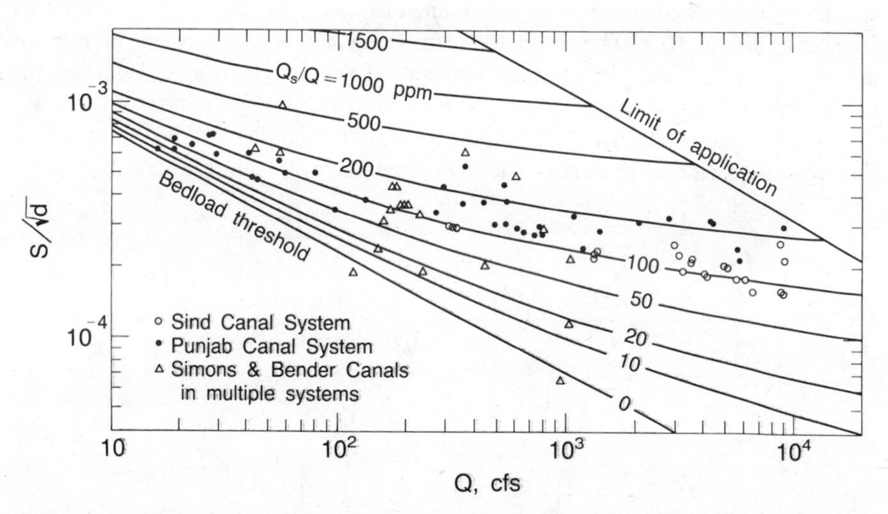


**Figure 10.8** Concentration of bed-material load as function of discharge, slope, and sediment size.

threshold for bed load and is represented by the equation

$$\frac{S_c}{d^{1/2}} = \frac{0.00238}{Q^{0.51}} \tag{10.28}$$

where $S_c$ is the critical channel slope corresponding to bed-load threshold, $d$ is in millimeters, and $Q$ is in cubic feet per second. The limit of application shown in these figures is at the upper limit of the dune bed defined by Eqs. 6.1 and 6.2 because stable alluvial canals must stay in the lower flow regime.

The graphical relationship for the width in Fig. 10.6 is represented by the following equation:

$$B = 4.17\left(\frac{S}{d^{1/2}} - \frac{S_c}{d^{1/2}}\right)^{0.05} Q^{0.5} \tag{10.29}$$

Examination of the exponents shows that the width is primarily a function of $Q$; its dependence on $S$ and $d$ is not very significant. The depth relationship shown in the figure can be represented by the following equation:

$$D = 0.055\left(\frac{S}{d^{1/2}} - \frac{S_c}{d^{1/2}}\right)^{-0.3} Q^{0.3} \tag{10.30}$$

Therefore, the depth is more dependent on $S$ and $d$ than is the width. These relations for $B$ and $D$ are in general agreement with previous regime formulas given in Sec. 10.1.

Because the analysis is based on minimum stream power, that is, best hydraulic efficiency, the hydraulic geometry so obtained must approach the best hydraulic section for rigid channels as the bed load approaches zero. For a trapezoidal channel, the best hydraulic section is given by $R = D/2$, which is represented by the bottom line in Fig. 10.6.

The Punjab Canal data, Sind Canal data, and Simons and Bender Canal data are plotted in Fig. 10.8, based on the measured discharge, slope, and median sediment size. The regime canals are characterized by low velocity, small bed load, and bed form in the lower flow regime of ripples or dunes. It is particularly interesting to note that all canals are plotted within the limit of application of the design chart which corresponds to the transition from the dune bed to the flat bed. For a canal with greater discharge or slope such that it is beyond the limit of application, field evidence reported by Punjab Irrigation Research Institute (1941) has shown that the straight alignment and stable geometry for such a canal could not be maintained with the normal maintenance.

► ***EXAMPLE 10.3.*** Design a stable alluvial canal to carry the discharge of 2100 cfs using the rational method. If the canal is to be constructed on a slope of 0.00018 and it has a median bed-sediment size of 0.34 mm, determine the stable width, depth, and concentration of bed-material load that should be admitted to the canal. This example is a different version of Example 10.1.

First, compute the parameter $S/d^{1/2} = 3.1 \times 10^{-4}$. With this value and $Q = 2100$ cfs, read the $B$ and $D$ values from Fig. 10.6.

$$B = 130 \text{ ft} \quad \text{and} \quad D = 6.7 \text{ ft}$$

The concentration of bed-material load is obtained from Fig. 10.8 as 200 ppm.

## 10.3 DESIGN OF STABLE ALLUVIAL CANALS IN A SYSTEM

For the purpose of water distribution, irrigation canals of different sizes are often connected in a distributary network, as shown schematically in Fig. 10.9. The trunk canal, which draws water from the river at the diversion structure, distributes its discharge into branches or distributary channels and then irrigation turnouts. Stable alluvial canals are characterized by a mobile bed that neither silts nor scours. When such canals are connected within the same system, their widths, depths, and slopes must be properly related under the specific distributions in water discharge and sediment load in order to maintain the approximate equilibrium or regime. Therefore, sediment transport problems are essential considerations in systems design of alluvial canals. This section describes canal headworks and stable geometries of canals connected in a system.

### Canal Headworks

A barrage (check dam) is usually used to maintain the river stage at the canal intake for the purpose of diversion (see Figs. 10.9–10.11). The inflow from a river is seldom sediment-free. Because of the difference in size, most rivers have a much higher sediment content per unit discharge than do the offtaking canals. Consequently, specific measures for desilting are often necessary. Several types of sediment excluder or ejector devices are in use, including vortex tubes and settling basins. The design methods for intake structures and desilting facilities are given in the ASCE Sedimentation Manual (1975) and by Dominy (1966) and Mercer (1971), among others.

The desilting works for the All-American Canal is shown in Fig. 10.11. Inflow to the canal passes through settling basins, each containing twenty-four 125-ft-diameter rotary scrapers. The levees forming the long sides of the basins constitute skimming weirs over which the desilted water discharges into the effluent

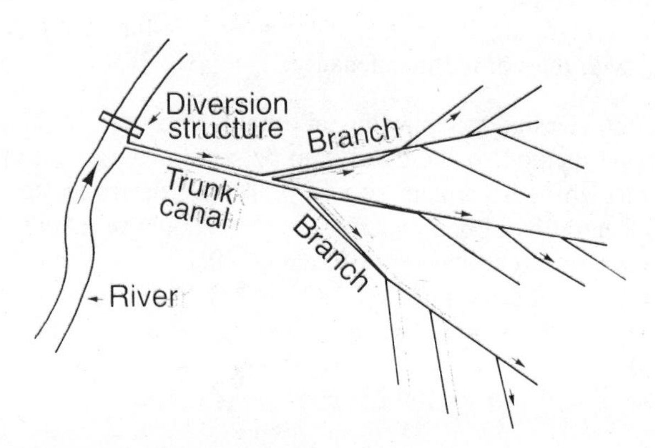


**Figure 10.9** Schematic of distributary canal system.

**Figure 10.10** Physical model of barrage and canal intake structure at Poona, India.

**Figure 10.11** Imperial Dam and desilting basins at the All-American Canal headworks (courtesy of Imperial Irrigation District).

channel leading to the All-American Canal. Bypass channels that carry sediment-laden flow directly to the canal provide the adjustment of sediment load admitted into the canal. Settled deposits in the basins are removed by the scraper blades and flushed into the sluice channel through the sludge pipes.

## Canals in a Distributary System

The stable geometry of an individual canal as a function of its discharge $Q$, slope $S$, and median sediment size $d$ is given in Fig. 10.6, the design chart for stable alluvial canals. Applications of this design chart has been extended by Chang (1985) to canals in a distributary system; this development is described below. The regime channel slope is controlled by the water discharge, bed load, and sediment caliber. To maintain the approximate equilibrium slope in the system, specific load must be admitted. To avoid sediment problems in systems design, the geometries and slopes of all canals in a system must be so selected that these canals neither erode nor silt. Under such an equilibrium condition, bed-material load admitted to the system must be in balance with the outflows. Bed-material load includes bed load and suspended load but not wash load. The ratio between bed load and suspended load may vary from one canal to the other with the change in flow condition.

To maintain sediment equilibrium, canals connected in a system should be designed to have equal concentration of bed-material load, explained in the following. Branches from a sediment-bearing canal could draw a greater or lesser concentration of sediment than does the parent canal. Adjustment of sediment distribution to the branches can be made at the bifurcation headworks. However, because water is the agent for sediment transport, the distribution of bed load or suspended load at the bifurcation should be more or less proportional to the discharge under the equilibrium condition. Of course, the distribution of bed-material load with discharge can be altered by disassociating the load from the water flow using structural control of the channel-bed elevation. But, for most canal systems without such structural sediment controls, the concentration of bed-material load remains similar among different branches. The natural made of sediment distribution at the bifurcations is more or less maintained when canals in a system are designed to have equal sediment concentration.

Concentration of bed-material load is given in Fig. 10.8. Along any line of equal concentration, one has

$$S \propto Q^{-1/6} d^{1/2} \tag{10.31}$$

where the exponent of $-1/6$ is the average slope of the contours. This relation bears strong similarity to Lacey's regime formula, Eq. 10.6, and Blench's formula, Eq. 10.14. The difference lies in the exponent for $d$, which is 0.833 in Lacey's formula and 5/12 in Blench's formula.

Three sets of canal data—the Punjab Canal data, the Sind Canal data, and the Simons and Bender Canal data—are used to compare with relationships shown in Fig. 10.8. These sand-bed canals are plotted using measured values of $Q$, $S$, and $d$. It is important to note that the Punjab Canal System, as well as the Sind Canal System, has several trunk canals that draw water from the river source with a similar sediment concentration. The Simons and Bender Canal data, on the other hand, are from different geographical locations in Colorado, Wyoming, and Nebraska; that is, they are from multiple canal systems.

For the Punjab Canal System as well as the Sind Canal System, the data of each set fall along a line of constant sediment concentration in Fig. 10.8. That is, canals of each system have an approximately equal sediment concentration. On the other hand, the Simons and Bender canal data, which are from different systems, are not correlated with constant sediment concentration.

In summary, because of the sediment problems in systems design, the geometries and slopes of all canals in the same system must be interrelated in order to maintain sediment equilibrium. Without the use of structural sediment controls, sediment equilibrium is maintained by designing canals with an equal sediment concentration. Under such a system design, the channel slope is inversely proportional to the one-sixth power of the discharge. Canals with greater discharges have higher velocities but milder slopes. The analytical discharge–slope relationship, Eq. 10.31, concurs with previous empirical formulas; thus, the analysis provides a better understanding of the intrinsic nature of such empirical formulas.

### Design Procedure

The design chart for stable alluvial canals in Fig. 10.6, when used in conjunction with Fig. 10.8, provides a simple graphical method for the design of canals in a system. When discharges of canals are given under a specified plan for water distribution, the slopes of these canals in the system should be chosen so that they fall along a line of constant sediment concentration in Fig. 10.8. The width and depth of each canal are then directly obtained from Fig. 10.6. Of course, the headworks of the canal system must be operated to supply the trunk canal with the proper sediment concentration in order to maintain the slope stability.

## 10.4 MATURING OF CANALS

The bed and bank materials of newly excavated canals are determined by the local soil profile but will gradually adjust to the sediment entering the channel which may be different in composition from that in the new canal. This gradual transition from the new canal to the final quasi-equilibrium is often referred to as the *maturing process*. One of the features in this process is the removal of silt and clay from the channel bed and the deposition of such cohesive material on the banks to form side berms. A mature canal typically has cohesive banks and a noncohesive bed depleted of silt and clay. Because of the adjustments in bed material and side slope, canals usually undergo changes in width, depth, and channel slope during the maturing process. The trend of changes caused by the adjustments in sediment size can be determined from Figs. 10.6 and 10.8. Steeper side slopes are associated with greater hydraulic efficiency, and the same inflow of water-sediment mixture will be transported at a milder slope and a greater depth. The impact from the change in channel slope can be quite serious in a long reach but can usually be absorbed by the drop structures. To avoid slope adjustments, the designer should consider in advance the sediment characteristics in the inflow water–sediment mixture. The inflow at the headworks can also be adjusted for an existing canal system to minimize changes.

## 10.5 HYDRAULIC GEOMETRY OF GRAVEL-BED STREAMS

Earlier studies by Leopold and Maddock (1953) and others have produced empirical relationships for the width and depth of gravel streams as a function of the discharge $Q$ (see Sec. 2.6). With $Q$ being either the mean annual discharge or bankfull discharge, the width was found to be proportional to $Q^{0.5}$ in these studies. This relationship, while valid for most gravel-bed streams, is in conflict with the observations for steep streams, which tend to be wider and shallower than streams of the same discharge at lower slopes. Such observations were made by Lane (1957), Leopold and Wolman (1957), and Charlton (1977), among others. This means that the stream width is not a function of $Q$ alone; instead, it also depends on the channel slope. The variation of width with slope becomes more evident as braiding develops at steeper slopes.

Earlier rational analyses of gravel channel geometry were based on the threshold theory. This theory holds that materials on the channel boundary are subject to flow conditions just sufficient to cause incipient motion of these particles. The analyses by Glover and Florey (1951), Lane et al. (1959), and Li et al. (1976) have obtained channel cross sections with width–depth ratios less than 15. According to Lane et al. (1959), the threshold channel section is actually the best hydraulic section; its width–depth ratio depends on the size of the bed material and its angle of repose. An increase in the angle of repose will mean a steeper bank slope and a smaller width-depth ratio. The threshold theory is only applicable to channels subject to scour but not silt. It cannot be applied to larger streams, whose channel beds are mobile at larger discharges of rather rare occurrence, although, under such circumstances, the shear stress on the mobile bed is usually slightly greater than the critical stress for incipient motion. Active gravel-bed streams have large width–depth ratios. For example, the Alberta gravel rivers reported by Kellerhals et al. (1972) have width–depth ratios between 15 and 100.

Kellerhals (1967) obtained the following empirical equation for regime width of gravel streams:

$$B = 1.8Q^{0.5} \tag{10.32}$$

where $B$ is the surface width in feet and $Q$ is the bankfull discharge in cubic feet per second. For the mean depth, he found that

$$\overline{D} = 0.166Q^{0.4}k_s^{-0.12} \tag{10.33}$$

where $\overline{D}$ is mean flow depth in feet and $k_s$ is Nikuradse's sand grain roughness in feet.

Using mostly different data, Parker (1979) obtained a similar empirical equation for the width:

$$B_* = 4.4(Q_*)^{0.5} \tag{10.34}$$

where $B_* = B/d$ is the surface width normalized by the median particle size $d$, $Q_* = Q/[(s-1)gd^5]^{1/2}$ is the dimensionless bankfull discharge, and $s$ is the specific gravity of sediment. In a rational approach, Parker (1978, 1979) used singular perturbation techniques to obtain a bed-stress distribution that allows a mobile bed but immobile banks at bankfull discharge. Three regime relations were thus rationally derived. To obtain the hydraulic geometry using this method, Eq. 10.34 for the width is used concurrently with the following respective equations for the depth and slope:

$$D_* = 0.253(Q_*)^{0.415} \tag{10.35}$$

$$S = 0.223(Q_*)^{-0.410} \tag{10.36}$$

where $D_* = \overline{D}/d$. These results were obtained with the assumption that the bankfull bed stress exceeds the critical stress by about 20%, resulting in a low, but nonzero, rate of sediment transport.

## Rational Method

A rational method was developed by Chang (1980b) for active gravel-bed streams with any degree of bed mobility. In this method, the channel cross section is assumed to be trapezoidal in shape, consisting of a mobile central region and immobile banks. If the bank slope $z$ is selected according to the angle of repose of gravel and other physical conditions, then the hydraulic geometry for a straight channel may be uniquely determined by its surface width $B$, mean flow depth $\overline{D}$, and channel slope $S$. For a set of water discharge and sediment load imposed upon the stream from its drainage basin, the three unknowns in width, depth and slope, are determined using three conditions, consisting of the flow-resistance equation, the bed-load equation, and the concept of minimum stream power. Note that the flow velocity is not treated as an additional unknown, since it may be computed readily using the continuity equation of flow for any set of water discharge, width, and depth. The flow-resistance equation and the bed-load equation employed in the development are described below, followed by the resulting regime relationships for the hydraulic geometry.

Unlike sand-bed streams, resistance to flow in gravel-bed streams is primarily due to grain roughness because dunes tend to be poorly developed. For fully developed turbulent flow, the equations for flow resistance given in Sec. 3.7 usually take a logarithmic or power form. Three resistance equations, Eqs. 3.41, 3.42, and 3.43, were tested in developing the rational method. Equation 3.43 by Bray was selected on the basis of best results.

For coarse gravel with a median diameter exceeding 16 mm, the sediment transport is almost exclusively bed load. Parker (1979) obtained a bed-load equation based on the experimental moving gravel data contained in the more than 6000 data sets in the compendium of Peterson and Howell (1973). The Einstein bed-load intensity $\Phi$ and flow intensity $\Psi$ were fitted by the eye to the relation

$$\Phi = 11.2\left(\frac{1}{\Psi} - 0.03\right)^{4.5} \Psi^3 \tag{10.37}$$

where

$$\Psi = \frac{q_s}{[(s - 1)gd^3]^{1/2}}, \qquad \frac{1}{\Psi} = \frac{\tau_0}{g(\rho_s - \rho)d} = \tau_* \tag{10.38}$$

where $q_s$ is the volumetric bed-sediment discharge per unit width, and $\tau_*$ is the dimensionless shear stress or the Shields stress. Equation 10.37 has its lower boundary at $\tau_* = 0.03$, which is taken as the threshold Shields stress for coarse gravel. This bed-load equation was intended for Shields stress in the range between 0.03 and 0.1. Although this range covers all the data that Parker used, it does not extend far enough into regions of steeper slopes. Adding the Zurich laboratory data by Meyer-Peter, et al. (1934), which has a median grain size of 28.6 mm, Chang (1980b) obtained the following relation for bed load

$$\Phi = 6.62\left(\frac{1}{\Phi} - 0.03\right)^{5} \Psi^{3.9} \tag{10.39}$$

This equation is shown in Fig. 10.12 together with Eq. 10.37 and Einstein's bed-load function for comparison. The three equations nearly coincide in the region shown. For this reason, they produce similar results for channel geometry.

The three independent conditions of flow resistance, bed load, and minimum stream power were solved following computing steps shown in Fig. 10.4. For a given set of water discharge, sediment load, grain size, bank slope, and so on, channel depth and width at the unique minimum slope were obtained in the computation. The computed width and depth at the minimum slope were used for the stable channel geometry. The results are summarized in Fig. 10.13.

The bank slope $z$ for the channel was selected and used as an input parameter. Since the bank slope affects the proportions of the bank region and the central region, it therefore influences the channel geometry. While the bank slope

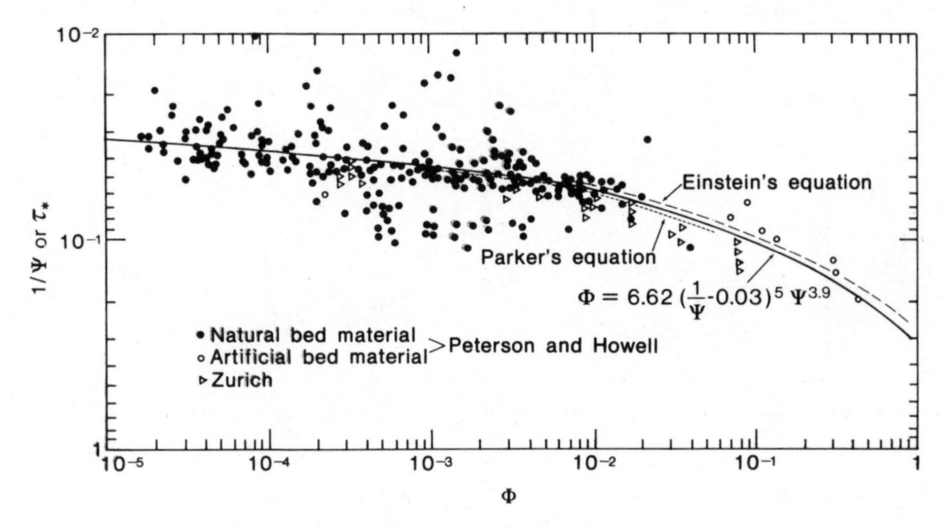


**Figure 10.12** Bed-load relations for gravel beds.

is always less than the angle of repose, it is also affected by other physical conditions. In general, small streams in confined valleys have steeper bank slopes than do large streams in broad flood plains. In this study, the $z$ value was selected to be 2.5 for the bankfull discharge of 10 cfs, 3 for $10^2$ cfs, 3.5 for $10^3$ cfs, 4 for $10^4$ cfs, and 4.5 for $10^5$ cfs. This selection of $z$ values simply represents a general trend for gravel-bed streams; it does not preclude other appropriate $z$ values to be used for given physical conditions. Other $z$ values, of course, will result in similar but somewhat different relations.

The graphical relations shown in Fig. 10.13 represent the rational regime relations obtained. These relations were first obtained using median diameter $d$ of 50 mm. For other gravel size between 16 mm and 200 mm, one obtains similar relations that are essentially only different in the vertical scale for channel slope. If the channel slope is multiplied by the factor $(50/d)^{1.15}$, relations for different gravel sizes may be approximated into one as shown in Fig. 10.13 (top and middle panels). Therefore, with $S(50/d)^{1.15}$ as the ordinate, these graphical relations are applicable to different grain sizes within the range specified.

The graphical relations are for active streams with nonzero bed load at the bankfull discharge. At the lower boundary of vanishing bed load, this method reduces to the threshold theory. At the threshold condition, the cross-sectional geometry obtained based on the rational method is that of the best hydraulic section, in agreement with the analysis by Lane et al. (1959). The threshold condition shown in Fig. 10.13 is represented by the equation

$$S_c = 0.000442d^{1.15}Q^{-0.42} \tag{10.40}$$

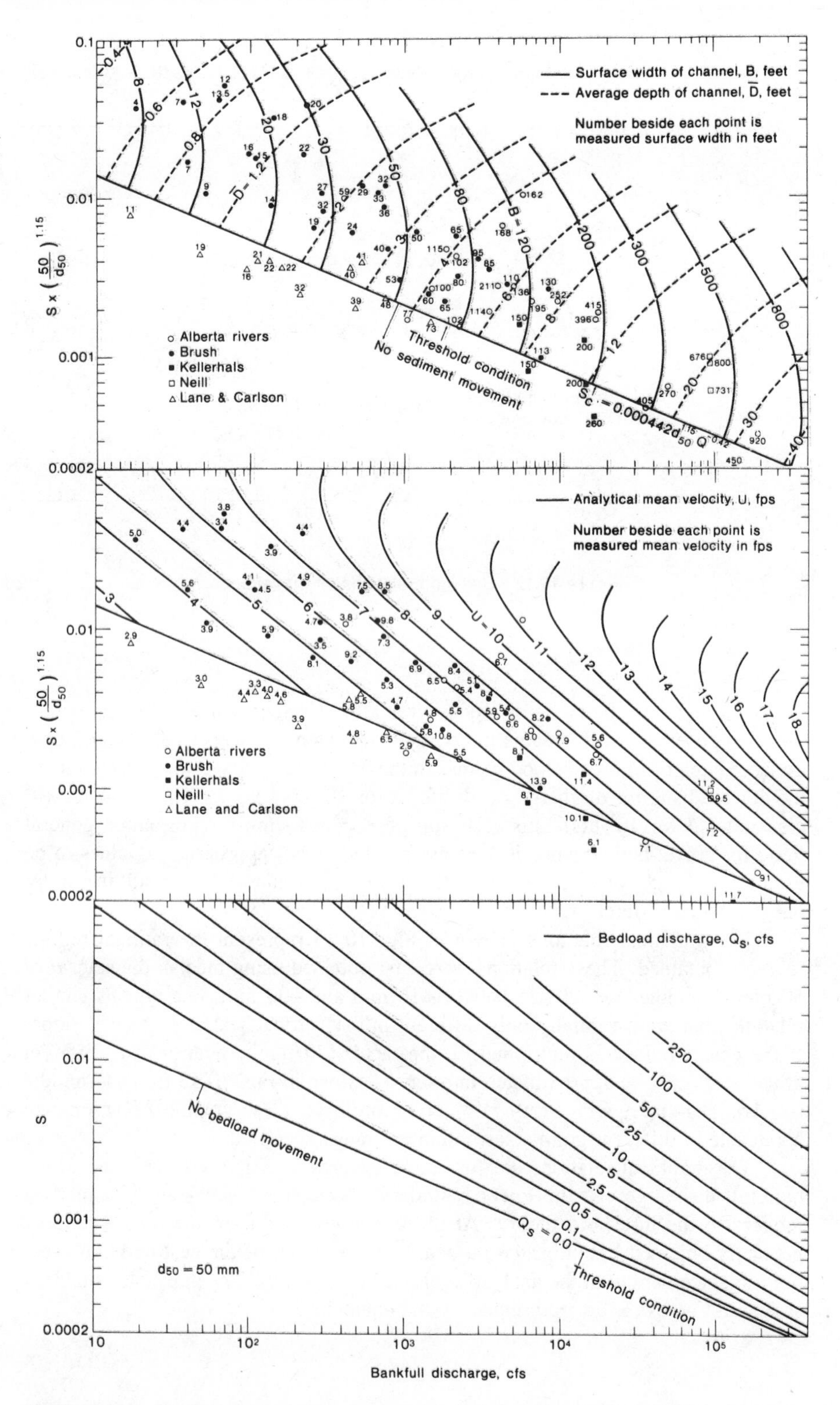


**Figure 10.13** Regime channel geometry, mean velocity, and bed load as functions of bankfull discharge, slope, and sediment size for gravel-bed streams.

where $S_c$ is threshold channel slope for impending sediment motion. Above this line, the gravel stream is considered active; below this line, the stream bed is basically immobile, although "through-put" load of finer grains may still exist.

For active gravel streams, the surface width $B$ shown in Fig. 10.13 varies with both bankfull discharge and channel slope. Along lines parallel to $S_c$, $B$ varies only with the discharge, that is,

$$B = CQ^{0.47} \tag{10.41}$$

This rational relationship for channel width is similar to the empirical relations given in Eqs. 10.32 and 10.34. The graphical relationship for width as a function of $Q$, $S$, and $d$ is represented by the general equation

$$B = \left[1.905 + 0.249\left(\ln \frac{0.001065d^{1.15}}{SQ^{0.42}}\right)^2\right]Q^{0.47}, \qquad S > S_c \tag{10.42}$$

This equation indicates that, except for steep slopes much greater than $S_c$, the width is essentially only a function of the discharge, consistent with other empirical relations. At steep slopes, however, the width increases rapidly with the slope, accompanied by a decrease in depth, indicating braiding tendency for the stream. It is generally agreed that braiding is associated with steep slopes and heavy sediment load.

For a given sediment size, the mean channel depth $\overline{D}$ is found to vary with the bankfull discharge and channel slope, as shown in Fig. 10.13. Along lines parallel to $S_c$, the depth varies in direct relation to the discharge, that is,

$$\overline{D} \propto Q^{0.42} \tag{10.43}$$

Note that this rational relation for depth is similar to the regime relations given by Eqs. 10.33 and 10.35. The general relation for depth as a function of $Q$, $S$, and $d$ is

$$\overline{D} = \left[0.2077 + 0.0418 \ln \frac{0.000442d^{1.15}}{SQ^{0.42}}\right]Q^{0.42}, \qquad S > S_c \tag{10.44}$$

The regime relation for velocity may be obtained from the discharge and the cross-sectional geometry. The results, as shown graphically in Fig. 10.13, (middle panel) indicate that the velocity for regime channel varies in direct relation with both the bankfull discharge and channel slope. On steep slopes, however, the velocity actually decreases with slope because of a rapid increase in channel width.

The analytical relationships are compared with several sets of data for gravel-bed streams in Fig. 10.13, including the Alberta Rivers data by Kellerhals et al. (1972), the Brush data (1961), the Kellerhals stable channel data (1967), the Neill data (1968), and the San Luis Canal data by Lane and Carlson (1953). The bank-full discharge is used as the dominant or "channel-forming" discharge, which covers a range from 17 to 192,000 cfs. The bed material has its median grain size in the range from 16 mm to 265 mm with most sizes between 30 and 70 mm. The rational relationships are in general agreement with the field data, although discrepancies can be observed. The analytical widths tend to be smaller for large streams but greater for small streams; the analytical depths tend to be greater for large streams but smaller for small streams. One explanation for this discrepancy is that bank formation of large streams in broad flood plains is often inhibited by the slow movement of gravel.

# 11

# ANALYTICAL RIVER MORPHOLOGY

River channel formation is a delicate adaptation to the imposed environmental conditions accompanied by compatibility with the flow and sediment transport processes. These fluvial processes are complicated by the flow–sediment interaction for which the channel roughness is not prescribed beforehand but, instead, is determined by the flow condition. Although few other subjects have received so much attention as river morphology, the complexity of this subject has inhibited mathematical formulation in the past. Modern developments in analytical river morphology stem from advances in the predictive methods for sediment transport, flow resistance, channel-width formation, and the dynamics of flow in curved channels. Such developments have extended the knowledge of river morphology by providing a better understanding of the underlying mechanics of previous empirical relations. This understanding is essential in mathematical modeling of rivers and in river control and regulation. The self-formed geometry of rivers in long-term equilibrium and its responses to changes are presented in this chapter from the analytical viewpoint. The scope includes analysis of river meanders, river morphology and thresholds, river channel changes pertaining to the adjustments in regime, and formation of alternate bars.

It is generally acknowledged by river researchers that the sediment-laden flow imposed upon the river from its drainage basin is the cause from which river channel formation follows as an effect. Holtorff (1983) compared the regime geometry to a feedback system in which the cause (the flow) and its effect (the channel) have interacted to reach a state of mutual compatibility. Channel characteristics, which consist of width, depth, bank slope, channel slope, meandering pattern, and so on, are dependent variables that represent the degrees of freedom that are not prescribed but need to be determined in the analysis. The number for the degrees of freedom varies with the type of alluvial river. The simplest alluvial river is a stable alluvial canal with a straight alignment described in Chapter 10, which has four degrees of freedom in its width, depth, bank slope, and channel slope. The velocity is a dependent variable but is not an additional degree of freedom because it is determinable from the discharge and channel geometry. A natural river has at least one more degree of freedom in its meandering pattern. But depending on the level of details desired, natural channels can possess nine degrees of freedom (Hey, 1978).

The analysis of river morphology, as shown in Fig. 11.1, consists of the following three major steps: (1) identification of independent, or controlling, variables, (2) application of physical relations for the fluvial processes, and (3) computation of the resulting channel geometry. This procedure for geometry determination has already been demonstrated in Sec. 10.2 for the case of stable alluvial canals with a straight alignment; it is applied to other types of rivers in this chapter. The difficulty lies in the fact that the channel geometry is not prescribed beforehand but instead needs to be determined for the given controlling variables. This is in contrast to the ordinary analysis for which the channel geometry is given and the flow and sediment transport are computed.

A river is the author of its own geometry. Under the long-term equilibrium, it is so adjusted that its ability to transport balances the water discharge and sediment load supplied from the drainage basin. The adjustments, which may include channel geometry, slope, meandering pattern, roughness, and so on, partially reflect changes in the river's resistance—that is, in power expenditure. This notion lends itself to the energy approach. In the following analyses, the hydraulic geometry is obtained such that the inflow quantities of water and sediment are carried with minimum power expenditure per unit channel length as well as minimum power for the river reach. The power expenditure of the reach is a minimum when the water-surface profile becomes a straight line through crossings and bends (see Sec. 9.2).

## 11.1 ANALYSIS OF RIVER MEANDERS

Meandering rivers are characterized by a streamwise changing geometry and curvature and a gradually-varied flow. The analysis by Chang (1984a) described in

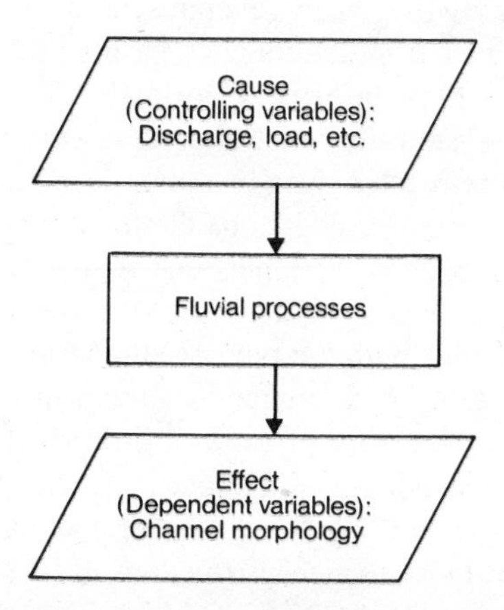


**Figure 11.1** Major steps for hydraulic geometry determination.

this section is for sand-bed meandering rivers; it incorporates the effects of secondary currents on the hydraulics of flow and sediment transport processes. The analysis will establish the maximum curvature for which a river does the least work in turning. This curvature, stated as the radius-of-curvature–channel-width ratio, $r_c/B$, has an average value of 3; it shows only minor variation within the meandering range. This aspect of meander morphology can be compared to a sinuous highway on a hillslope. The meandering river and the sinuous highway can be quite similar in planform. The maximum curvature for the highway is specified in the design criteria in order to make the driving easy. It is interesting to find that the self-formed meander geometry also limits its maximum curvature to avoid the unnecessary energy loss in turning.

A meander curvature increases the power expenditure due to secondary currents, but this is matched by a compensating increase in the sediment efficiency associated with the transverse bed slope. This analysis will also demonstrate how uniform utilization of power (that is, linear water-surface profile) and equal sediment discharge are maintained through river meanders. In the following, the variables for river meanders, applicable physical relations, the method of analysis, and results of analysis are described. The analytical results will also be employed to elucidate the nature of power transformation in river meanders.

### Variables for Meander Analysis

The river flow to be analyzed is shown schematically in Fig. 11.2, which also defines some of the notation employed. Three representative cross sections are selected: Section I is at the point of inflection or crossing; Section K is at the bend apex, which coincides approximately with the zone of maximum curvature; and Section J is located somewhere between these two cross sections. Each cross-sectional shape, approximated by straight lines as shown in Fig. 11.2, is uniquely defined by the surface width $B$, center depth $D$, bank slope $z$, and transverse bed slope $S_t$.

For the purpose of river analysis, variables for regime conditions have been identified as independent variables, dependent variables, and constraints. Water discharge and sediment inflow as well as their respective properties, which are determined in the watershed, are independent variables for the river. Dependent variables in the analysis include the flow velocity, channel width, flow depth, channel slope, and radius of curvature. Note that channel roughness and transverse bed slope in the curved channel are not additional dependent variables because they may be computed based on other variables.

Valley slope is treated as another independent variable because the time scale for its formation is much greater than that for regime channel geometry. Bank slope is another dependent variable. Although a steeper bank slope is hydraulically more efficient, it nevertheless is constrained by the erodibility of bank materials. Cohesive banks can be steeper than noncohesive ones because of the soil erodibility and the angle of repose. Vegetation can decrease the erodibility of a bank; it

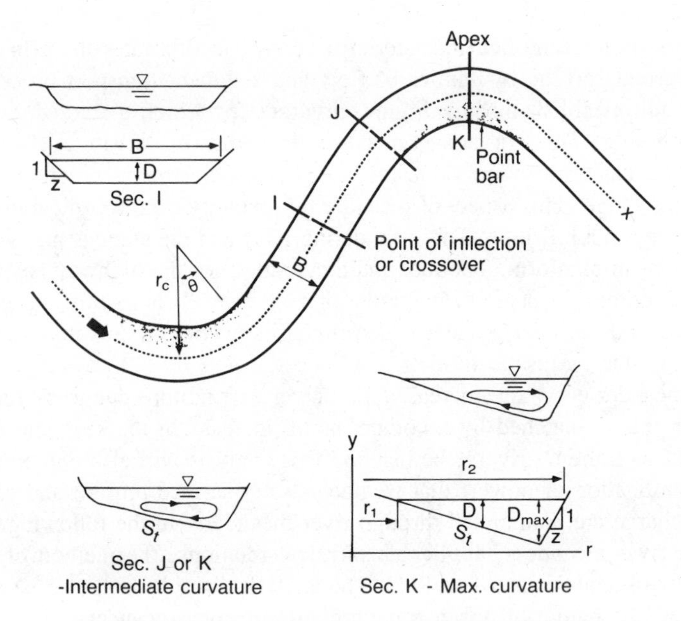


**Figure 11.2** Definition sketch of meander geometry.

can also trap sediment in a depositional environment. The effect of vegetation may be such that it inhibits the development of the self-formed channel geometry, especially in smaller streams. Nonetheless, effects of vegetation and heterogeneity of the bed material cannot be considered in the present analysis.

## Applicable Physical Conditions

The dependent variables or unknowns for the given independent variables and constraints are determined using the following physical conditions: (1) flow-continuity equation, (2) sediment-load equation, (3) flow-resistance equation, (4) minimum stream power per unit channel length, and (5) minimum stream power for a river reach. Physical relations pertaining to curved channels are also employed.

For steady, incompressible flow, the continuity equation is given by $Q = AU$. The bed-load equation is used in the analysis because the bed load is responsible for molding the geometry of rivers. Two equations, the DuBoys equation (Eq. 7.5) and the Einstein–Brown equation (Eqs. 7.26 and 7.27), are used separately; they produce closely similar results for the regime channel geometry except in the small region of vanishing bed load. The flow-resistance equation provides the stage or the depth of flow for a discharge. Engelund's method (see

Sec. 6.5) which considers alluvial bed forms, is employed herein. This method, being a shear-stress relationship, accounts for the transverse variation in resistance at a section.

River flow in the curved reach is characterized by the helical motion, or transverse circulation, due to the difference of centripetal forces between the upper and lower layer of the flow. The transverse circulation is associated with features of the curved reach such as the cross-sectional shape, power expenditure, erosional and depositional pattern, and migration pattern. These features are described in terms of the transverse bed slope, flow resistance and power expenditure, and sediment transport in a meander.

Several analytical models have been developed for the prediction of the transverse bed slope in a curved, erodible bed channel as described in Sec. 8.4. In the present analysis, the transverse bed slope is based on Eq. 8.55, developed by Odgaard. The transverse bed profile varies with the distribution in grain size, such variations are generally small enough so that the profile can be approximated by a straight line, as assumed in the present analysis.

In a curved channel, the power expenditure per unit channel length consists of two components: that associated with streamwise resistance and that due to transverse circulation, that is,

$$\gamma QS = \gamma QS' + \gamma QS'' \tag{11.1}$$

or

$$S = S' + S'' \tag{11.2}$$

where $S$ is the total energy gradient, $S'$ is the streamwise energy gradient, and $S''$ is the transverse energy gradient. The streamwise energy gradient or power expenditure, which is related to the alluvial bed forms, is computed using Engelund's method for flow resistance.

The river channel is self-formed, therefore, meander geometry must reflect the gradual flow development in the curved channel. A factor that limits meander curvature is the tendency for flow separation in sharp bends. Because of separation-induced sediment deposition, separated zones are generally eliminated in a self-formed channel, as is the sharp bend. In the absence of flow separation, the transverse power loss is attributed to the following major causes: (1) internal fluid friction due to transverse circulation and (2) boundary shear associated with transverse shear. The transverse loss due to these causes is included in Eq. 8.74, which is used in the present analysis. For a given channel configuration, Eq. 8.77 shows that the transverse energy gradient is a direct function of the square of the Froude number $\mathbf{F}$, the square of the depth–radius ratio, and the channel roughness; it decreases with the width–depth ratio. The transverse energy gradient can be expressed in the following functional form:

$$S'' = F\left[\mathbf{F}^2, \left(\frac{D}{r_c}\right)^2, \frac{B}{D}, f, \text{cross-sectional shape}\right] \tag{11.3}$$

Transverse currents in curved channels have important effects on sediment transport. Onishi et al. (1976) measured sediment rates in curved channels. They found that, for a given mean depth and flow velocity, the sediment rate per unit width in the meandering channel with the $r_c/B$ value of 3.7 was greater than that in the straight channel, which, in turn, was greater than that in the meandering channel, with an $r_c/B$ value of 7.3. This suggests that a meandering channel can be more efficient than a straight one, in that a given discharge can transport a given load with a smaller energy gradient or less power expenditure. They also provided analysis to show that nonuniformity of the streamwise velocity distribution associated with the transverse bed slope can increase the sediment load.

Bed load in the streamwise direction is computed herein using the streamwise boundary shear $\tau_0'$, which is related to the streamwise energy gradient as

$$\tau_0' = \frac{\gamma D S' \, dr}{[(dr)^2 + (dz)^2]^{1/2}} \tag{11.4}$$

where $D$ is the local depth, $r$ is the radial coordinate, and $z$ is the vertical coordinate. Because of the nonuniformity in shear distribution, total bed load is obtained by integrating the unit load over the channel width.

## Method of Analysis

Cross sections along the meander are shown in Fig. 11.2. Section I at the crossover is approximated as a trapezoid. For sections J and K in the curved channel, the base of the trapezoid is replaced by the transverse bed slope. For a set of channel width and center depth, the maximum transverse bed slope occurs when the sloping bed extends to the water surface on the convex bank. Since the transverse bed slope is proportional to the channel curvature (see Eq. 8.55), the maximum transverse bed slope occurs under the maximum channel curvature (or min $r_c$) for the given width and center depth.

For a given set of independent variables and constraints, the physical conditions are employed to obtain the dependent variables. As a first step, physical conditions consisting of the flow-continuity equation, sediment-load equation, flow-resistance equation, and those pertaining to curved channels are solved to obtain the velocity, depth, slope, radius of curvature, and so on, for various assumed widths. Figure 11.3 is a flow diagram showing the major steps of computation. In the diagram, section I (crossover) and sections J and K in the curved channel are computed separately. The sediment-load equation and flow-resistance equation are related to the streamwise shear which is computed using the streamwise energy gradient according to Eq. 3.4. For section I, in the absence of channel curvature, the total energy gradient $S$ is also the streamwise energy gradient $S'$. But for sections J and K, $S$ is the sum of $S'$ and $S''$ in which $S''$ is computed using Eq. 8.74.

The flow diagram includes two different channel curvatures: the maximum and an intermediate channel curvature. The transverse bed slope at the maximum

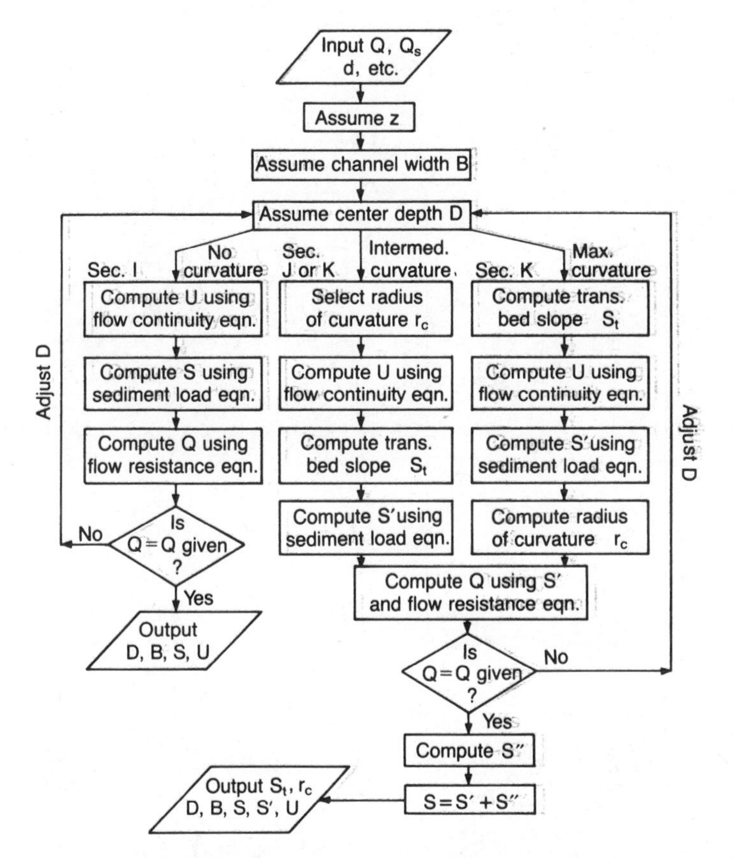


**Figure 11.3** Flow diagram showing major steps of computation.

curvature is determined from the geometric relationship using the assumed $B$ and $D$. Then the radius of curvature $r_c$ is computed using Eq. 8.55; it represents the minimum $r_c$ or the maximum curvature for the given conditions. For the case of intermediate curvature, an intermediate $r_c$ is first selected, then the transverse bed slope is computed using Eq. 8.55.

For the water discharge of 1000 cfs, sediment load of 0.05 cfs, and sediment size of 0.3 mm, sample results for sections I, J, and K are obtained as given in Fig. 11.4, in which the computed variables are shown as a function of the width. From the condition of minimum stream power per unit channel length, the stable width at each section is obtained where $S$ is a minimum. Other variables such as $B$, $D$, $U$, $S$, $S'$, $S_t$ and $r_c$ are obtained at the same point.

Whether these sections of different channel geometries can exist together along the same channel reach is now investigated. Because of the condition of minimum stream power for a river reach of gradually varied flow, these sections can occur along the same reach if their minimum energy gradients are equal. The only remaining variable that can be adjusted in order to satisfy this condition is the

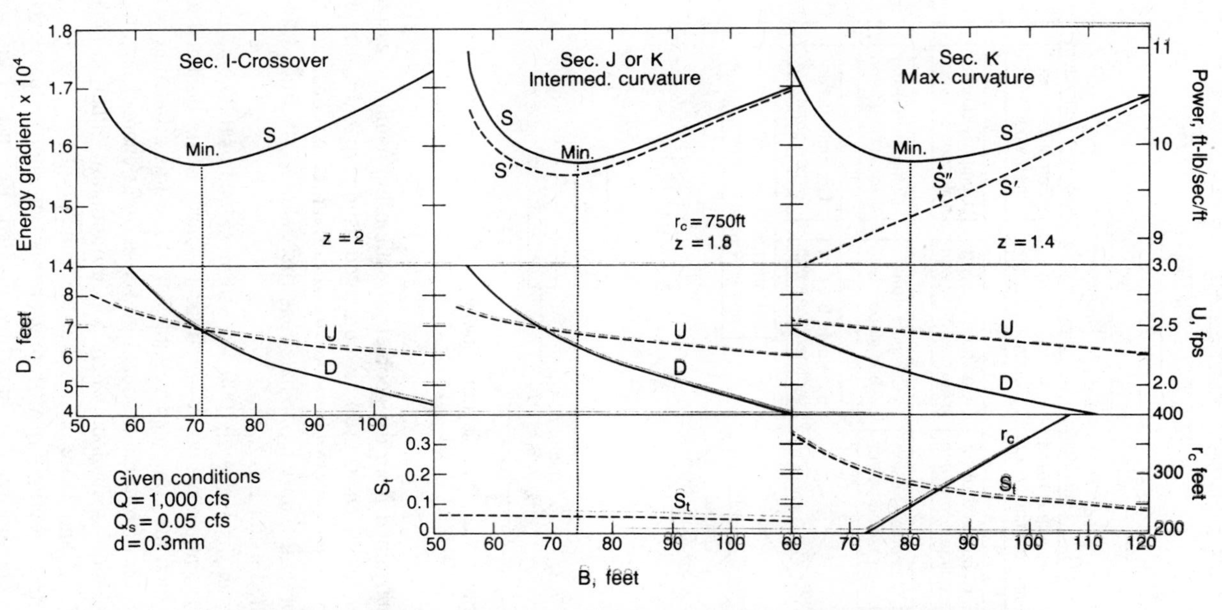


**Figure 11.4** Sample results: Variations of computed variables with channel width.

bank slope represented by $z$. Although a steeper bank slope usually decreases the energy gradient, it is, however, constrained by bank erodibility. For the sample case shown in Fig. 11.4, the slope angle of 36° is assumed for section K under the maximum curvature. This slope, which gives the $z$ value of 1.4, is the approximate angle of repose for sand. With the $z$ values of 1.8 for the intermediate curvature and 2 for the crossing, the minimum energy gradients of these cases are equalized, as shown in Fig. 11.4. Therefore, they can coexist along the same reach of gradually-varied flow under the dynamic equilibrium. They can usually coexist because meanders provide the adjustment in channel pattern so that a river of flatter channel slope may exist on a steeper valley slope.

### Results of Analysis

Analytical results are first given in Fig. 11.4, in which the dependent variables as computed are shown as a function of the channel width for the representative cross sections I, J, and K. The geometric variables for these sections, obtained at the minimum total energy gradient of 0.000157, are tabulated as follows:

| Variable (1) | Section I Crossing (2) | Section J or K Intermediate Curvature (3) | Section K Maximum Curvature (4) |
|---|---|---|---|
| Surface width $B$ (ft) | 71.2 | 74.2 | 79.8 |
| Center depth $D$ (ft) | 6.9 | 6.3 | 5.5 |
| Maximum depth $D_{max}$ (ft) | — | 8.2 | 10.3 |
| Mean velocity $U$ (ft/sec) | 2.45 | 2.41 | 2.43 |
| Transverse bed slope $S_t$ | 0 | 0.06 | 0.16 |
| Radius of curvature $r_c$ (ft) | — | 750 | 242 |

The foregoing sample results show that, despite the variation in cross-sectional shape along the meander, the channel width stays approximately the same. The average width for this case is 75 ft, and the average center depth is 6.2 ft. Under the maximum channel curvature, the value of $r_c/B$ is obtained as 3.2. Because it is obtained using conditions of minimum stream power, it therefore represents the maximum curvature for which a river does the least work in turning. Since the power expenditure is equal for all three cases, a river also does the least work for a lesser channel curvature or in the absence of curvature.

Different cases are computed in order to develop a general relationship for meanders. The maximum curvature in terms of $r_c/B$ is found to be a function of the channel slope, discharge, sediment size, and average width–depth ratio, as shown graphically in Fig. 11.5. The lower limit of this relationship at $S_c$ corresponds to the condition of vanishing bed load; the upper limit is at $S_A$, beyond which braided channels may develop. As shown in Fig. 11.5, the value of $r_c/B$

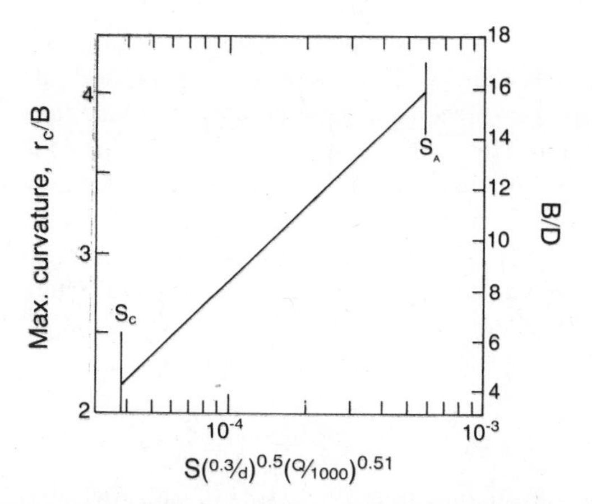


**Figure 11.5** Relationship of maximum curvature with slope, discharge, sediment size, and width–depth ratio.

ranges from 2.2 to 4.0; it varies in direct proportion to the channel slope or to the width–depth ratio. The median value is about 3. Although this ratio is remarkably constant, it does indicate the possibility of more acute bends for streams of flatter slopes, which are also associated with a smaller width–depth ratio.

The median value for $r_c/B$ of about 3, as analytically determined herein, is close to the median value of 2.7 determined from measurements by Leopold et al. (1964). Their data are not plotted, because there is a lack of other information on channel slope and sediment size. A source of the statistical variations is the heterogeneity that often exists in natural rivers. The heterogeneity, which includes such factors as vegetation, cohesive sediment content, variation in sediment size, chance placement of rocks, fallen trees, and so on, may inhibit the development of self-formed meander geometry.

## Power Transformation in Meanders

That the changing geometry along river meanders is associated with significant power transformation may be explained by the sample results shown in Fig. 11.4. While all sections have the same minimum total energy gradient $S$ or equal power expenditure per unit channel length, the streamwise energy gradient $S'$ is less in the curve because of the presence of transverse currents which means greater $S''$. The streamwise sediment load depends on the streamwise shear, which is related to $S'$; the effect of smaller $S'$ in the curve is compensated by the increased sediment efficiency associated with the nonuniformity in bed profile. Thus, constant streamwise sediment load is maintained throughout the river reach of meander bends and crossings. Therefore, it may be stated that while transverse circulation causes increased power expenditure, the increase is matched by an increase in sediment efficiency contributed by the circulation.

The adjustment in channel geometry in meanders is also associated with significant power transformation, as shown in Fig. 11.4 by section K under the maximum curvature. An increase in channel curvature (decrease in $r_c$) improves the sediment efficiency (decrease in $S'$), but this is also accompanied by a rapid increase in the transverse loss $S''$, which is a direct function of $(D/r_c)^2$ (see Eq. 11.3). The stable geometry at minimum $S$ represents the optimal condition under the different forms of power expenditure. The meander geometry is adjusted in just such a way that minimum power is maintained along the channel reach.

## 11.2 POWER APPROACH TO RIVER MORPHOLOGY AND THRESHOLDS

The scientific and engineering literature is replete with regime relationships for alluvial rivers. The quasi-equilibrium channel geometry is usually related to the slope, discharge, sediment properties, and so on (see Sec. 2.6). Such relationships, however, are by no means continuous because there exist several apparent thresholds or discontinuities between pattern states. Therefore, regime adjustment involving a small change in slope may lead to a large change in channel pattern if the slope is close to a critical value. It is clear that a large change in channel pattern is accompanied by a significant change in channel geometry. In addition, it appears that there exists some threshold for the silt–clay (sediments smaller than 0.063 mm) content in the channel perimeter because Schumm (1960, 1963, 1968, 1977) has found certain distinct channel features associated with the variation in channel silt–clay content.

The development of regime relationships for alluvial rivers is complicated by the existence of thresholds, yet the thresholds and different morphological features separated by them are obviously important parts of our knowledge about river engineering. This section presents an analysis of river morphology and thresholds for sand-bed rivers using a power approach. The discontinuity in flow resistance (that is, in power expenditure) between lower and upper flow regimes (see Sec. 6.2) is found to be related to the threshold in channel geometry, channel pattern, and silt–clay content. Certain quantitative relationships pertaining to channel geometry and channel patterns in different regions separated by thresholds are presented. The analysis is also compared with channel morphology and silt–clay content using field data.

Brice (1983) classified alluvial rivers into four major types as shown in Fig. 2.4. In the order of increasing channel slope, they are: sinuous canaliform rivers, sinuous point-bar rivers, sinuous braided rivers, and nonsinuous braided rivers. The threshold conditions for which these channel types occur are analyzed herein. It will be demonstrated that morphological features predicted by the analysis are, to a large extent, coherent with those of the four major types.

The foregoing analysis in Sec. 11.1 is limited to sinuous canaliform rivers, characterized by flat slopes and flow resistance in the lower flow regime. For steeper rivers, there exist possibilities for flow resistance of both lower and upper flow regimes. The sudden change in flow resistance between these two flow regimes should be matched by responses in river morphology. The following analysis by Chang (1985) will demonstrate that the adjustment in flow resistance and its associated power transformation account, in part, for the characteristic geometry associated with each channel type.

## Analytical Basis

The sketch of a point-bar river with riffles and pools is shown in Fig. 11.6. It represents a sinuous point-bar river if it is nonbraided; otherwise, it is a sinuous braided river. Transverse flow distribution is uneven at the pool section, with most of the flow concentrated in the pool. The effective width at the pool section is defined as that portion of the total width which provides power expenditure of the section in excluding the low-velocity spaces at the channel margins from the calculations. This definition for effective width is equivalent to that given by Cherkauer (1973). He noted that the effective width more closely approaches the total channel width in riffles than in pools. Thus, although the total width may be constant, effective width oscillates from narrow pools to wide riffles.

For the purpose of this analysis, each channel cross section is approximated as a trapezoid or, for braided channels, as a number of trapezoids separated by islands. If a section is in a curved channel, the base of the trapezoid is replaced by the transverse bed slope $S_t$. For a given channel width $B$ (or effective width) and center depth $D$, the maximum transverse bed slope occurs when the sloping bed extends to the water surface on the convex bank (see Fig. 11.6). Because $S_t$ is

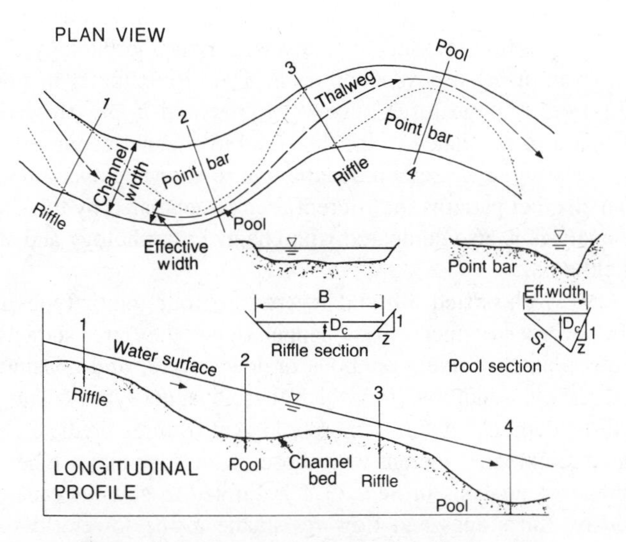


**Figure 11.6** Schematic of sinuous point-bar river.

directly related to the channel curvature (see Eq. 8.55), the maximum $S_t$ occurs under the maximum curvature (or minimum $r_c$) for the given condition.

Variables and physical conditions used in the analysis are outlined in Sec. 11.1. In the following, the case of straight and nonbraided channel is first considered; then the effects of channel curvature and braided channels on power expenditure and on channel geometry are analyzed.

## Straight and Nonbraided Case

Given a set of independent variables and constraints, the physical conditions are employed to obtain the dependent variables. As a first step, three of the physical conditions, represented by the flow-continuity equation, flow-resistance relation, and bed-load equation, are solved to obtain the mean velocity, depth, and energy gradient at a series of incremental widths. Procedures for solving the three unknowns, using the three physical conditions, are given in Fig. 11.3 under the case of no channel curvature. Some sample results are shown in Figs. 11.7 and 11.8, in which each graphical relation showing the variation of energy gradient, S (or power expenditure per unit channel length $\gamma QS$) with channel width (or effective width) for the given set of independent variables is called a *stream power*

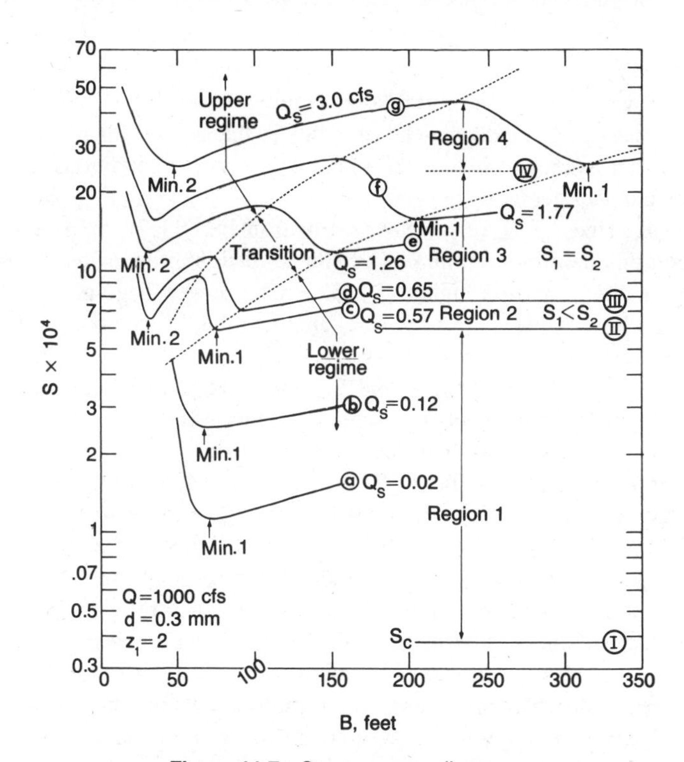


**Figure 11.7** Stream power diagrams.

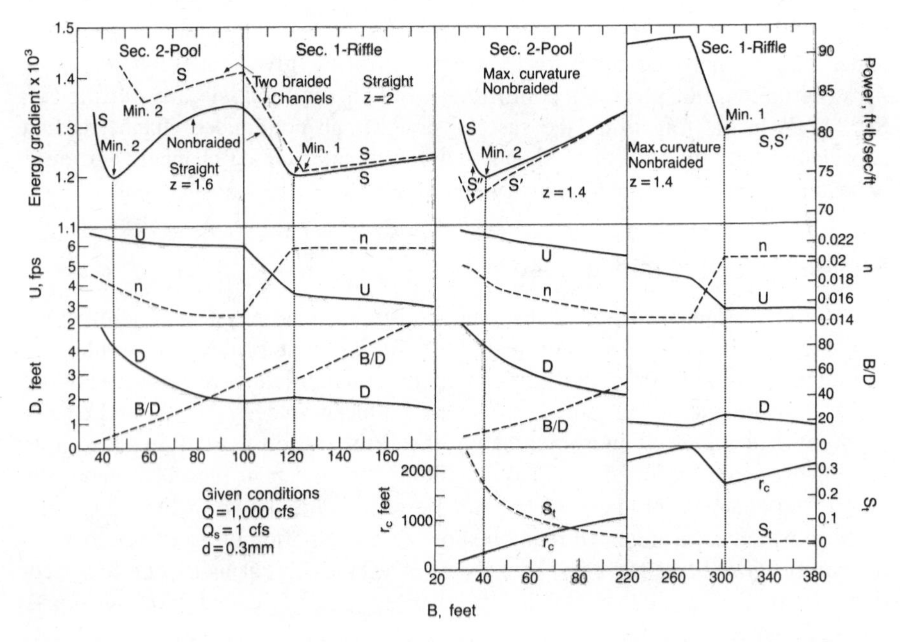


**Figure 11.8** Sample results: Variations of computed variables with channel width or effective width.

*diagram*. The water discharge of 1000 cfs, sediment size of 0.3 mm, and various bed loads are used as the independent variables in the sample calculations. These diagrams are first examined before the remaining physical conditions are applied to obtain the channel geometry.

For small values of $Q_s$ (e.g., curves a or b in Fig. 11.7), the stream power diagram has a unique minimum in $S$. The alluvial bed is in the lower flow regime with ripples or dunes. With a higher $Q_s$, the stream power diagram has two minima, one in the lower flow regime (Min. 1) and the other in the upper flow regime (Min. 2), as shown by curves c–g in Fig. 11.7. The minimum energy gradients at Min. 1 and Min. 2 are designated $S_1$ and $S_2$, respectively. From computed results shown in Fig. 11.7, the channel represented at Min. 1 (riffle) is wider and shallower and has a lower velocity than the channel at Min. 2 (pool).

The computed results in Fig. 11.7 show greater sediment efficiency, that is, less power expenditure, associated with a higher degree of channel roughness in terms of Manning's $n$. The existence of Min. 1 is attributed to the greater channel roughness of the lower flow regime. As the width decreases from this point, the flow begins to transform into the upper flow regime through the transition. Because this development is associated with a significant decrease in roughness (see Fig. 6.3), it is also accompanied by rapid increase in power expenditure.

From the condition of minimum stream power per unit channel length, the channel width at the minimum energy gradient $S_1$ or $S_2$ is a stable width. Other variables associated with this energy gradient and width, including the depth, velocity, and width–depth ratio, are directly obtained as results of the computation. Such values for the sample calculation shown in Fig. 11.8 for the straight and nonbraided case are tabulated as follows:

| Variable | Section 1 Riffle | Section 2 Pool |
|---|---|---|
| Minimum energy gradient $S$ | 0.00119 | 0.00119 |
| Effective width $B$ (ft) | 121 | 44.5 |
| Center depth $D$ (ft) | 2.1 | 4.2 |
| Average velocity $U$ (ft/sec) | 3.7 | 6.5 |
| Effective width–depth ratio $B/D$ | 57.6 | 10.6 |

If a stream power diagram has a unique minimum in $S$, a unique channel geometry is obtained. However, if a diagram has two minima, the same water discharge and bed load may be transported by two stable channel geometries. From the condition of minimum stream power for a channel reach, gradually-varied flow of the riffle and pool pattern may develop along the reach if power expenditures per unit channel length for these sections are equal so that the water-surface profile is a straight line. If they are unequal, the lesser of the two is more stable.

Whether a stream power diagram may have two equal minima is now being investigated. The minimum energy gradients, $S_1$ and $S_2$, are also functions of the bank slope using the trapezoidal approximation for the cross-sectional shape. Under the same independent variables, the value of $S_1$ or $S_2$ decreases as the bank slope steepens; it reaches a minimum when the bank slope angle equals 60° or $z = 1/3^{1/2}$. Note that this is the bank slope for the best hydraulic section. Although a steeper bank slope within the limit of $1/3^{1/2}$ is associated with greater efficiency, the bank slope nevertheless is constrained by the stability of bank materials. In consideration of possible cohesive materials on channel banks, the maximum bank slope is assumed to be 60°. Within this constraint for bank slope, the values of $z$ for these sections may be adjusted so as to equalize $S_1$ and $S_2$, if possible. In the present analysis, the 2–1 bank slope is assumed for the riffle section, that is, the value of $z$ is 2, whereas it is adjusted within the range of $1/3^{1/2}$ to 2 for the pool section. For stream power diagrams c and d in Fig. 11.7, $S_2$ is greater than $S_1$ even when the maximum bank slope is used for $S_2$. Therefore, channel geometry represented at Min. 1 is more stable. For diagrams e and f, $S_1$ and $S_2$ can be equalized if a steeper bank slope is used for $S_2$. For diagram g, $S_2$ is equal to $S_1$ when the same 2–1 bank slope is used for both $S_1$ and $S_2$. Therefore, multiple channel geometries may coexist in the same reach for rivers represented by stream power diagrams e–g in Fig. 11.7.

### Effects of Channel Curvature on Regime

Channel curvature contributes to the transverse bed slope and increased power expenditure associated with transverse currents; it also affects sediment transport, as described in Sec. 11.1. For a set of independent variables $Q$, $Q_s$, and $d$, the physical conditions of flow continuity, flow resistance, bed-load transport, and those pertaining to curved channels are employed to solve for the center depth, mean velocity, energy gradient, and transverse bed slope $S_t$ (or $r_c$) at a series of incremental widths. Major steps of computation are shown in Fig. 11.3 under the cases of maximum curvature and any intermediate curvature. Note that the energy gradient at a curved channel section, $S$, is partitioned into the components of $S'$ and $S''$, where $S'$ is due to streamwise resistance and $S''$ is associated with the transverse loss.

Sample results for the pool section at the maximum channel curvature are shown in Fig. 11.8. With the $z$ value of 1.4 for the bank slope, the minimum energy gradient of this section is equal to that for the straight case. Other variables, including the effective width, center depth, and velocity, are similar for both cases. Similar results have also been obtained for the pool section under intermediate channel curvatures. Therefore, it may be concluded that the pool section can be equally stable under different channel curvatures within its maximum value. Under the maximum curvature, the radius of curvature for this case is obtained to be 2.9 times the width of the riffle section that more closely approximates the surface width (see Fig. 11.6). For different sets of independent variables, the ratio of radius of curvature to channel width under the maximum curvature is found to be nearly constant, with a value of about 3. Because it is obtained using conditions of minimum stream power, it therefore represents the maximum curvature for which a river does the least work in turning.

The conclusion regarding the effects of channel curvature on the pool does not apply to the riffle section. Because of the smaller depth and lower velocity, power expended through transverse currents is less at the riffle section (see Eq. 8.77). But, under the maximum curvature or any intermediate curvature, the change in channel roughness due to transverse bed slope is such that it reduces the sediment efficiency, thereby increasing the power expenditure, as shown in Fig. 11.8. This result concurs with previous findings by Onishi et al. (1976) for wide curved channels. It may therefore be concluded that, for a wide riffle section, the straight channel pattern is more stable.

### Effects of Braided Channels on Regime

The water discharge and bed load are subdivided among the anabranches at a braided river section. Formation of braided channels reflects, in part, a river's adjustment in power expenditure that, in turn, affects the channel stability. The effects of braided channels on channel stability are shown in Fig. 11.8 by the variations of energy gradient $S$ (or power expenditure $\gamma QS$) with channel width or effective width. Graphical relations as shown are for two equal braided channels, and the total width is the combined widths of the anabranches. It shows that for the pool section, which has a small width–depth ratio, braiding results in significant increase in power expenditure. Therefore, the nonbraided channel is more stable for the pool. For the riffle section having a large width–depth ratio, the braided condition is associated with slightly greater power expenditure and channel width. Since braiding has little effect on the power expenditure for riffles, the braided and nonbraided patterns are approximately equal in stability. Wide streams, in reality, are usually braided because of high sediment rates, bank erosion, and physical heterogeneity (Leopold et al., 1964).

## 11.3 CHANNEL GEOMETRY, CHANNEL PATTERNS, AND THRESHOLDS

Based on characteristics of the stream power diagrams in Fig. 11.7, rivers are classified by Chang (1985a) into four regions (regions 1–4) separated by thresholds. Results pertaining to regime channel geometry for these regions are shown in Fig. 11.9. Whereas this graphical relationship is obtained using a 2–1 bank slope (for Min. 1), similar relationships using other values may be obtained. In constructing this figure, different sets of independent variables $Q$, $Q_s$, and $d$ in the sand-size range were used to produce the width, depth, and slope, as illustrated by the sample given in Fig. 11.8. Then, channel width and average center depth are plotted in Fig. 11.9 as functions of $Q$, $S$, and $d$. Note that in the case of riffle and pool patterns, channel width as shown is the width of the riffle section because it more closely approximates the channel width (see Fig. 11.6).

Several thresholds or discontinuities for channel geometry designated as lines I–IV are also shown in Figs. 11.7 and 11.9. From the bottom, the first threshold (line I) that corresponds to the critical slope for bed load, $S_c$, is represented by the equation

$$\frac{S_c}{d^{1/2}} = 0.00238Q^{-0.51} \tag{11.5}$$

where $d$ is in mm and $Q$ is in cfs. If $Q$ is in $m^3/sec$, the constant becomes 0.000386. This equation is obtained in the computation using the critical shear for bed load developed by Straub (Eq. 7.9). The equation for the second threshold (line II) in Fig. 11.9 is represented by

$$\frac{S}{d^{1/2}} = 0.05Q^{-0.55} \tag{11.6}$$

and the third threshold (line III) is given by

$$\frac{S}{d^{1/2}} = 0.047Q^{-0.51} \tag{11.7}$$

where $Q$ is in cfs. If $Q$ is in $m^3/sec$, the multiplier in Eq. 11.6 becomes 0.00704 and it is 0.00763 in Eq. 11.7. There may also be a fourth threshold, represented by the dashed line (line IV) at the lower boundary for steep braided streams. It is

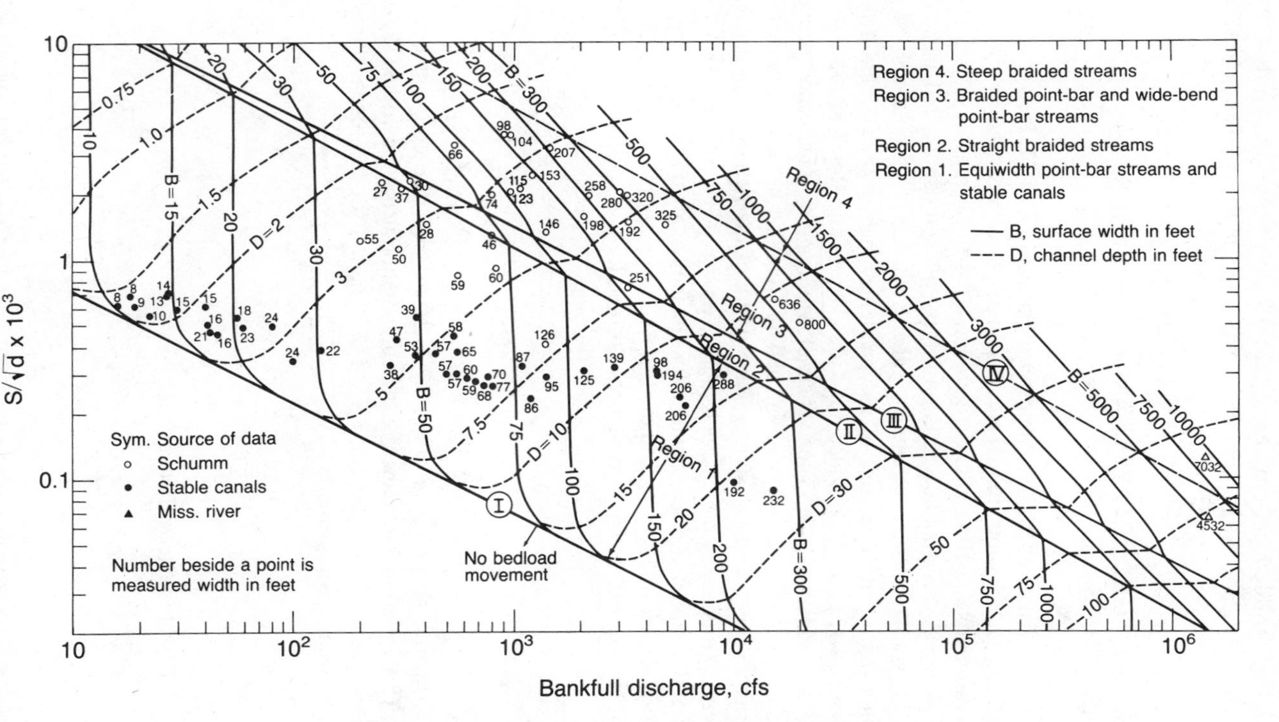


**Figure 11.9** Regime channel geometry for sand-bed rivers.

not a sharp discontinuity, and its precise location is not determined in this analysis.

In the following, regime channel geometry and its associated channel pattern and channel type for each region are described.

## Region 1

This region has its lower threshold at incipient motion for bed load (line I) and its upper threshold at line II in Figs. 11.7 and 11.9. A regime channel in this region is characterized by a flat slope, low velocity, small bed load, and flow resistance in the lower regime of ripples and dunes. The channel is relatively deep and narrow, with a width–depth ratio generally ranging from 4 to 20. It is interesting to find that all human-made stable canals are within this region. Each of these canals has constant width and depth, as well as long straight reaches. Natural channels in this region are usually meandering in channel pattern. It has been shown in Sec. 11.1 that, for such channels, the channel width and center depth stay approximately the same along the meandering channel. Therefore, they are sinuous canaliform rivers.

Graphical relations for channel geometry shown in Fig. 11.9 are also expressed in mathematical forms. The surface width varies with $Q$, $S$, and $d$, that is,

$$B = 3.49\left(\frac{S}{d^{1/2}} - \frac{S_c}{d^{1/2}}\right)^{0.02} Q^{0.47} \tag{11.8}$$

where the constant 3.49 is replaced by 5.68 if $B$ is in m and $Q$ is in m$^3$/sec. This equation denotes that width is primarily a function of the discharge; its dependence on $S$ and $d$ is not significant because of the small exponent. The following equation for the center depth is obtained from the figure

$$D = 0.51Q^{0.47} \exp\left[-0.38\left(\frac{S}{S_c} - 1\right)^{0.4}\right] \tag{11.9}$$

where the constant 0.51 is replaced by 0.83 if $D_c$ is in m and $Q$ is in m$^3$/sec. Empirical relations given in Sec. 10.1 for stable canals are in general agreement with these relations.

Although most natural rivers in region 1 have a meandering channel pattern, a straight channel pattern for which the channel slope equals the valley slope may also be maintained as long as constant inflows of $Q$ and $Q_s$ are in balance with the channel's capacity for transport. Field evidence from stable alluvial canals indicate that the straight channel pattern can be maintained with periodic minor repairs. However, if the discharge or slope of a canal is large enough such that the canal crosses the threshold into region 2 or 3, then rapid bank erosion starts to occur, as experienced from the India–Pakistan canals.

## Region 2

This small region, represented by stream power diagrams c and d, is between lines II and III in Figs. 11.7 and 11.9; each diagram has two unequal minima, of which $S_1$ is smaller and, thus, is the global minimum. This means that the channel geometry represented at Min. 2 is less stable and is therefore unlikely to develop. In other words, rivers in this region are more likely to assume a unique channel geometry represented at Min. 1. Figures 11.7 and 11.9 show that the channel geometry is sensitive to the slope in that an increase in slope is accompanied by a rapid increase in channel width and a rapid decrease in depth. Because the width–depth ratio is usually large, rivers in this region are quite often braided. A river with a unique channel geometry, which has a large width–depth ratio, is usually straight in channel pattern because meandering development is inhibited by the large width.

The existing literature describing rivers possessing these characteristics is limited. This may be because this region is small in range, especially at lower bankfull discharges. The most interesting study was perhaps done by Lane (1957). He classified a group of rivers as straight, braided rivers, including 12 river reaches of the upper Mississippi, the lower Nile, the lower Yangtze, and so on. With bankfull discharges ranging from 40,000 to 500,000 cfs, these rivers all fall within or near region 2 (Chang, 1979b). These data points are not plotted in Fig. 11.9 because of the lack of measured channel geometry.

## Region 3

This region is between lines III and IV in Figs. 11.7 and 11.9. A stream power diagram in this region (e.g., e or f in Fig. 11.7) has two minima, of which $S_1$ is for the riffle section and $S_2$ is for the pool. For the same bank slope, $S_1$ is less than $S_2$ and thus represents a more stable condition. However, $S_1$ and $S_2$ can be equalized because $S_2$ can be associated with a steeper bank slope. Under this situation, the riffle and pool sections may coexist in the same reach while maintaining a straight water-surface profile. Both channel width and depth are sensitive to the slope, in sharp contrast to rivers in region 1. Since an increase in slope is accompanied by rapid increase in width and decrease in depth, rivers in this region may be braided; the extent of braiding is in direct relation to the slope.

Graphical relations for channel width and depth in Fig. 11.9 are also expressed by equations. The width $B$ is a function of both channel slope and bankfull discharge, that is,

$$B = 33.2Q^{0.93}\left(\frac{S}{d^{1/2}}\right)^{0.84} \tag{11.10}$$

where 33.2 is replaced by 278 if $B$ is in m and $Q$ is in m$^3$/sec. The equation for the average center depth, $D$, in English units is

$$D = \left(0.015 - 0.025 \ln Q - 0.049 \ln \frac{S}{d^{1/2}}\right)Q^{0.45} \tag{11.11a}$$

and in SI metric units

$$D = \left(-0.112 - 0.0379 \ln Q - 0.0743 \ln \frac{S}{d^{1/2}}\right) Q^{0.45} \qquad (11.11b)$$

The unit of $d$ is mm in both equations. The width–depth ratio is generally between 25 and 100 for rivers in region 3.

When the discharge is less than the bankfull discharge, a river in this region may move into region 2 or 1. Therefore, it may have quite different channel characteristics at lower discharges.

The relationship between channel slope and valley slope determines the channel pattern. As a general rule, the channel slope for alluvial rivers cannot be steeper than the valley slope under the equilibrium state. If $S_1$ is equal to the valley slope, then the river has a straight pattern. With a unique channel geometry that is wide and shallow, it is usually a braided river. The extent of braiding increases with the channel slope. If $S_1$ is flatter than the valley slope, then $S_1$ and $S_2$ can be equal to allow the coexistence of riffle and pool sections along the same channel reach. As described previously, a river does less work in turning through the pool than through the riffle, and, therefore, channel bends usually form at pools. The resulting alternating riffle-pool pattern allows the channel to stay in dynamic equilibrium on a steeper valley slope. A river of these characteristics is a sinuous point-bar river or a sinuous braided river. The latter, which have steeper slopes, are usually wider. The large width associated with a braided pattern may inhibit the concentration of flow for the formation of pools; therefore, pools may be absent in certain reaches of such wide rivers. This is why sinuous braided rivers are less sinuous than sinuous point-bar rivers.

### Region 4

Rivers in this region are similar to those in region 3 in that the stream power diagram may have two equal minima for riffles and pools but those in this region are steeper in slope. No clear division between these two regions is delineated, but an approximate line of separation (line IV) is shown in Fig. 11.9. Since the width–depth ratio represented at Min. 1 is very large, rivers in this region are usually highly braided. The width–depth ratio is usually greater than 100. While there exists another possible stable channel geometry represented at Min. 2 for the pool, this geometry can be equally stable only if it is nonbraided. Because concentration of flow in the pool is usually physically prevented by the large width, the pool section is usually missing in such rivers. In the absence of the pool section, the braided channel pattern is usually straight although some of the anabranches may be sinuous.

### Comparison with River Data

Different data for sand-bed rivers are used for comparison with the analysis. These include stable canal data from Punjab Canals and All American Canals compiled by Blench and Simons (1974), river data of the midwestern United States compiled by Schumm (1968), and the Mississippi River data.

In addition to channel geometry, the midwestern river data contain information on meander sinuosity (ratio of valley slope to channel slope) and silt–clay content in the channel perimeter. These data, as plotted in Fig. 11.10, indicate a distinct discontinuity near the second threshold, with rivers above this line having less sinuosity and silt–clay content. With no discontinuity in the size of bed sediment across line II or III, the average value of silt–clay content is 5% for those rivers above line II and is 26% for those near or below this threshold. This discontinuity correlates remarkably well with the sudden change in channel geometry, as shown in Fig. 11.9. Whereas the silt–clay content should be related to the supply of such materials from the watershed, it also appears to be related to the channel geometry and flow characteristics. The variation of silt–clay content shown in Fig. 11.10 is consistent with the findings by Schumm (1977), because the lower silt–clay content is associated with steeper and wider rivers.

Channel width data are plotted in Fig. 11.9 for comparison with the prediction. At the same discharge, rivers that are plotted above line II are considerably wider, and the discontinuity of measured widths is in general agreement with the analysis.

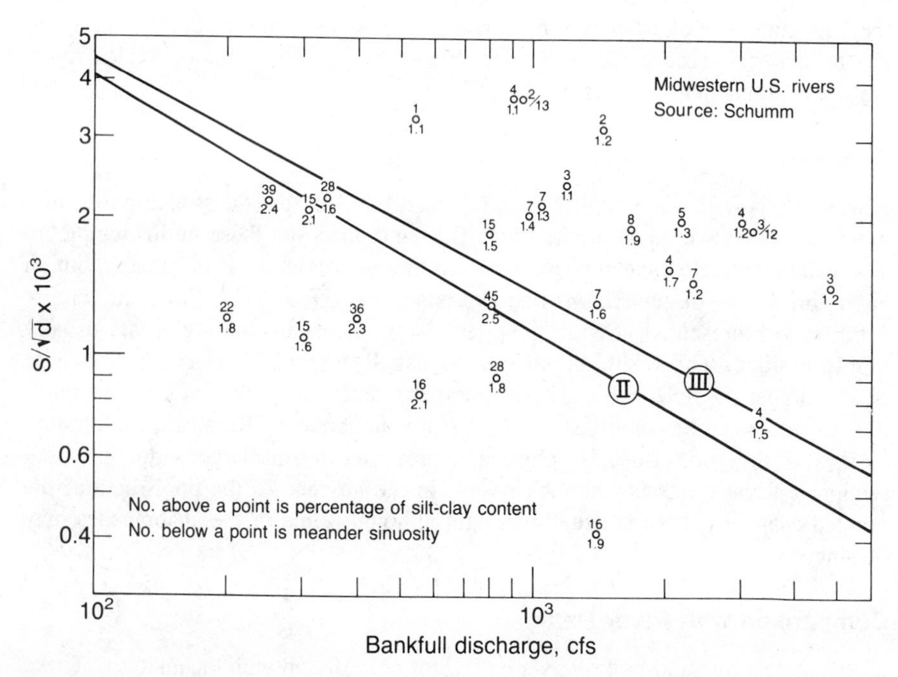


**Figure 11.10** Variations of channel silt–clay content and meander sinuosity near thresholds II and III.

## 11.4 RIVER CHANNEL CHANGES: ADJUSTMENTS OF EQUILIBRIUM

In response to changes in water and sediment discharges, the river channel adjusts in order to establish the dynamic equilibrium. Adjustments of equilibrium as used herein refer to river channel changes from one quasi-equilibrium state to the other. Channel morphology exhibited during the course of variation (i.e., the transient behavior) is not included. However, transient models do exist, such as the conceptual process-response model by Hey (1979) and the mathematical model by Chang (1982, 1984b) which is described in Part IV. The time scale for adjustment varies with the channel's sensitivity to change. Significant adjustment in stable humid-zone rivers may require hundreds of years. But in semiarid areas, the scale for river adjustment can be much shorter, close to that for the random response of a catastrophic event.

Earlier methods of analysis are basically qualitative in nature; some of them are described in Sec. 2.8. In this section, a quantitative method for river channel's adjustments of equilibrium by Chang (1986) is presented and illustrated by examples. This method is based on the quantitative relationships among the variables of water discharge, bed-material discharge, slope, sediment size, and channel width and depth for sand-bed rivers under dynamic equilibrium given in Sec. 11.3. In response to changes of certain variables, the directions and magnitudes of adjustments for the others may be determined using this method.

The adjustment in equilibrium may sometimes involve a dramatic transformation in channel character, or channel metamorphosis, because of certain thresholds separating channels of distinct characters. Such thresholds are intrinsic to the relationships of this method. It is therefore useful to identify potential dramatic changes of a river channel induced by a plan of river regulation or river control. Applications of this method are illustrated by several examples (Chang 1986). The cause and effect of channel adjustment for each case are elucidated.

In the graphical relationship of Fig. 11.9, rivers of distinct morphological features are classified into four regions (regions 1–4) separated by thresholds I–IV. It should be noted that these thresholds, particularly IV, are not sharp discontinuities. From the bottom, threshold I corresponds to the critical slope for bed-load movement. Thresholds II and III stem from the discontinuity in flow resistance between lower and upper flow regimes.

Rivers in region 1 are characterized by a flat slope, small width and width–depth ratio, and flow resistance in the lower regime. The width is insensitive to the slope and is essentially only a function of bankfull discharge. Rivers in region 2 have fairly flat slopes but are steeper than those in region 1. The regime width is fairly wide and highly sensitive to the slope. Rivers in region 3 have slopes ranging from moderately steep to fairly steep, characterized by the alternating riffles and pools at bankfull stage. The regime width is very sensitive to the slope. Rivers that exist on less steep slopes are usually less braided but more sinuous; they are classified as sinuous point-bar rivers. Those on steeper slopes are usually more braided and less sinuous; they are sinuous braided rivers. Rivers in region 4 exist on steep slopes, characterized by a highly braided but fairly straight channel pattern.

Because of the sensitivity of channel width to slope and discharge in regions 2–4, a change in slope or discharge may be associated with a large change in width. Braided channels start to occur in region 2, and the chances for being braided escalate with the increase in slope and width–depth ratio. This trend is essentially coherent with the meandering–braiding threshold proposed by Leopold and Wolman (Eq. 2.8), which generally falls within region 3 in Fig. 11.11a.

## Quantification of Lane's Relation

The variables $Q$, $Q_s$, $S$, and $d$ are interrelated in Lane's qualitative relation (Eq. 2.30) for river adjustments. A quantitative relationship involving these vari-

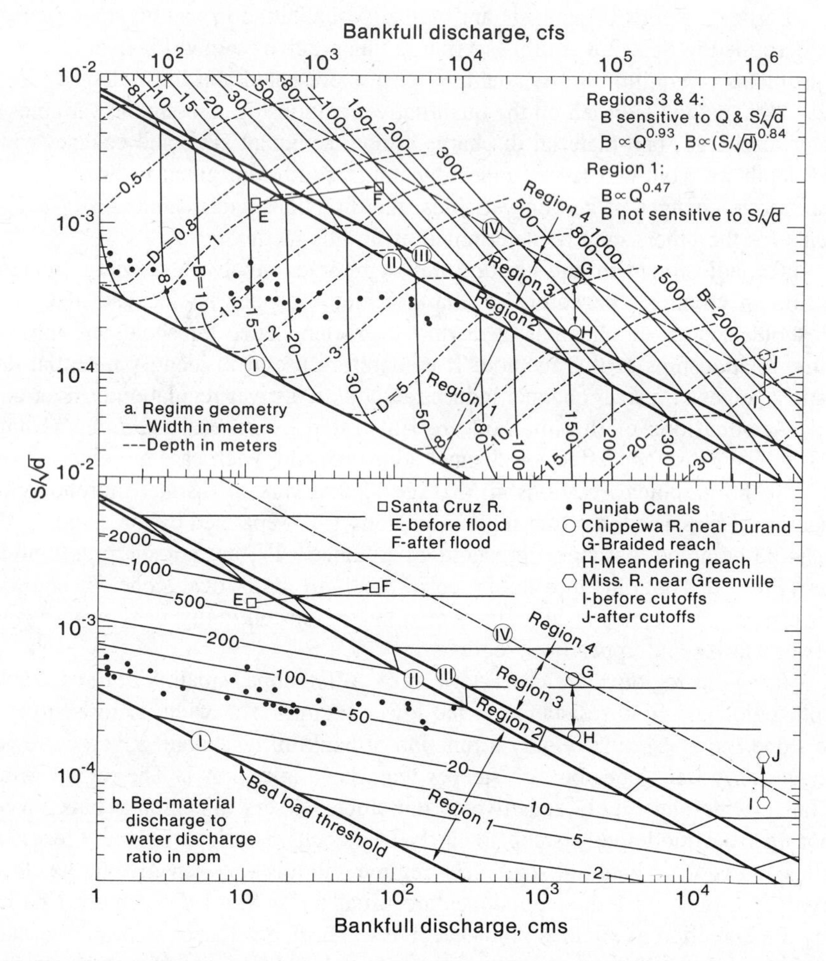


**Figure 11.11** Regime relationships of sand-bed rivers with sample cases of river channel changes.

ables may be obtained from the relationship in Fig. 11.11a because it includes the variables $Q$, $S$, $d$, $B$, and $D$ from which the sediment discharge can be computed. The bed-material load $Q_s$ has been computed using the Engelund–Hansen formula, and the concentration of bed-material load $C_s$ in parts per million by weight is shown in Fig. 11.11b. Because the variables $Q$, $Q_s$, $S$, and $d$ are interrelated, this relationship represents the quantification of Lane's relation.

This graphical relationship for $C_s$ has discontinuities along lines II and III, across which dissimilarities exist. In region 1, the graphical relationship is represented by the equation

$$\frac{S}{d^{1/2}} = 5.27 \times 10^{-5}(C_s)^{0.58}Q^{-0.17} \tag{11.12}$$

where $Q$ is in cubic meters per second and $d$ is in millimeters. In regions 3 and 4, the equation is

$$\frac{S}{d^{1/2}} = 7.28 \times 10^{-6}(C_s)^{0.87} \tag{11.13}$$

From Eqs. 11.12 and 11.13 and for constant $Q$ and $d$, the value of $Q_s$ increases with $S^{1.73}$ in region 1 but it varies with $S^{1.15}$ in regions 3 and 4. This dissimilarity is explained by the accompanying variation in channel geometry because, at steep slopes, the increase in $Q_s$ with $S$ is slowed down by the widening development.

From the quantitative relations among the variables $Q$, $Q_s$, $S$, $d$, $B$, and $D$ and their thresholds shown in Fig. 11.11, the adjustments of certain variables in response to changes of the others may be determined. When this method is compared with the qualitative method of channel metamorphosis in Table 2.1, the directions of morphological responses to particular combinations of changing discharge and sediment yield from this method are coherent with the qualitative method. Herein, applications of the quantitative method are illustrated by the following examples: (1) instability problem for alluvial canals, (2) response of ephemeral rivers to flood, (3) downstream changes of the Chippewa River, and (4) response of the Mississippi River to cutoffs. These examples should also demonstrate, for river control and regulation, proper practices by which adverse river adjustments can be avoided.

### Instability Problem for Alluvial Canals

The regime method originated from engineers working on the India–Pakistan canals. An important contribution by Lacey is the relation for the stable canal width, which may be written in the form (see Eq. 10.5)

$$B = CQ^{0.5} \tag{11.14}$$

This equation indicates that stable width of such canals is essentially only a function of discharge and independent of channel slope. Field evidence has shown that the stability of canals designed following this relationship can be maintained only within a certain limit of slope and discharge. When the limit is exceeded, channel instability, demonstrated by bank erosion and channel widening, usually occurs. This channel adjustment may be elucidated by Fig. 11.11, in which the stable Punjab canals are located entirely within region 1. The geometric relationships represented in region 1 are consistent with those of stable alluvial canals. With a greater $Q$ or $S$ or both, a canal near line II may cross this threshold into region 2 or 3. This adjustment is associated with rapid increase in width and decrease in depth, inconsistent with the regime relations. Therefore, canals designed using the regime method cannot be expected to remain stable beyond region 1. This analysis reveals the underlying mechanics for this type of canal instability.

## Response of Ephemeral River to Flood

Ephemeral rivers, characteristic of semiarid regions, are more sensitive to changes. In response to a major flood, channel adjustment may be dramatic in character and is often swift in occurrence. For the Santa Cruz River near Tucson, the adjustment of equilibrium from a small meandering stream (see Fig. 11.12) to a wide braided channel (see Fig. 11.13) is illustrated herein. The Santa Cruz River is a tributary of the Gila River that was studied by Burkham (1972); channel changes of these two rivers were generally similar. The small meandering stream shown in Fig. 11.12 was formed during episodes of low flow. It had a width of

**Figure 11.12** Santa Cruz River in form of small meandering stream before 1983 flood.

**Figure 11.13** Wide braided channel of Santa Cruz River after 1983 flood.

15 m, a depth of 1 m, and a channel slope of 0.0024 on a valley slope of 0.0028. During the major flood of October 1983, parts of this river channel went through a complete transformation in character. The post-flood channel shown in Fig. 11.13 had a much larger width of about 75 m and it was less sinuous. The bankfull depth, on the other hand, remained similar. The median diameters of bed sediment were 3 mm and 2.5 mm for the respective channels before and after the event. This case history of river channel changes also demonstrates that the adjustment in width can be an order of magnitude greater than aggradation or degradation changes.

Channel metamorphosis of the Santa Cruz River, demonstrated basically by dramatic widening, is explained herein using Fig. 11.11. Points E and F shown in the figure, plotted according to the bankfull discharge, $S$, and $d$, represent the respective quasi-equilibrium conditions before and after the adjustment. The pre-flood channel at E is shown in region 1 but near thresholds II and III. In response to the major flood, it crosses these thresholds into region 3. This adjustment, as predicted in Fig. 11.11a, is associated with a large increase in width but a small change in depth, consistent with measurements.

## Downstream Changes of the Chippewa River

The Chippewa River of Wisconsin, as shown in Fig. 11.14, was studied by Schumm and Beathard (1976) as a case of demonstration for the threshold concept in geomorphic approach. Based on their data, downstream changes of the Chippewa River and the appropriate means of river control are studied using the

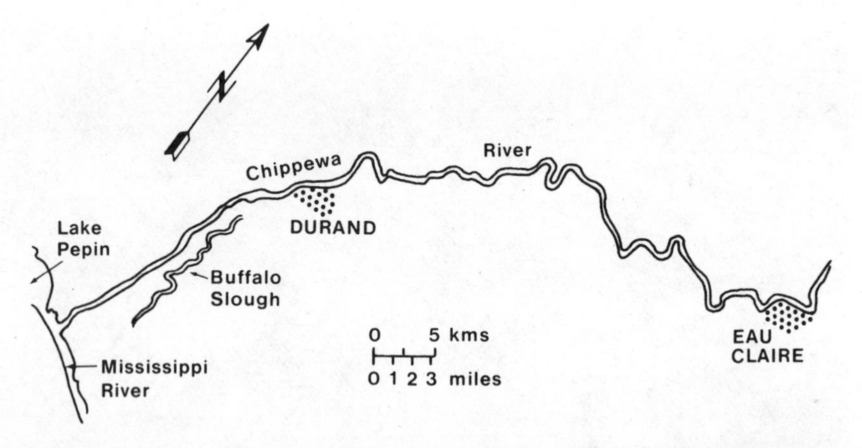


**Figure 11.14** Chippewa River near Durand, Wisconsin (after Schumm and Beathard 1976).

quantitative method. Downstream of Durand to its confluence with the Mississippi, the Chippewa River is braided, characterized by a wide and shallow channel with shifting sand bars. This lower reach has a bankfull width of about 305 m, a median grain size of 0.53 mm, a channel gradient of 0.000333, and a valley slope of 0.000347. The sinuosity, defined as the ratio of valley slope to channel slope, is thus obtained as 1.04, indicating a nearly straight river pattern. Upstream of Durand to Eau Claire, the Chippewa River abruptly changes to a meandering pattern with a sinuosity of 1.45. It has a bankfull width of 195 m, a channel slope of 0.000278, and a valley slope of 0.000403. It is interesting to note that the upper meandering reach has a flatter channel slope but a steeper valley slope than the lower braided reach. The median size of bed material in the upper reach is coarser than 2 mm, composed of coarse granules to cobble size material. Bankfull discharge for the river is given as 1645 cms.

From the given bankfull discharge, channel slopes, and median sediment sizes, the lower braided reach is plotted in Fig. 11.11 at point G and the upper meandering reach is plotted at point H. The analytical channel widths shown for these two points (380 m and 210 m, respectively) compare reasonably well with measurements. Both G and H are in region 3, where channel width and depth at the same $Q$ are sensitive to the variation of $S/d^{1/2}$. The width–depth ratio for G is 80 as shown in the figure and is 30 for H. The lower braided reach is associated with a much greater width–depth ratio than the upper meandering reach.

As shown in Fig. 11.11b, these two river reaches represented at G and H have different bed-material loads, their ratio being 2.3. This imbalance in sediment, according to Schumm and Beathard (1976), may have been supplied by bank erosion along the lower reach which proceeded since the development of the new downstream channel following abandonment of the old Buffalo Slough, as shown in Fig. 11.14. This sinuous remnant of the Chippewa River had a lower sediment rate than the present channel. This increased sediment rate has also resulted in a noticeable growth of its delta into the Mississippi River valley, where it has not been completely adjusted to the river regime.

An appropriate means of channel stabilization and sediment-load reduction in this case, as suggested by Schumm and Beathard, is the development of a sinuous channel for the lower Chippewa. If this plan for river control is implemented, a sinuous channel so developed will have a flatter slope. The associated river adjustment, as predicted in Fig. 11.11, will involve the formation of a narrower but deeper channel, accompanied by a reduction in sediment discharge. The channel slope may also be reduced by the use of drop structures.

### Response of the Mississippi River to Cutoffs

Artificial cutoffs are often employed to straighten a river for the intended purposes of navigation improvement and flood-level reduction. The cutoff program of the lower Mississippi River was one of the largest construction programs of this type ever attempted. The potamology investigations by Winkley (1977) for the U.S. Army Engineer District, Vicksburg, reveals that a complete knowledge of cutoffs is necessary prior to any future work of this type.

Cutoffs increase channel slope and alter the quasi-equilibrium of the river. The effects must be absorbed by the river in order to establish the new equilibrium. If a river is in or near regions 2, 3, and 4 as shown in Fig. 11.11, an increase in slope should be accompanied by significant increases in width and width–depth ratio. This type of channel adjustment is illustrated using the lower Mississippi River between the Arkansas River junction and Greenville, Mississippi; a portion of this river reach is shown in Fig. 11.15. The sinuosity of this river reach was reduced from about 3 to 1.4 by cutoffs constructed from 1933 to 1937. The river channels of 1933 and 1975 are plotted in Fig. 11.11 as points I and J according to the respective measured values of $S/d^{1/2}$ and the bankfull discharge of 31,000 cms (1.1 million cfs).

Because both I and J are in region 3, where channel width is sensitive to the slope, steepening of the channel gradient due to cutoffs has resulted in substantial widening and even braiding. As a result of this development, an extensive levee system was constructed in order to maintain this unnatural channel alignment, as shown in Fig. 11.15. The predicted bankfull width of 1350 m for the 1933 channel shown in Fig. 11.11a compares favorably with the measured average width of about 1310 m. The predicted width of 2800 m for the 1975 channel is greater than the measured value of about 2000 m, probably because the levee system has inhibited the width development. Without the levee control, on the other hand, the development of a wider channel should be accompanied by a shallower depth and more braiding, detrimental to navigation. Apparently, such channel adjustment was not expected beforehand and it certainly calls for more in-depth study before any future plan is made.

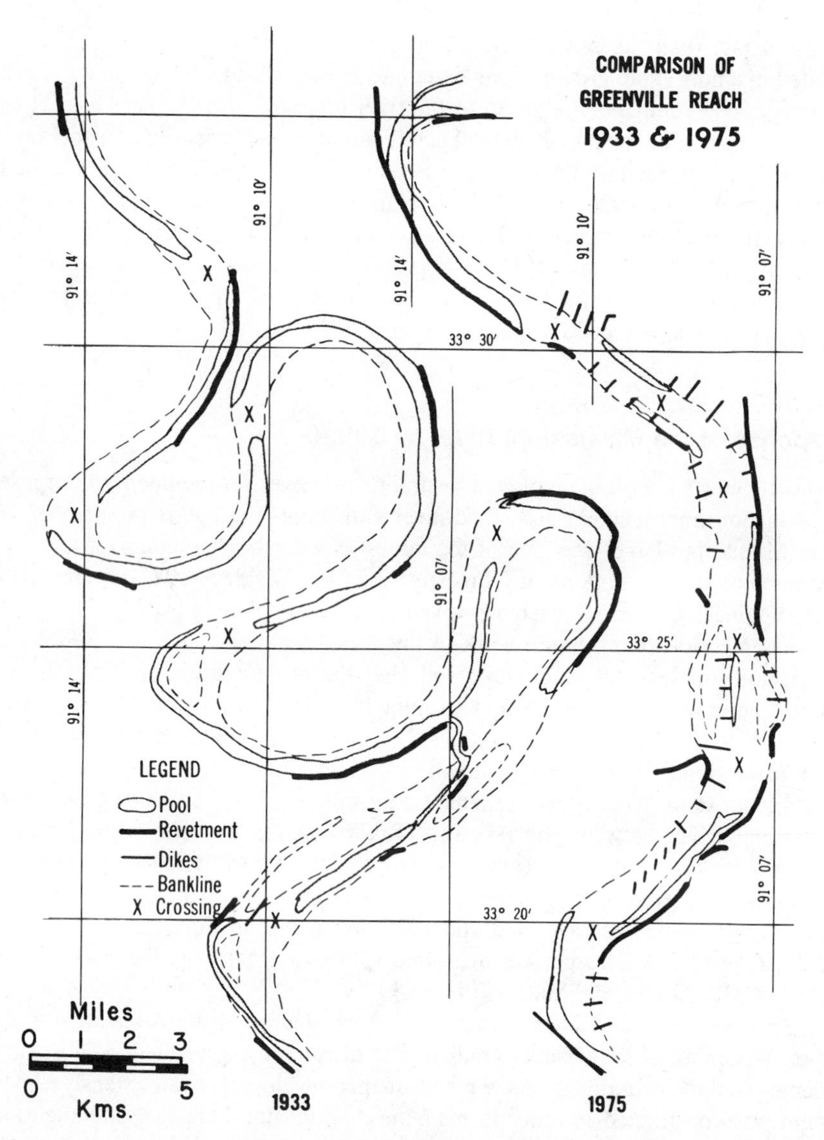


**Figure 11.15** Mississippi River near Greenville, Mississippi in 1933 and 1975. Levee system in 1975 channel was constructed to control channel widening while maintaining navigation depth.

## 11.5 FORMATION OF ALTERNATE BARS

Alternate bars are often observed in channelized alluvial rivers at lower stages of flow. An example of alternate bar formation in the Colorado River near Needles, California reported by Leopold (1982) is shown in Fig. 11.16. In this case, con-

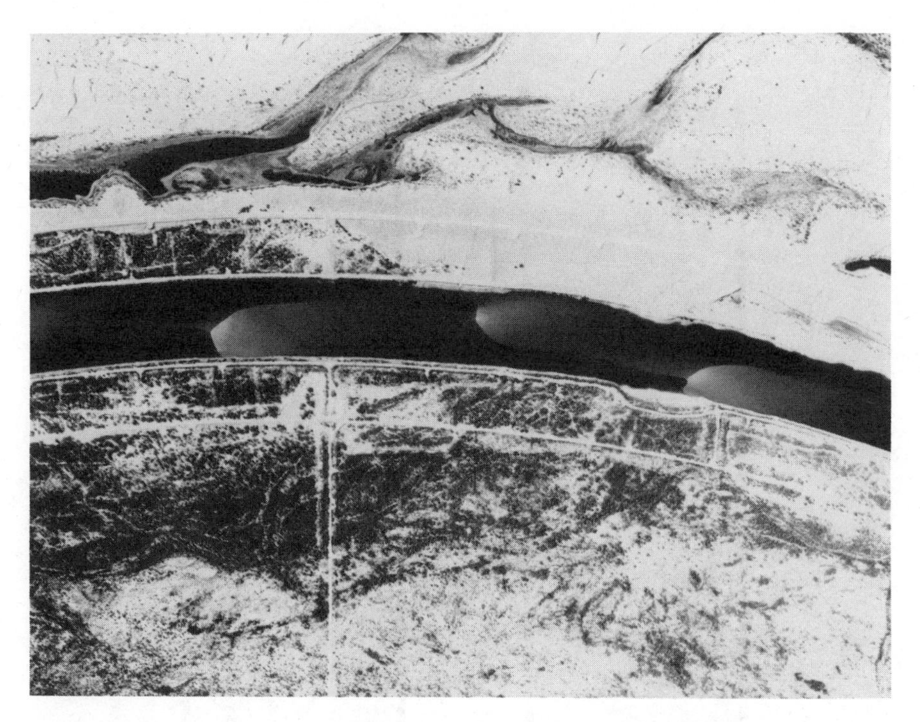

**Figure 11.16** Alternate bar formation in Colorado River near Needles, California (courtesy of L. B. Leopold).

trolled discharge and bank stabilization led to a series of bars with crests near alternate sides of the channel. Such bed features are missing at higher stages of flow. Researchers have generally agreed on the significant relationship between alternate bar formation and river meandering. In contrast to free meanders, alternate bars are confined meanders within the width constraint. In the case of erodible banks, free meanders may develop as alternate bars disappear.

Following earlier attempts by European researchers, Kinosita (1962) started the study of alternate bars in Japan, where land reclamation from flood plains resulted in the channelization of many rivers, especially in urban areas. In channels with alternate bars, substantial scour at bank toes, to a depth equivalent to several percentages of the channel width, often necessitates maintenance work. Because of the engineering and ecological significance, experimental studies of alternate bars have been made by Kinosita (1962), Chang et al. (1971), Inokuti et al. (see Sukegawa, 1971), and Jaeggi (1984), among others. Samples of bed topography with alternate bars obtained by Chang et al. are given in Fig. 11.17. Analytical and experimental studies have also been carried out on different aspects of alternate bars, such as the cause of occurrence (Sukegawa, 1970; Parker, 1976), criteria for formation (Kinosita, 1962; Sukegawa, 1972; Jaeggi, 1984), meander wavelength (Chang et al. 1971; Parker, 1976; Ikeda 1984), and scour

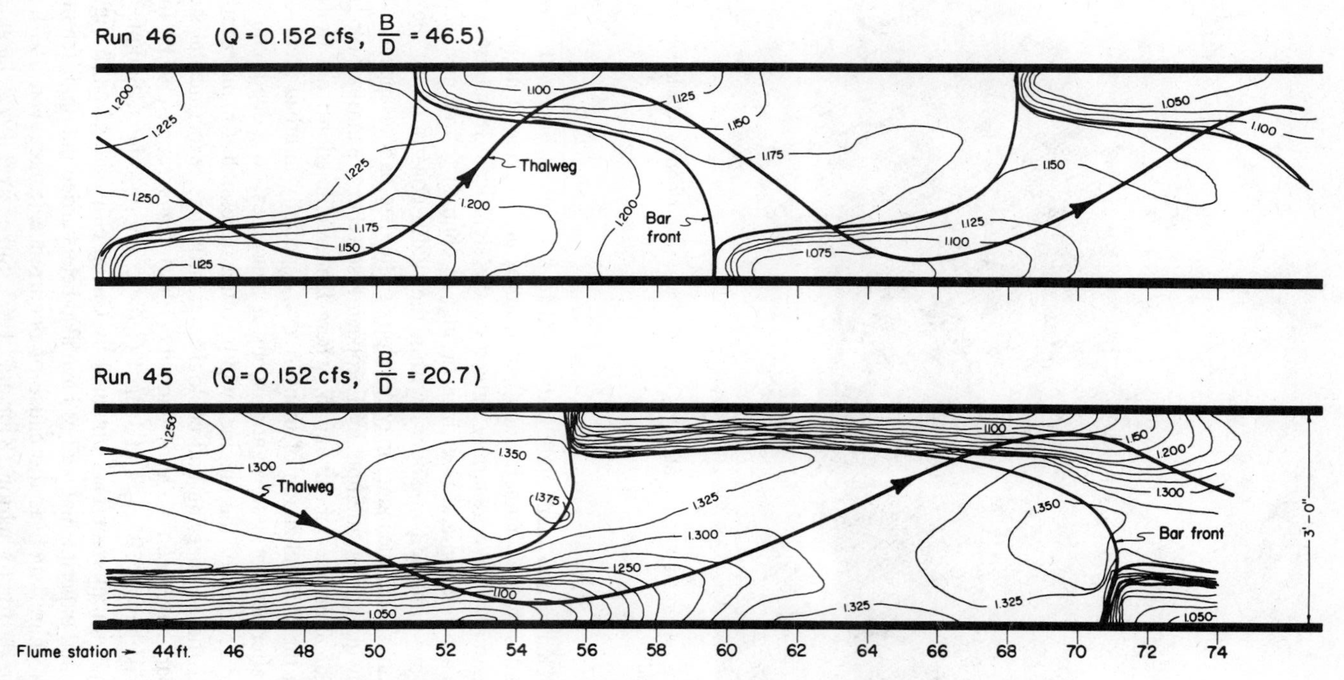


**Figure 11.17** Samples of bed topography with alternate bars in laboratory flume.

depth (Jaeggi, 1984; Ikeda, 1984). Details of certain findings are reviewed in this section.

## Empirical Studies

Alternate bar formation is within a specific range whose limits may be defined by the variables of flow and sedimentation. The lower limit of this range is often delineated using the Shields stress (see Sec. 5.1), defined as

$$\tau_* = \frac{\tau_0}{(\gamma_s - \gamma)d} \tag{11.15}$$

where $\tau_*$ is the Shields stress, $\tau_0$ is the boundary shear stress, and $d$ is median bed-material size. On the basis of boundary shear stress, alternate bars, as well as other bed forms, do not develop if the Shields stress is less than the critical Shields stress $\tau_{*c}$ for incipient motion of the bed material. Therefore, the lower limit for alternate bar formation is given by

$$\frac{\tau_*}{\tau_{*c}} = 1 \tag{11.16}$$

For rivers with nonuniform bed materials, alternate bar formation, however, is inhibited by the development of a stable armor layer. To account for the greater shear stress required of coarser materials, Jaeggi (1984) developed a more general expression than Eq. 11.16, defining the lower limit of alternate bar formation as

$$\frac{\tau_*}{\tau_{*c}} = \left(\frac{d_{90}}{d}\right)^{0.67} \tag{11.17}$$

where $d_{90}$ is the grain size exceeding 90% of bed material by weight. This implies that a higher value of $\tau_*/\tau_{*c}$ is required for graded materials to form bars.

More studies have been made on the upper limit of alternate bar formation (see, e.g., Kinosita, 1962; Sukegawa, 1972; and Jaeggi, 1984). Empirical approaches, based on dimensional analysis, were employed by these researchers to determine the relevant dimensionless parameters and their values distinguishing channel flows with and without alternate bars. Each criterion was established using experimental data obtained for rectangular flumes. Sukegawa (1972) developed the following criterion for alternate bar formation:

$$\frac{B}{R} \geq \frac{1}{125}\left(\frac{\tau_*}{\tau_{*c}}\right)^2 \frac{1}{S} \tag{11.18}$$

where $B$ is channel width, $R$ is the hydraulic radius, and $S$ is channel slope. By introducing new parameters in the dimensional analysis, Jaeggi (1984) obtained the upper limit for alternate bar formation:

$$\frac{\tau_*}{\tau_{*c}} = 2.93 \ln \frac{\gamma BS}{(\gamma_s - \gamma) d \tau_{*c}} - 3.13 \left( \frac{B}{d} \right)^{0.15} \tag{11.19}$$

When compared with experimental data, Eqs. 11.18 and 11.19 provide fairly good approximations for the upper limit of alternate bar formation. These equations, applicable to sediments of variable densities, demonstrate the important role of channel width or width–depth ratio on the criterion for alternate bar formation.

## Analytical Criteria for Alternate Bar Formation

An analysis of alternate bar formation was made by Sukegawa (1970). He used the linear stability analysis, for which the instability of the bed against a disturbance was hypothesized as the primary cause of alternate bar formation. This type of analysis has also been applied to explain the meandering of rivers.

A rational criterion for alternate bar formation was developed by Chang (1985c) for straight alluvial channels with smooth rigid banks. This method is based on the relative magnitudes of the stable width of stream flow $B_s$ and the channel width between the banks $B$. The formation of alternate bars is attributed to meandering development within the confined channel, which becomes possible if the stable width is less than the confinement width, that is,

$$B_s < B \tag{11.20}$$

The stable width is equivalent to the regime width a stream tends to assume for a set of discharge, slope, and sediment characteristics. In the case of alluvial canals, for example, the regime width has been found to be proportional to the square root of the discharge (see Secs. 10.1 and 10.2), that is,

$$B_s \propto Q^{0.5} \tag{11.21}$$

Similar regime relationships have also been established for the river width (see Sec. 11.3).

While the stable width varies with the discharge, the channel width is fixed by the rigid banks. For the purpose of the present discussion, the slope of the stream flow at the stable width is called the *stream slope*; the average longitudinal slope of the channel bed between rigid banks is termed the *valley slope*. Depending on the magnitude of $B_s/B$, three different flow patterns may develop as follows.

If, at a higher discharge, the stable width, which is directly related to the discharge, is greater than the channel width, then the stable width is constrained by the rigid banks. In such a situation, meanders or alternate bars do not form due to the lack of freedom for stream slope adjustment under the equilibrium condition. So long as the stable width is constrained, stream bed adjustment from one equilibrium slope to the other occurs through scour or fill but not through meandering development.

The second case is when the stable width is less than the channel width. This may result from either a lower discharge or a greater channel width. In this case, there exists some limited freedom for stream slope adjustment within the confined channel. Meandering develops if the stream assumes a flatter slope by wandering back and forth on the steeper valley slope between the banks. Alternate bars can be considered large-scale bed forms, which reflect meandering development in a confined channel. Whether meandering will develop, if such a freedom exists, is not included in this criterion. That meandering usually develops if the freedom exists is explained in Sec. 12.4 on the cause of meandering development.

For the third case, the stable width is much smaller than the confinement width. Such a condition exists, for example, if the confinement width is very large. Since there is little or no constraint on stream channel formation, meander waves are more fully developed. Such small alluvial streams in river valleys are called *underfit streams* (Dury, 1964). The development of free meanders means the disappearance of alternate bars; therefore, there should be another limit separating alternate bars from free meanders, but this feature has not been seriously examined.

### Stable Width

The stable width relationship is determined using the method described in Sec. 10.2. Among the physical conditions used in the analysis, the flow-resistance relationship is based on Engelund's method given in Sec. 6.5, and the bed load is computed using the Einstein–Brown formula described in Sec. 7.1. These relations may also be replaced by other valid formulas for flow resistance and bed load. Closely similar results can be obtained when other valid formulas are used.

Different sets of stable width and depth are computed corresponding to various sets of $Q$, $Q_s$, $S$, $d$, and $\gamma_s$ at 10°C. For each case, the Shields stress is computed according to Eq. 11.15, and the critical Shields stress is obtained from the Shields diagram. From the computed results, the stable width, $B_s$, has been correlated to other variables. For a given bed material, $B_s$ has been found to be a direct function of $\tau_*/\tau_{*c}$ and $Q$. For different bed materials, $B_s$ is a function of $Q/[(s-1)d/1.65]^{0.3}$ in addition to $\tau_*/\tau_{*c}$, where $s$ is the specific gravity of sediment. The general graphical relationships for $B_s$ and $B_s/D$ are shown in Fig. 11.18; they are represented by the following mathematical expressions:

$$B_s = 6.62\left[1 + 1.3\left(\frac{\tau_{*c}}{\tau_*}\right)^{0.4} \ln \frac{\tau_*}{\tau_{*c}}\right]\left[\frac{1.65}{(s-1)d}\right]^{0.15} Q^{0.48} \tag{11.22}$$

and

$$\frac{B_s}{D} = 2\left(1 + 2.2 \ln \frac{\tau_*}{\tau_{*c}}\right) \tag{11.23}$$

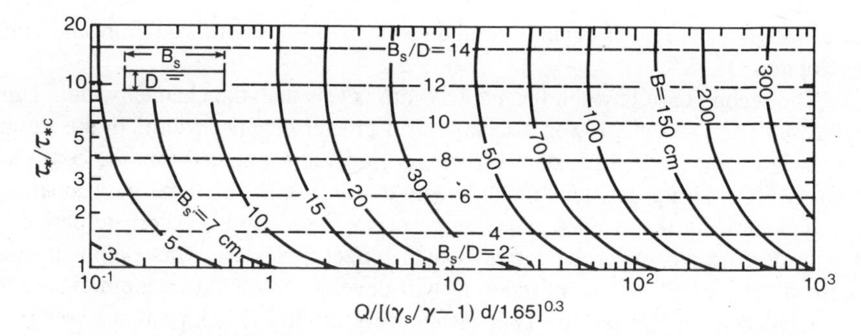


**Figure 11.18** Stable width and width–depth ratio as functions of Shields stress ratio, discharge, specific gravity, and median size of sediment.

where $B_s$ and $D$ are in centimeters, $Q$ is in liters per second, and $d$ is in millimeters. Equation 11.22 shows that $B_s$ is directly proportional to $Q$ raised to the power of 0.48. This rational result is remarkably similar to the empirical regime relationships. The rational geometry approaches the best hydraulic section for rigid channels at the bed-load threshold, where $\tau_*/\tau_{*c}$ has the value of unity. For a rectangular channel, the best hydraulic section has a width–depth ratio of 2, located at the bottom line in Fig. 11.18. Above this bottom line, the Shields stress is greater than the critical value, indicating a mobile bed.

## Comparison with Experimental Data

This rational criterion for alternate bar formation, Eq. 11.22, is now compared with three sets of experimental data, obtained by Chang et al. (1971), Inokuti et al. (see Sukegawa, 1972), and Jaeggi (1984), respectively. All experiments were carried out in straight, rectangular flumes with a mobile bed. Chang et al. were primarily interested in the meander wavelength; therefore, alternate bars were produced in each run. The other experiments, on the other hand, covered cases with and without alternate bars. For these data sets, the discharge is in the range of 0.2–64.3 liters/sec, the width 15–91.4 cm, the particle size 0.45–4 mm, and the specific gravity for the bed material 1.05–2.65.

The data of each run is shown in Fig. 11.19, where a solid data point indicates the presence of alternate bars and where a hollow point shows otherwise. Any data point falling below the lower limit indicates an immobile bed and hence is not plotted. In Fig. 11.19, the criterion for alternate bar formation is for the stable width to be less than the flume width. The stable width of each data point can be determined in the figure from the analytical curves for the stable width. A point located to the left of the flume-width curve has an analytical stable width less than the flume width. Under this situation, alternate bars can usually be expected to form according to this criterion. On the other hand, a data point located on the right side of the flume-width curve has a stable width greater than the flume width; alternate bar formation is not predicted. Those solid points indicating alter-

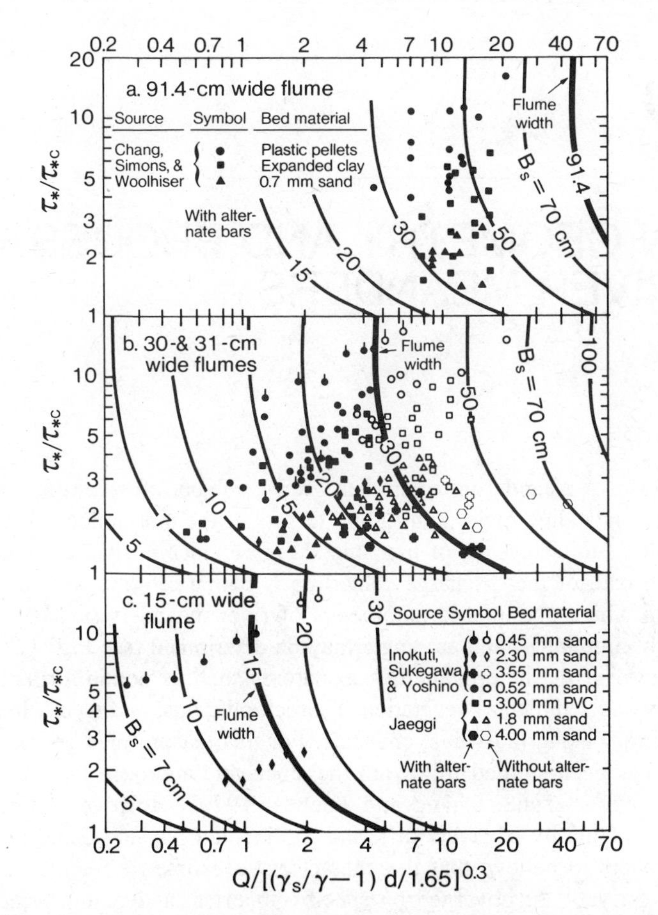


**Figure 11.19** Comparison of stable width for each data point with flume width. Criterion for alternate bar formation is for stable width of streamflow to be less than flume width.

nate bars are generally located to the left of the flume-width curve, while those hollow points designating no alternate bars are generally to the right. There are exceptions to these generalizations; stable widths for such points are usually close to the flume width.

# 12

# PLAN GEOMETRY AND PROCESSES OF RIVER MEANDERS

The planform of meandering rivers has been a subject of scientific investigation because understanding major controlling factors of the morphological features and their responses to changes is of fundamental importance in river control and regulation. The path of the discharge centerline is herein referred to as the *meander path*, from which planimetric parameters of river meanders such as arc length, wavelength, amplitude, and arc angle may be determined (see Fig. 12.1).

Because of the heterogeneities in nature, such as nonuniform river valley topography and sediments, vegetation, bedrock outcrops, and so on, meander size and shape vary along individual channels. For this reason, plan geometry of river meanders has been studied following the stochastic approach (see, e.g., Thakur and Scheidegger, 1968; Chang and Toebes, 1970). However, the distinctive planimetric regularity of rivers differing in size and physiographic province has led researchers to believe that the characteristic geometric features common to meandering rivers are governed by specific theories or dynamic principles. For example, stability theories have been applied to delineate the origin of meandering. Analyses by Ikeda et al. (1981) and Parker et al. (1983) using a dynamic

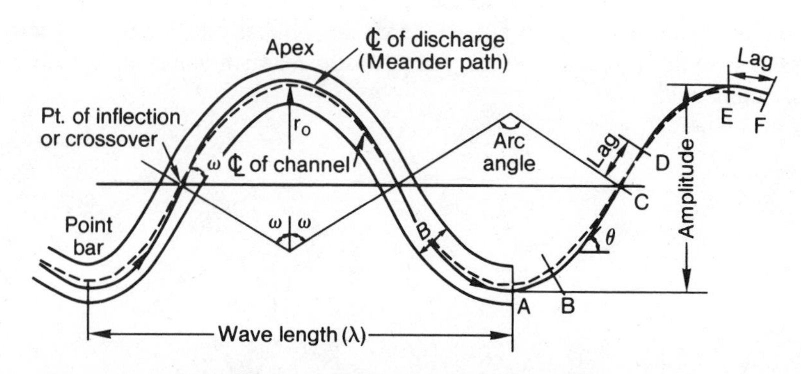


**Figure 12.1** Definition sketch of river meanders.

equation for longitudinal velocity and a kinematic equation for bank erosion have provided the mechanistic founding of the meander path. An expression for the wavelength was derived as

$$\frac{\lambda}{D} = 8\left(\frac{\pi B}{fD}\right)^{1/2} \tag{12.1}$$

where $\lambda$ is the wavelength, $f$ is the friction factor, $B$ is the width, and $D$ the depth.

Earlier mathematical models for regular meander paths, as summarized by Ferguson (1973), define meander paths as sine wave, circular arcs, Fargue's spiral, von Schelling's curve, or the sine-generated curve; the last three models are generally similar. Each model is specified by one scale parameter (arc length) and one shape parameter (arc angle, sinuosity, or maximum curvature) at the bankfull stage. In a more recent development, the meander path is related to the streamwise variation in helical motion or secondary currents. Some of these models are described in the following sections.

## 12.1 SINE-GENERATED CURVE

The origin of the sine-generated curve is traced to the most probable path between fixed points. With a fixed number of steps, von Schelling (1951) assumed the normal probability distribution for the change of direction at the end of each step. For a path that is a continuous curve, he showed that a general condition for the most probable path is for the variance to be a minimum, that is,

$$\sum \frac{\Delta s}{r^2} = \text{a minimum} \tag{12.2}$$

where $s$ is the curvilinear coordinate along the meander path, $\Delta s$ is the increment in $s$, and $r$ is the radius of curvature.

The theory of minimum variance was offered by Langbein and Leopold (1966) as a cause for meandering because meanders that achieve the minimum variance are more stable than a straight alignment. From the most probable path obtained by von Schelling, Langbein and Leopold derived the following equation for a regular meander path:

$$\theta = \omega \sin \frac{2\pi s}{M} \tag{12.3}$$

where $\theta$ is the angle that the meander path at a given point makes with mean downpath direction, $\omega$ is the maximum angle a path makes with mean downpath direction, and $M$ is the meander arc length, defined as the distance measured along the meander path between repeating points. Equation 12.3 is referred to as the *sine-generated curve*; it contains a scale parameter $M$ and a shape parameter $\omega$, both of which need to be specified in application. With properly selected values of $M$ and $\omega$, the sine-generated curve fits the shape of river meanders quite well, as illustrated in Fig. 12.2. Because the scale parameter and shape parameter need to be specified, the sine-generated curve does not provide the wavelength and sinuosity by itself. In other words, the mechanics underlying the formation of river meanders is not given by the curve.

## 12.2 MEANDER PATH BASED ON STREAMWISE VARIATION OF HELICAL MOTION

The path of meandering rivers can be determined by associating this path with the pattern of river flow, which is characterized by helical motion or secondary currents in meander bends. Since the river channel is self-formed, the equilibrium planform is so adjusted that it is compatible with the flow pattern. The channel curvature of the small stream shown in Fig. 12.3 may be considered a reflection of the flow curvature. Along a meandering channel consisting of bends and crossovers, each bend apex, which approximately coincides with the zone of maximum curvature, is preceded and followed by transition curves. Helical motion is at the maximum strength near the apex and is weak near the crossover. Therefore, helical motion peaks, decays, reverses, and then grows between two consecutive apexes. The decay and growth of helical motion, as explained by Rozovskii (1957), is due to the interaction of centripetal force and turbulent shear that the flow has to overcome in transforming from the helical pattern into parallel flow and vice versa.

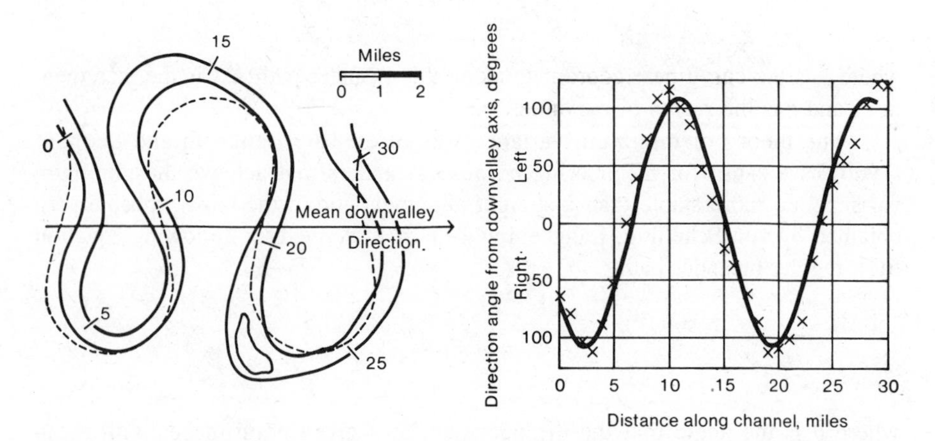


**Figure 12.2** (Left) Map of channel compared to sine-generated curve (dashed curve) and (Right) plot of actual channel direction against distance (crosses) and a sine curve (full line) for Mississippi River at Greenville, Mississippi (Langbein and Leopold, 1966).

**Figure 12.3** Meandering stream in Death Valley, California. The channel curvature may be considered a reflection of the flow curvature.

The meander path model (Chang, 1984) presented in this section is based on the equation of motion for transverse circulation in channels. A distinctive feature of this model is that the scale parameter is analytically determined from the channel length required for the decay and growth of the transverse circulation; therefore, only the shape parameter needs to be specified in its application. For a specified arc angle or curvature at the apex, this model provides the shape and scale of the meander path, from which other planform parameters such as arc length, wavelength, and amplitude can be directly obtained. This model is applicable only to nonbraided rivers with approximately uniform width and mean depth.

## Analytical Basis

A plan view of river meanders to be analyzed is shown schematically in Fig. 12.1. In this discussion, the meander path is distinguished from the channel path. The former is called as the *discharge centerline* or the *flow path*, and the latter is called the *channel centerline*. The meander path, being the discharge centerline, has zero net transverse discharge; that is, the inward and outward transverse flows at different depths across this line are in balance. Under this condition, the transverse flow may be considered to be a circulation, which, by definition, has no net lateral discharge. The flow path is generally in line with the locus of the high velocity. For nonbraided meandering streams, the depth of flow and mean flow velocity along the meander path are more or less uniform; they are assumed constant in the model. A more sophisticated model considering streamwise depth variation is described in Chapter 13.

A schematic showing the streamwise variation of the transverse velocity at the water surface along the meander path is given in Fig. 12.4a, in which the magnitude of this velocity is in direct proportion to the path curvature. The equation of motion for the transverse velocity in a wide curved open channel under the steady condition has the form (see Eq. 8.79)

$$u\frac{dv}{ds} = \frac{u^2}{r} - gS_r + \frac{\partial}{\partial z}\left(\varepsilon\frac{\partial v}{\partial z}\right) \tag{12.4}$$

where $v$ is the transverse velocity, $u$ is the longitudinal velocity, and $S_r$ is the transverse water-surface slope. In the present analysis, this equation is evaluated along the discharge centerline at the water surface. The surface velocities for $u$ and $v$ are designated $\tilde{u}$ and $\tilde{v}$, respectively.

The longitudinal velocity profile is given by Eq. 8.30, from which the surface velocity is related to the mean velocity $U$ as follows:

$$\tilde{u} = \frac{1 + m}{m}U \tag{12.5}$$

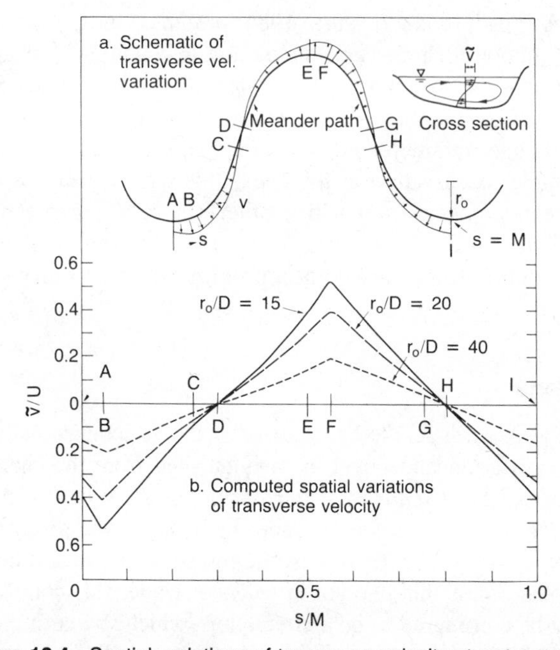


**Figure 12.4** Spatial variations of transverse velocity at water surface.

The transverse water-surface slope $S_r$ is given by Eq. 8.83. The last term in Eq. 12.4 is the Reynolds stress and is given by Eq. 8.81.

### Phase Lag

In a meandering river, the discharge centerline and the channel centerline are out of phase, as shown schematically in Figs. 12.1 and 12.4 by the phase lag. The points of interest are designated A to I. Observations in meandering rivers generally indicate that the maximum curvature of flow (at B or F) tends to be located downstream of the apex (at A or E), which is the location of the maximum channel curvature. At the point of inflection designated C, where the channel curvature is zero, the curvature of flow and transverse circulation are not completely zero but they reach zero at D, which is downstream of C.

Physically, the phase lag is attributed to the lag of channel's response to the flow in that channel formation tends to fall behind the pattern of river flow. Typically, the feedback mechanism lags somewhat behind the initial perturbation. Analytical determination of the phase lag is described in a later paragraph.

In the following, the centrifugal acceleration (second term in Eq. 12.4) and transverse water-surface slope are assumed to depend on the channel curvature $r_c$, but the transverse velocity of flow is related to the flow curvature $r$ along the meander path. Substituting Eq. 12.5 for $u$, Eq. 8.81 for the Reynolds stress, and Eq. 8.83 for $S_r$ into Eq. 12.4 yields

$$\frac{d\tilde{v}}{ds} + F_2(f)\tilde{v} = \frac{F_3(f)}{r_c} \tag{12.6}$$

where

$$F_2(f) = \frac{\kappa}{D}\left(\frac{f}{2}\right)^{1/2}\frac{m}{1+m}, \qquad F_3(f) = \left(\frac{1+m}{m} - \frac{m}{1+m}\right)U \tag{12.7}$$

and $ds = r\,d\theta$ is evaluated along the meander path.

### Meander Path Equation

A general equation that covers both growth and decay of transverse velocity along the meander path is first derived, from which the meander path equation is then obtained. At a point of maximum transverse velocity such as B or F in Figs. 11.1 and 11.4, we have

$$\frac{d\tilde{v}}{ds} = 0 \tag{12.8}$$

Therefore, Eq. 12.6 becomes

$$F_2\tilde{v}_0 = \frac{F_3}{r_{c0}} \tag{12.9}$$

where the subscript zero refers to the point of maximum transverse velocity at B or F. Subtracting Eq. 12.9 from Eq. 12.6 and adding $d\tilde{v}_0/ds = 0$ yields

$$\frac{d(\tilde{v}_0 - \tilde{v})}{ds} + F_2(\tilde{v}_0 - \tilde{v}) = \frac{F_3}{r_{c0}} - \frac{F_3}{r_c} \tag{12.10}$$

This equation, in which $\tilde{v}_o - \tilde{v}$ is the dependent variable and $r_{c0}$ and $r_c$ are functions of $s$ alone, is a linear differential equation of the first order. For the homogeneous linear equation consisting of only the left-hand side, the variables are separable and the solution is

$$\tilde{v}_0 - \tilde{v} = C \exp\left(-\int F_3\, ds\right) = C\exp(-F_3 s) \tag{12.11}$$

where $C$ is a constant of integration. The particular solution of Eq. 12.10 is obtained using the variation of parameters. Now substitute in the nonhomogeneous equation (Eq. 12.10) the expression

$$\tilde{v}_0 - \tilde{v} = \phi \exp(-F_2 s)$$

in which $\phi$, a function of $s$, has replaced the constant $C$. Equation 12.10 becomes

$$\frac{d\phi}{ds} \exp(-F_2 s) = \frac{F_3}{r_{c0}} - \frac{F_3}{r_c}$$

Therefore,

$$\phi = C' + \int \left(\frac{F_3}{r_{c0}} - \frac{F_3}{r_c}\right) \exp(F_2 s)\, ds$$

where $C'$ is a constant. The solution of the general linear equation is therefore

$$\tilde{v}_0 - \tilde{v} = \left[c + \int \left(\frac{F_3}{r_{c0}} - \frac{F_3}{r_c}\right) \exp(F_2 s)\, ds\right] \exp(-F_2 s) \tag{12.12}$$

From the boundary condition at the initial point for $s$, where $\tilde{v}$ equals the initial transverse velocity $\tilde{v}_i$, the constant $c$ is evaluated to be $\tilde{v}_0 - \tilde{v}_i$. Equation 12.12 becomes

$$\tilde{v} = \tilde{v}_0 - \left[(\tilde{v}_0 - \tilde{v}_i) - \int \left(\frac{F_3}{r_{c0}} - \frac{F_3}{r_c}\right) \exp(F_2 s)\, ds\right] \exp(-F_2 s) \tag{12.13}$$

This is the general equation for the spatial variation of transverse surface velocity along the meander path. Now, substituting Eq. 8.85 into Eq. 12.13 yields

$$\frac{F_1}{r} = \frac{F_1}{r_0} - \left[\left(\frac{F_1}{r_0} - v_i\right) - \int\left(\frac{F_3}{r_{c0}} - \frac{F_3}{r_c}\right) \exp(F_2 s)\, ds\right] \exp(-F_2 s) \qquad (12.14)$$

where

$$F_1 = \frac{D}{\kappa}\left[\frac{10}{3} - \frac{1}{\kappa}\frac{5}{9}\left(\frac{f}{2}\right)^{1/2}\right] U \qquad (12.15)$$

and $r_0$ is the minimum radius of path curvature which occurs at B or F. This radius of curvature is assumed to be equal to the radius of channel curvature at the apex. Equation 12.14, which relates meander path curvature $r$ to curvilinear coordinate $s$ and other quantities, is the meander path equation.

## Method of Computation

For a given radius of channel curvature at the apex ($r_0$) and other required quantities ($D$, $U$, and $f$), the meander path is obtained by solving Eq. 12.14 using the method of finite differences. The geometry of the path is computed at small increments of curvilinear distance, $\Delta s$. The path from B to F is a complete cycle of decay and growth for the transverse velocity, which changes sign at point D. The computation starts at B, but the location of this point measured by the distance from A (i.e., the phase lag) is an unknown. The meander path is a function of the phase lag; therefore, different phase lags will produce different meander curves. In the case of regular meander paths for which repeating points along the curve possess the same geometry, the phase lag is obtained by trial and error so that such regularity of each path is maintained. In the trial-and-error procedure, different assumed values for the lag are introduced in the computation, the phase lag that produces the regular repeating pattern for the path so obtained is found to be about one-fifth of the distance between an apex and its next crossover. The radii of channel curvature, $r_c$ and $r_{c0}$, in Eq. 12.14 are computed using the respective radius of path curvature at the lag distance downstream.

At each increment $\Delta s$, the value of $r$ is evaluated using Eq. 12.14; the transverse velocity $\tilde{v}$ is then computed using Eq. 8.85. Other geometric parameters are obtained from

$$\theta_{j+1} = \theta_j + \frac{\Delta s}{r_j} \qquad (12.16)$$

$$x_{j+1} = x_j + \Delta s \cos \theta_j \qquad (12.17)$$

and

$$y_{j+1} = y_j + \Delta s \sin \theta_j \qquad (12.18)$$

where $x$ is the horizontal coordinate, $y$ is the vertical coordinate, and subscripts $j$ and $j + 1$ are $s$-coordinate indices. The accuracy of computation depends on the size of $\Delta s$. From the experience of this analysis, the computed results approach constancy if $\Delta s < D/4$.

## Plan Geometry of River Meanders

Sample regular meander paths computed using the mathematical model are shown in Fig. 12.5; the corresponding spatial variations in transverse velocity are given in Fig. 12.4b. The meander arc length, which is measured between repeating points along the path, is found to be a function of the depth and channel roughness, independent of channel curvature or arc angle. The arc-length–depth ratio, $M/D$, is inversely related to channel roughness. In other words, the dimensionless channel length, $M/D$, required for growth and decay of transverse velocity depends on the roughness. Its independence of the channel curvature is because the effects of opposite curvatures are mutually canceled between two repeating points. The independence of arc length on arc angle as determined herein concurs with the findings of Hey (1976), which was based on data from the Wye and Tweed rivers in Great Britain.

From the regular meander path model, a graphical relationship for the plan geometry of river meanders is obtained as shown in Fig. 12.6. The ratio $M/D$ is shown as a unique function of the friction factor $f$. The wavelength $\lambda$ is the straight-line distance of the arc length; the ratio $\lambda/D$ is a function of $M/D$ and arc angle. The sinuosity, defined as the ratio of arc length to wavelength, or $M/\lambda$, is a unique function of the arc angle and is independent of the wavelength.

An analytical path from the regular meander path model is compared with the sine-generated curve having equal wavelength and sinuosity, as shown in Fig. 12.7. The analytical path is very similar to the sine-generated curve, which fits the actual shape quite well and better than other alternatives. Whereas the sine-generated curve is a symmetrical curve, the analytical path is slightly skewed. A comprehensive study of the inherent asymmetry of meander curves was made by Carson and LaPointe (1983). The sine-generated curve, as well as other existing curves, has two unknowns $M$ and $\omega$, which need to be specified by a scale parameter and a shape parameter. However, the scale parameter (arc length) is analytically determined in the meander path model, therefore, only the shape parameter needs to be specified.

The shape parameter may be specified by the arc angle, sinuosity, or the maximum curvature at the apex, which are all uniquely related. In self-formed river meanders, the maximum curvature of a bend is constrained by the development of flow separation at sharp turning angles. Under such situations, sediment deposition induced in the eddy zone downstream of the point bar is accompanied by erosion on the opposite bank; thus the sharp bend is eliminated through this process of river channel formation. The maximum equilibrium curvature has been obtained in Sec. 11.1 using the energy approach; it represents the maximum curvature for which a river does the least work in turning. This curvature, represented

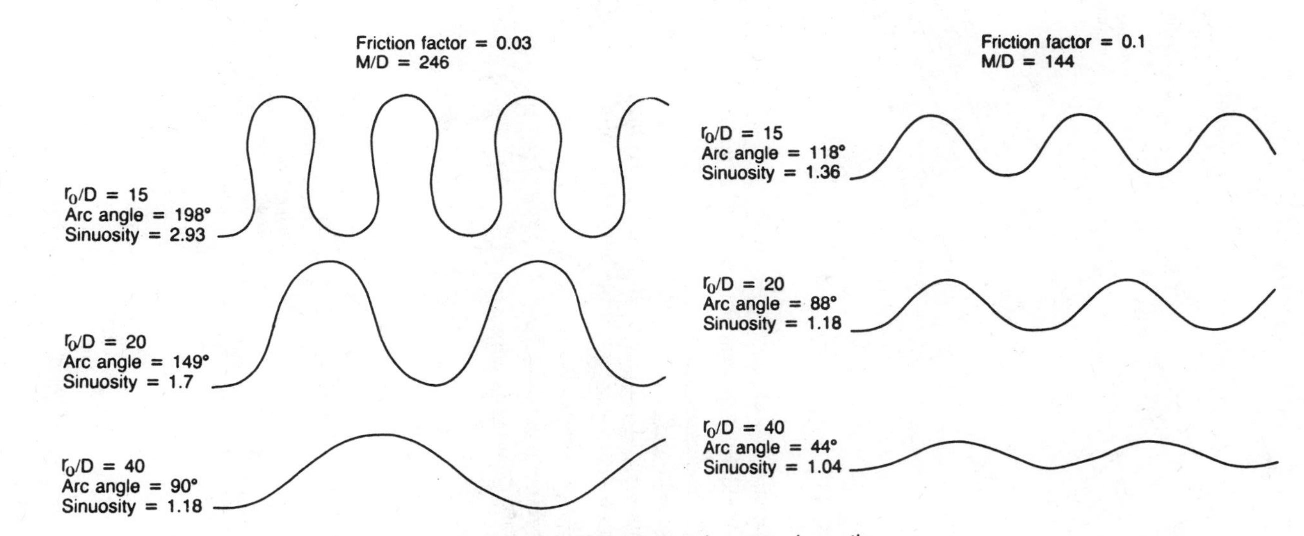


**Figure 12.5** Sample regular meander paths.

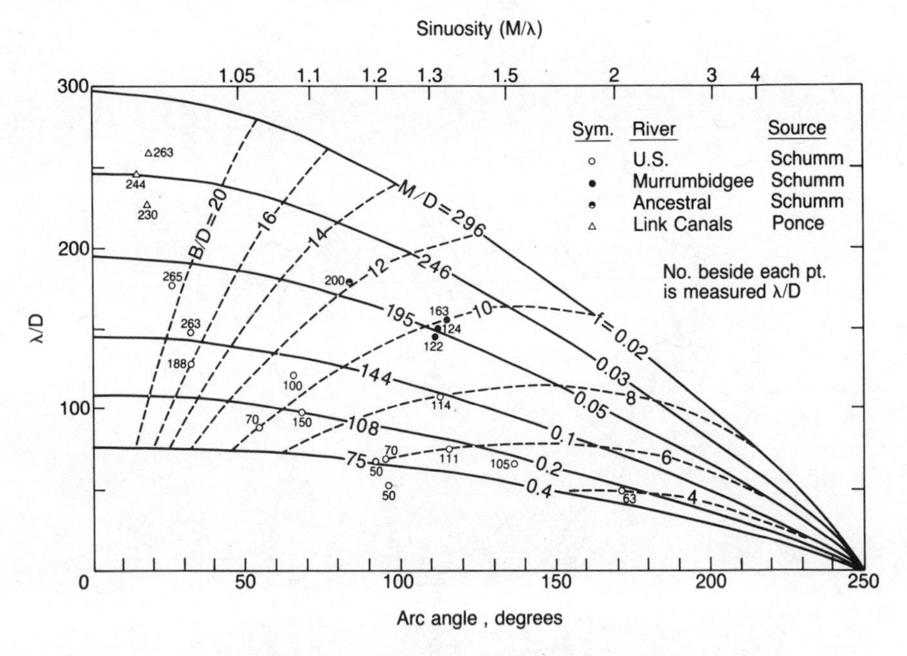


**Figure 12.6** Plan geometry of river meanders.

as the radius–channel-width ratio, $r_0/B$, has the average value of about 3 for meandering streams. Its minor variation with the width–depth ratio shown in Fig. 11.5 follows the relationship

$$\frac{r_0}{B} = 2.2 + 0.15\left(\frac{B}{D} - 4\right) \tag{12.19}$$

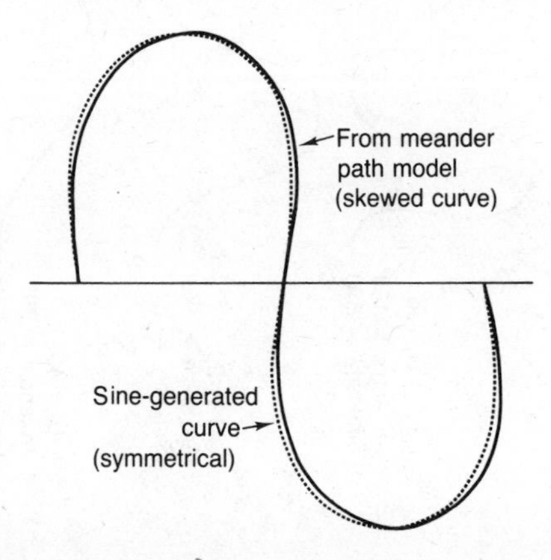


**Figure 12.7** Comparison of two different curves having equal wavelength and sinuosity.

It indicates that the maximum curvature decreases with an increase in width–depth ratio.

River meanders for which maximum bend curvatures are established are defined herein as fully-developed meanders. Plan geometry for this type of meandering river has apparently received more attention in the past. For a given $B/D$, the corresponding $r_0/B$ is obtained from Eq. 12.19, then the arc angle and sinuosity are uniquely determined from the regular meander path model. The resulting relationship is plotted in Fig. 12.6 using dashed lines. It shows that, for such rivers, the ratio $\lambda/D$ is directly related to $B/D$ but inversely related to $f$ and that the sinuosity increases with a decrease in $B/D$ *or* $f$. In other words, rivers with a small width–depth ratio or channel roughness, or both, can be more sinuous. The graphical relationship of Fig. 12.6 is similar to the relationship of Eq. 12.1 developed by Ikeda et al. (1981).

River data pertaining to the plan geometry of sinuous nonbraided streams at the bankfull stage are used for comparison with the analysis. Since planimetric parameters such as wavelength and arc length are usually related to channel width or discharge in most studies, published data seldom contains such information as channel depth, roughness, and arc angle that are required in this model. Therefore, river data useful for this study are rather limited.

River data compiled by Ponce (1976) and Schumm (1968) are used for comparison with the analytical predictions. The data reported by Ponce were measured from three of the Link Canals in Pakistan; these canals are essentially straight sand-bed channels with a meandering thalweg. A large sample of wavelengths was collected from each canal to determine the predominant (mode) wavelength of the thalweg. The data for meandering rivers compiled by Schumm include 12 rivers in the midwestern United States, three Murrumbidgee River sections, and one ancestral-river section. These rivers have flat slopes, nonbraided channel pattern, and width–depth ratios less than 22. These data are plotted in Fig. 12.6 using the measured $B/D$ and $f$. The measured value for $\lambda/D$ is written beside each data point for comparison with the analytical prediction.

## 12.3 PROCESSES GOVERNING MEANDER BEND MIGRATION

Meander bend may migrate in lateral and downvalley directions and may grow or subside. Major processes of flow and sediment transport that affect meander bend migration are described in this section. An adequate understanding of such processes, needless to say, is essential for river control and for mathematical modeling of river channel changes. The process that characterizes the flow through meander bends is the streamwise variation of the helical-motion strength to which so many phenomena of river meanders are related. In this section, processes governing lateral and downvalley migration and the associated curvature changes are described on the basis of the streamwise growth and decay of helical motion. The changing curvature of the meander path and its phase lag with the channel curvature have important effects on meander morphology, and they should be in some way related to the variation in helical strength.

An equation for transverse velocity profile along the depth given in Sec. 8.2 allows no net transverse flow and is assumed valid along the flow path. Such a relation signifies the unique relation between the helical strength (represented by $v$) and the curvature of the flow path.

The helical strength, or the flow curvature, varies in the streamwise direction. The variation is described by Eq. 12.4, which shows that streamwise variation of helical strength, transverse velocity, or flow curvature is under the influence of the centrifugal acceleration from the channel curvature, the transverse-water surface slope, and the internal turbulent shear. The internal shear is the resistance that the flow has to overcome in changing it curvature. Because of such resistance, the flow curvature does not adjust immediately to the channel curvature. This process is considered by deVriend and Struiksma (1983) to be the delayed response of flow curvature to channel curvature and is responsible for the phase lag depicted schematically in Fig. 12.8.

The flow pattern in Fig. 12.8 denotes that the migration rate varies along the channel. It should be related to the processes of flow and sediment transport in addition to the bank material and vegetation. The effects of flow on sediment transport and mode of migration may be analyzed by the flow pattern in relation to the channel pattern. The tractive force, or shear, exerted by the flow on the channel boundary has longitudinal and cross-stream components. The longitudinal direction, as used herein, is along the channel path; the cross-stream direction is normal to the channel path. As long as there is an angle between flow path and channel path, the primary flow and secondary circulation have components in both the longitudinal and cross-stream directions.

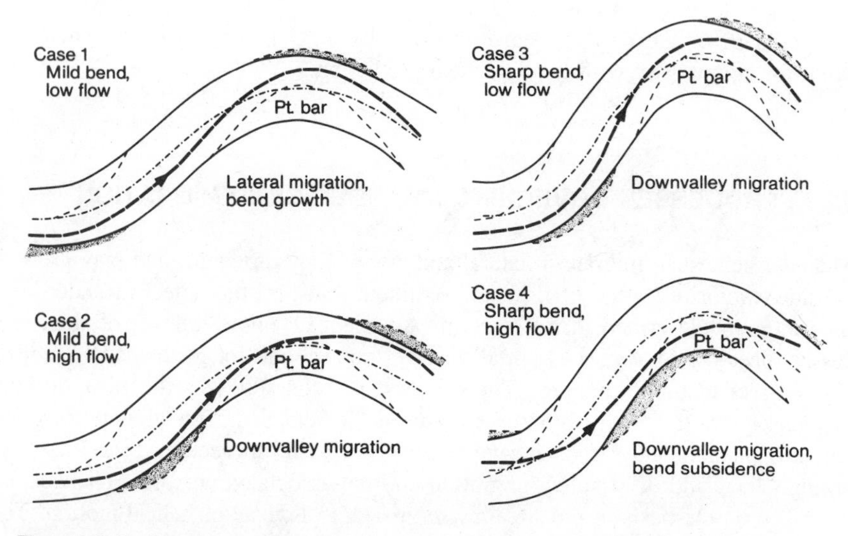


**Figure 12.8** Common modes of meander bend deformation in relation to flow pattern.

The longitudinal shear is essentially contributed by the primary flow; it generally causes bank erosion at places where the flow path hugs the bank, as exemplified by the shaded bank areas in Fig. 12.8. Deposition may be expected along banks away from the flow path.

The angular difference between the flow path and channel path also contributes to cross-stream discharge, which is described in Sec. 8.9 and depicted in Fig. 8.14. The secondary circulation exerts a downward action on the concave bank and moves sediment from the concave bank to the convex bank. Depending on the relative transport rate due to the cross-stream component of the primary flow and the secondary circulation, the sediment may move either away or toward a bank. In addition to the foregoing factors, meander bend migration is also affected by the wave action and the mode of bank failure. These topics are described by Thorne (1981) and Grissinger and Little (1986), among others.

## Meander Bend Migration and Flow Path

The variation of helix strength or flow curvature is a function of the flow condition, with which the phase lag between the flow path and channel path also varies. According to Eq. 12.14, the flow path adjusts more rapidly to the channel curvature at a shallow depth, or low flow, but the adjustment is slow at high flow. As the turning of the flow path is inhibited at a large depth, there is a greater difference in their curvatures and a greater phase lag. Depending on the pattern of flow path in relation to the channel path, meaner bend changes may be manifested in lateral migration or downvalley migration (or both) and the possibility of either bend growth or subsidence, as illustrated in Fig. 12.8 by the four cases for different combinations of channel curvature and flow condition. In case 1 (mild curvature and relatively low flow), the helical motion develops rapidly and thus the flow pathline is more in line with the channel configuration. Since the maximum flow curvature is close to the apex, it indicates the tendency of lateral migration and growth of the bend. In case 2 (mild curvature and high flow), the change in flow curvature is slow and the maximum flow curvature is thus farther downstream from the apex, denoting downvalley migration. For cases 3 and 4 (sharp channel curvature), there is greater difference between the channel curvature and the flow curvature. In case 3 (low flow), the flow pattern points to downvalley migration. In case 4 (high flow), the large difference between the flow curvature and channel curvature is associated with a large phase lag and high velocities near the convex bank. This relationship between flow pattern and channel pattern signifies the tendency of downvalley migration together with reduction in bend curvature.

As described in Secs. 5.7 and 8.8, Ippen and Drinker (1962) measured the distribution of boundary shear stresses in the curved reaches of smooth trapezoidal channels under conditions of subcritical flow. A series of tests was conducted to determine the effects of variation in discharge and bend geometry on the shear pattern. The locations of the shear maxima were generally found to be consistent with the path of highest velocity. The relative curvature $B/r_c$ is measure of the channel curvature in relation to its size. At large relative curvatures, high shear is found near the inside bank in the curve and near the outside bank below the curve exit. This shear pattern is consistent with the flow pattern depicted as case 4 in Fig. 12.8. At small relative curvatures, the increased stresses appear along the outer bank, in the downstream portion of the curve, consistent with the flow pattern in case 2.

In summary, the complicated flow and sediment processes affecting meander bend migration can be described from the viewpoint of streamwise variation in helical motion, which also reflects the pattern of flow development. The flow pattern lags behind the channel pattern because of the resistance that the flow has to overcome in its gradual development. Depending on the relationship between flow pattern and channel pattern, meander bend can migrate laterally or downvalley, and its curvature may grow or subside, as illustrated by examples in Fig. 12.8. Secondary or transverse circulation normal to the primary flow direction contributes to transverse sediment transport but not transverse discharge with respect to the primary flow direction. The angle between the primary flow and channel centerline contributes to cross-stream discharge, variation in superelevation, and redistribution of the longitudinal shear.

## 12.4 ON THE CAUSE OF RIVER MEANDERING

The cause of river meandering has been a subject of persistent debate in scientific and engineering literature. Earlier explanations of what causes river meandering have been extensively reviewed by Yang (1971), Chitale (1973), Callander (1978), and Hickin (1983), among others. Proposed causes have included the most probable path (minimum variance), secondary currents, dynamic instability, silt–clay content, and so on.

The present discussion focuses on the cause from which meandering follows as an effect. Since meandering is a feature of river channel formation, determining the cause of meandering requires identifying the cause of river channel formation. In technical terms, the cause refers to the independent or controlling variables that are imposed upon the river channel. The effect that results from the cause is described by the dependent variables, which include the channel geometry and meandering pattern. It is clear that any explanation of the cause of meandering must be derived from the independent variables and their interactions with the dependent variables, as illustrated in Fig. 11.1. Interdependency of the variables and their roles as dependent variables at some times and independent variables at other times must also be clearly understood.

Any thesis on the origin of meandering must explain why the channel slope must be flatter than the alluvially formed valley slope. This slope relationship is the necessary and sufficient condition for the occurrence of meanders. The ratio of valley slope to channel slope is defined as *meander sinuosity*.

## Theses on the Cause of Meandering

Certain doctrines on the origin of meandering for regime rivers are reviewed herein. Two major doctrines: the most probable path and secondary currents are included in the following discussion; others, such as dynamic instability, the Coriolis force, bank erosion, statistical analysis, and so on, are not but the same reasons apply.

The theory of minimum variance was offered as a possible cause of meandering because meanders that achieve the minimum variance are more stable than a straight alignment. From the most probable path obtained by von Schelling (1951), Langbein and Leopold (1966) derived the sine-generated curve, given by Eq. 12.3, for the meander path. Because the controlling variables of water and sediment inflows are not included in deriving this equation, it does not explain the cause of meandering. Since the shape parameter needs to be prescribed in the equation, the sine-generated curve does not predict the channel slope or the sinuosity of the meander path.

Development of secondary currents has been proposed as a possible cause of meandering, derived from the finding that secondary currents can occur in a straight channel with or without sediment motion (Tanner, 1960). Secondary currents undoubtedly play a very important role in the morphology of river meanders because many meander features, except sinuosity, can be modeled based on secondary currents. This is exemplified by the meander path model given in Sec. 11.2. In using this model, the shape parameter (sinuosity) has to be prescribed as an independent variable. In other words, the sinuosity or channel slope is not determined from secondary currents; the sinuosity of the river defines the helical strength instead.

## Cause of River Meandering

The cause of river meandering is explained herein by the fluvial processes involved in the formation of the alluvial valley and the river channel, as well as the interrelationship of their gradients. The explanation is based on the independent variables of river channel formation, from which meandering follows as a dependent variable (i.e., an effect) and comprises the following four stages: (1) formation of the channel slope, (2) relationship between channel slope and valley slope, (3) formation of the valley slope, and (4) formation of river meanders. This explanation elucidates why certain rivers have a distinctly meandering pattern with a channel slope that is significantly flatter than the valley slope.

1. *Formation of the Channel Slope*. For the graded-time span, the channel gradient is under the control of the water and sediment inflow rates and their characteristics. From Lane's relationship (Eq. 2.30) of balance, the channel slope $S$ can be written as

$$S = F(Q_s, Q, d) \tag{12.20}$$

where $Q_s$ is the sediment load, $Q$ is the water discharge, and $d$ is the median sediment size. This functional relationship implies that the slope of the river channel is the effect whereas $Q$, $Q_s$, and $d$ constitute the cause. Lane's relationship has been quantified, as given in Fig. 10.13 for gravel-bed rivers and by Eqs. 11.12 and 11.13 for sand-bed rivers.

2. *Relationship Between Channel Slope and Valley Slope*. Because the values of $Q$ and $Q_s$ are determined in the drainage basin, the channel slope produced by these inflow quantities is usually different from the valley slope, which is another independent variable for the river during the graded-time span. Considering all possibilities, the channel slope could be steeper than, equal to, or flatter than the valley slope. The first case can be eliminated because the channel slope cannot physically be steeper than the valley slope under dynamic equilibrium or regime. It may therefore be concluded that for rivers in regime, the channel slope must be equal to or flatter than the valley slope. It follows that regime rivers may be straight or meandering in channel pattern. Since the channel slope is primarily determined by the values of $Q$, $Q_s$, and $d$, it is possible for rivers of identical slopes to have different sinuosities depending on each valley slope. The following discussion of fluvial processes illustrates that the channel slope can be quite flatter than the valley slope.

3. *Formation of the Valley Slope*. Alluvial valleys are usually formed under different hydrologic or geologic regimes during a very long-term geologic time span and are affected by changes in climate, surface vegetation, paleohydrology, catastrophic floods, tectonic events, forest fires, channel piracy, overgrazing, human-made factors, and so on. Under different regimes represented by separate combinations of the inflow quantities of water and sediment, the river channel develops different slopes. If the channel slope, during any episode in geologic time, is steeper than the valley slope, then deposition occurs to steepen the valley slope. The valley slope may also be reduced depending on the water–sediment mixture supplied. Therefore, the valley slope, for the geologic time span, is a dependent variable, whereas the climate (or paleohydrology) and geology of the basin are independent variables. Examples of valley formation include (1) the paleochannels of the Murrumbidgee River reported by Schumm (1968) and (2) geomorphological response to an extreme flood in the Honda Valley, Spain reported by Harvey (1984).

4. *Formation of Meanders*. Meandering rivers with channel gradients significantly flatter than valley gradients are explained by the different fluvial processes involved in forming a river channei and an alluvial valley. Each undergoes periods of erosion and deposition because of changes in water discharge and sediment load. However, the geomorphological responses of channels and valleys to erosion and deposition are different. During deposition, which may be caused by overloading of sediment, stream adjustment is usually characterized by widening and braiding, while the alluvial valley is built up uniformly. Most active depositional valley formation occurs during major historical hydrologic events, but an alluvial valley may also be filled over a longer time span through evolutionary

adjustments. Whether it is through active braiding or gradual shifting of the stream course, deposition is usually rather evenly distributed on the valley floor. Therefore, deposition favors alluvial plain formation. During episodes of erosion that result from a deficit in sediment supply, the stream channel seeks to develop a flatter slope. In sharp contrast to deposition, morphological response to erosion is characterized by incision of the stream into the alluvial plain. Therefore, erosion favors formation of a more confined river channel. Incision or degradation of a river channel is inhibited by coarsening of the bed material due to hydraulic sorting; it may also be prevented when an armored layer develops. Reduction of channel slope through incision would require tremendous degradation. For these reasons, a river channel usually adjusts by developing a flatter slope through meandering. River channel formation is associated with the dominant, or bankfull, discharge, which has a return period of no more than a few years (see Sec. 2.2). Given the length of the geologic time span and its many hydrologic and geologic regimes, the valley slope usually results from historical conditions different from those forming the existing channel. Because the valley slope is steepened during any episode in geologic time when the channel slope is steeper than the valley slope, the valley slope can be quite steeper than the current channel slope. This explains why certain rivers have a distinctly meandering pattern.

A river in a broad alluvial plain has sufficient freedom to develop a slope much flatter than the valley slope by the combination of channel meanders and valley meanders, as exemplified in Figs. 2.5 and 2.6. The smooth serpentine meander path is compatible with the gradual flow development of secondary currents. For highly sinuous channels, the curvature at an apex, defined by the radius-of-curvature–channel-width ratio, has an average value of about 3. It represents the maximum curvature for which a river does the least work in turning while maintaining streamwise uniformities in sediment transport and energy expenditure (see Sec. 11.1). Sharper turning angles, which cause greater energy loss (or a steeper energy gradient), are normally eliminated in the process of meander formation.

In the foregoing explanation for river meandering, consideration was given to the alluvial-valley formation in geologic time. Meandering represents subsequent changes in slope and pattern reflecting the river's adjustment to altered discharge and load. While meandering formation is a long-term development, this process may be shortened considerably by human interference. For example, major dams reduce the discharge and depletes the bed load for the downstream channel. This change in regime often results in the formation of a more sinuous channel. A sand pit, such as that shown in Fig. 12.9, has the same effect as a dam on channel morphology. In response to the depletion of bed load, the small stream coming out of the borrow pit established a sinuous pattern in addition to incision.

In summary, the cause and effect (i.e., independent and dependent variables) must be identified in interpreting the fluvial processes. In the graded-time span for river channel formation, the regime channel slope is a dependent variable controlled by the water and sediment inflows from the drainage basin, and it is the minimum slope needed to transport such inflow quantities. The valley slope,

**Figure 12.9** Meandering and incision of stream below sand pit in response to sediment deficit.

which is formed during the much longer geologic time span, is an independent variable for the river during the graded time span. The river slope cannot physically be steeper than the valley slope for a river in regime. It may either be equal to or flatter than the valley slope. For distinctly meandering rivers, the channel slope is significantly flatter than the valley slope. This is because the alluvial valley is usually formed during the very long-term span under various hydrologic and geologic conditions different from those shaping the current channel. In the geologic time span, the valley slope is built up during episodes of deposition whenever the channel slope exceeds the valley slope. But the valley slope is not easily reduced during periods of erosion. Because degradation is inhibited by sediment sorting, the morphological response to sediment deficit is to form a meandering channel, which, in turn, reduces erosion. For these reasons, the present channel can be quite flatter than the alluvially formed valley.

## REFERENCES FOR PART III

ASCE, *Sedimentation Engineering*, Manuals and Reports on Engineering Practice, No. 54, Vita A. Vanoni, ed., 1975.

Blench, T., "Regime Theory for Self-Formed Sediment-Bearing Channels," *Trans. ASCE*, **117**, pp. 383–408, 1952.

Blench, T., "Regime Theory Design of Canals with Sand Beds," *J. Irrig. Drainage Div. ASCE*, **96**(IR2), pp. 205–213, June 1970.

Blench, T. and Simons, D. B., "Regime Data, Flume Data, and Regime Basics," International Commission on Irrigation and Drainage, New Delhi, India, 1974.

Brice, J. C., "Planform Properties of Meandering Rivers," *River Meandering*, Proceedings of the Conference Rivers '83, New Orleans, Louisiana, October 24-26, 1983, pp. 1–15.

Brownlie, W. R., "Compilation of Alluvial Channel Data: Laboratory and Field," Report No. KH-R-43B, W. M. Keck Laboratory, California Institute of Technology, November 1981.

Brush, L. M., "Drainage Basins, Channels, and Flow Characteristics of Selected Streams in Central Pennsylvania," *USGS Professional Paper 282-F*, 1961.

Burkham, D. E., "Channel Changes of the Gila River in Safford Valley, Arizona 1846–1970," *USGS Professional Paper 655-G*, 1972, 24 pp.

Callander, R. A., "River Meandering," *Annual Review of Fluid Mechanics*, **10**, pp. 129–158, 1978.

Carson, M. A. and LaPointe, M. F., "The Inherent Asymmetry of River Meander Planform," *J. Hydrol.*, **91**, pp. 41–55, 1983.

Chang, H., Simons, D. B., and Woolhiser, D. A., "Flume Experiments on Alternate Bar Formation," *J. Waterways Harb. Coast. Eng. Div. ASCE*, **97**(WW1), pp. 155–165, February 1971.

Chang, H. H., "Geometry of Rivers in Regime," *J. Hydraul. Div. ASCE*, **105**(HY6), pp. 691–706, June 1979a.

Chang, H. H., "Minimum Stream Power and River Channel Patterns," *J. Hydrol.*, **41**, pp. 303–327, 1979b.

Chang, H. H., "Stable Alluvial Canal Design," *J. Hydraul. Div. ASCE*, **106**(HY5), pp. 873–891, May 1980a.

Chang, H. H., "Geometry of Gravel Streams," *J. Hydraul. Div. ASCE*, **106**(HY9), pp. 1443–1456, September 1980b.

Chang, H. H., "Mathematical Model for Erodible Channels," *J. Hydraul. Div. ASCE*, **108**(HY5), pp. 678–689, May 1982.

Chang, H. H., "Analysis of River Meanders," *J. Hydraul. Eng. ASCE*, **110**(1), pp. 37–50, January 1984a.

Chang, H. H., "Modeling of River Channel Changes," *J. Hydraul. Eng. ASCE*, **110**(2), pp. 157–172, February 1984b.

Chang, H. H., "Regular Meander Path Model," *J. Hydraul. Eng. ASCE*, **110**(10), pp. 1398–1411, October 1984c.

Chang, H. H., "River Morphology and Thresholds," *J. Hydraul. Eng. ASCE*, **111**(3), pp. 503–519, March 1985a.

Chang, H. H., "Design of Stable Alluvial Canals in a System," *J. Irrig. Drainage Eng. ASCE*, **111**(1), pp. 36–43, March 1985b.

Chang, H. H., "Formation of Alternate Bars," *J. Hydraul. Eng. ASCE*, **111**(11), pp. 1412–1420, November 1985c.

Chang, H. H., "River Channel Changes: Adjustments of Equilibrium," *J. Hydraul. Eng. ASCE*, **112**(1), pp. 43–55, January 1986.

Chang, T. P. and Toebes, G. H., "A Statistical Comparison of Meander Planforms in the Wabash Basin," *Water Resour. Res.*, **6**(2), pp. 557–578, April 1970.

Charlton, F. G., "An Appraisal of Available Data on Gravel Rivers," Report No. INT 151, Hydraulic Research Station, Wallingford, England, August 1977.

Cherkauer, D. S., "Minimization of Power Expenditure in a Riffle-Pool Alluvial Channel," *Water Resour. Res.*, **9**(6), pp. 1613–1628, December 1973.

Chitale, S. V., "Theories and Relationships of River Channel Patterns," *J. Hydrol.*, **19**, pp. 285–308, 1973.

Davies, T. R. H. and Sutherland, A. J., "Extremal Hypotheses for River Behavior," *Water Resour. Res.*, **19**(1), pp. 141–148, February 1983.

Dominy, F. E., "Design of Desilting Works for Irrigation System," *J. Irrig. Drainage Div. ASCE*, **92**(IR4), pp. 1–26, December 1966.

deVriend, H. J. and Struiksma, N., "Flow and Bed Deformation in River Bends," *River Meandering*, Proceedings of the Conference Rivers '83, New Orleans, Louisiana, October 24-26, 1983, pp. 810–828.

Dury, G. H., "Principles of Underfit Streams," *USGS Professional Paper 452-A,* 1964.

Ferguson, R. I., "Regular Meander Path Models," *Water Resour. Res.*, **9**(5), pp. 1079–1086, August 1973.

Gilbert, G. K., "Report on the Geology of the Henry Mountains," U.S. Geographical and Geological Survey of the Rocky Mountain Region, 1880.

Glover, R. E. and Florey, Q. L., "Stable Channel Profiles," *Hydrology, No. 325*, U.S. Bureau of Reclamation, 1951.

Grissinger E. H. and Little, W. C., "Similarity of Bank Problems on Dissimilar Streams," Proceedings of the Fourth Federal Interagency Sedimentation Conference, **2**, pp. 5.51-5.60, 1986.

Hancu, S. and Batuca, D., "Morphological Equations Based on Variational Principles," Proceedings of the First International Symposium on River Sedimentation, **1**, Beijing, 1980.

Harvey, A. M., "Geomorphological Response to an Extreme Flood: a Case from Southeast Spain," *Earth Surface Processes and Landforms*, **9**, pp. 267–279, 1984.

Hey, R. D., "Geometry of River Meanders," *Nature*, **262**, pp. 482–484, August 5, 1976.

Hey, R. D., "Determinate Hydraulic Geometry of River Channels," *J. Hydraul. Div. ASCE*, **104**(HY6), pp. 869–885, June 1978.

Hey, R. D., "Dynamic Process-Response Model of River Channel Development," *Earth Surf. Processes*, **4**, pp. 59–72, 1979.

Hickin, E. J., "River Channel Changes," *Modern and Ancient Fluvial Systems*, J. D. Collins and J. Lewin, eds., Int. Asso. Sediment. Spec. Pub. **6**, Blackwell Scientific Pub., Oxford, England, pp. 61–83, 1983.

Holtorff, G., "The Evolution of Meandering Channels," Proceedings of the Second International Symposium on River Sedimentation, Nanjing, China, 1983, pp. 692–705.

Ideka, S., "Prediction of Alternate Bar Wavelength and Height," *J. Hydraul. Eng. ASCE*, **110**(4), pp. 371–386, April 1984.

Ikeda, S., Parker, G., and Sawai, K., "Bend Theory of River Meanders. Part 1. Linear Development," *J. Fluid Mech.*, **112**, pp. 363–377, 1981.

Inglis, C. C., "Meanders and Their Bearing on River Training," The Institute of Civil Engineers, Maritime and Waterways Engineers Division, London, pp. 3–54, 1974.

Ippen, A. T. and Drinker, P. A., "Boundary Shear Stresses in Curved Trapezoidal Channels," *J. Hydraul. Div. ASCE*, **88**(HY5), pp. 143–179, September 1962.

Jaeggi, M. N. R., "Formation and Effects of Alternate Bars," *J. Hydraul. Eng. ASCE*, **110**(2), pp. 142–156, February 1984.

Keller, E. A., "Areal Sorting of Bed-Load Material: the Hypothesis of Velocity Reversal," *Geol. Soc. Am. Bull.*, **82**, pp. 753–756, March 1971.

Kellerhals, R., "Stable Channels with Gravel-Paved Beds," *J. Waterways Harb. Div. ASCE*, **93**(WW1), pp. 63–84, February 1967.

Kellerhals, R., Neill, C. R., and Bray, D. I., "Hydraulic and Geomorphic Characteristics of Rivers in Alberta," *River Engineering and Surface Hydrology Report*, Research Council of Alberta, Alberta, Canada, No. 72-1, 1972.

Kinosita, R., "Study on the Channel Evolution of the Isikari River," Bureau of Resources, Department of Science and Technology, Japan, 1962 (in Japanese).

Kirkby, M. J., "Maximum Sediment Efficiency as a Criterion for Alluvial Channels," *River Channel Changes*, K. J. Gregory, ed., John Wiley & Sons, New York, 1977, pp. 429–442.

Kondap, D. M. and Garde, R. J., "Application of Optimization Principles in the Study of Stable Channels," Proceedings Inter. Workshop on Alluvial River Problems, University of Roorkee, Roorkee, India, March 1980.

Lacey, G., "Stable Channels in Alluvium," *Proc. Inst. Civ. Eng.*, London, **229**, 1930.

Lacey, G., "Uniform Flow in Alluvial Rivers and Canals," *Proc. Inst. Civ. Eng.*, London, **237**, 1935.

Lacey, G., "Flow in Alluvial Channels with Sand Mobile Beds," *Proc. Inst. Civ. Eng.*, London, **9**; discussion, **11**, 1958.

Lane, E. W. and Carlson, E. J., "Some Factors Affecting the Stability of Canals Constructed in Coarse Granular Materials," International Association for Hydraulic Research, Minneapolis, Minnesota, September 1953.

Lane, E. W., "A Study of the Shape of Channels Formed by Natural Streams Flowing in Erodible Material," U.S. Army Engineering Division, Missouri River, Corps of Engineers, M.R.D. Sediment Series No. 9, Omaha, Nebraska, 1957.

Lane, E. W., Lin, P. N., and Liu, H. K., "The Most Efficient Stable Channel for Comparatively Clean Water in Noncohesive Material," Report CER59HKL5, Colorado State University, Fort Collins, Colorado, April 1959.

Langbein, W. B., "Geometry of River Channels," *J. Hydraul. Div. ASCE*, **90**(HY2), pp. 301–312, March 1964.

Langbein, W. B. and Leopold, L. B., "River Meanders—Theory of Minimum Variance," *USGS Professional Paper 422-H*, 1966.

Langhaar, H. L., *Energy Methods in Applied Mechanics*, John Wiley & Sons, New York, 1962.

Leopold, L. B., "Water Surface Topography in River Channels and Implications for Meander Development," in *Gravel-Bed Rivers*, R. D. Hey, J. C. Bathurst, and C. R. Thorne, eds., John Wiley & Sons, New York, pp. 359–383, 1982.

Leopold, L. B. and Maddock, T. Jr., "The Hydraulic Geometry of Stream Channels and Some Physiographic Implications," *USGS Professional Paper 252*, 1953.

Leopold, L. B. and Wolman, M. G., "River Channel Patterns; Braided, Meandering and Straight," *USGS Professional Paper 282-B*, pp. 45–62, 1957.

Leopold, L. G., Wolman, M. G., and Miller, J. P., *Fluvial Processes in Geomorphology*, W. H. Freeman, San Francisco, 1964, 522 pp.

Li, R-M., Simons, D. B., and Stevens, M. A.,"Morphology of Cobble Streams in Small Watersheds," *J. Hydraul. Div. ASCE*, **102**(HY8), pp. 1101–1117, August 1976.

Mackin, J. H., "Concept of the Graded River," *Geol. Soc. Am. Bull.*, **59**, pp. 463–512, May 1948.

Mercer, A. G., "Diversion Structures," in *River Mechanics*, **II**, H. W. Shen, ed., P.O. Box. 606, Fort Collins, Colorado, 1971.

Meyer-Peter, E., Favre, H., and Einstein, H. A., *Neuere Versuchsresultate uben den Geschiebetrieb*, SBZ, Zurich, Switzerland, 1934.

Neill, C. R., "A Reexamination of the Beginning of Movement for Coarse Granular Bed Materials," Report No. INT 68, Hydraulics Research Station, Wallingford, United Kingdom, June 1968.

Onishi, Y., Jain, S. C., and Kennedy, J. F., "Effects of Meandering in Alluvial Streams," *J. Hydraul. Div. ASCE*, **107**(HY7), pp. 899–917, July 1976.

Parker, G., "On the Cause and Characteristic Scales of Meandering and Braiding in Rivers," *J. Fluid Mech.*, **76**(3), pp. 457–480, 1976.

Parker, G., "Self-formed Rivers with Equilibrium Banks and Mobile Bed: Part II. The Gravel River," *J. Fluid Mech.*, **89**(1), pp. 127–148, 1978.

Parker, G., "Hydraulic Geometry of Active Gravel Rivers," *J. Hydraul. Div. ASCE*, ASCE, **105**(HY9), pp. 1185–1201, September 1979.

Parker, G., Diplas, P., and Akiyama, J., "Meander Bends of High Amplitude," *J. Hydraul. Eng. ASCE*, **109**(10), pp. 1323–1337, October 1983.

Peterson, A. W. and Howell, R. F., "A Compendium of Solids Transport Data for Mobile Boundary Channels," Report. No. HY-1973-ST3, Department of Civil Engineering, University of Alberta, Edmonton, Alberta, Canada, January 1973.

Ponce, V. M., "Boundary Flow Interaction in Straight Alluvial Channels," Ph.D. Thesis, Colorado State University, Fort. Collins, Colorado, 1976.

Punjab Irrigation Research Institute, "Report for the Year Ending April, 1941," Lahore, Punjab, Superintendent of Government Printing, 1943, 234 pp.

Rammette, M., "A Theoretical Approach on Fluvial Processes," Proceedings of the International Symposium on River Sedimentation, Beijing, China, April 1980.

Rozovskii, I. L., "Flow of Water in Bends of Open Channels," The Academy of Sciences of the Ukrainian SSR, 1957, translated from Russian by the Israel Program for Scientific Translations, Jerusalem, Israel, 1961 (available from Office of Technical Services, U.S. Department of Commerce, Washington, D.C., PST Catalog No. 363, OTS 60-51133).

Schumm, S. A., "The Shape of Alluvial Channels in Relation to Sediment Type," *USGS Professional Paper 352-B,* 1960.

Schumm, S. A., "Sinuosity of Alluvial Rivers on the Great Plains," *Geol. Soc. Am. Bull.* **74**, pp. 1089–1100, September 1963.

Schumm, S. A., "River Adjustments to Altered Hydrologic Regimen — Murrumbidgee River and Paleochannels, Australia," *USGS Professional Paper 598,* 1968.

Schumm, S. A., *The Fluvial System*, John Wiley & Sons, New York, 1977, 338 pp.

Schumm, S. A. and Beathard, R. M., "Geomorphic Thresholds: An Approach to River Management, " *Rivers 76*, **1**, Third Symposium of the Waterways, Harbors and Coastal Engineering Division, ASCE, 1976, pp. 707–724.

Simons, D. B. and Albertson, M. L., "Uniform Water Conveyance Channels in Alluvial Material," *J. Hydraul. Div. ASCE*, 86(HY5), pp. 33–71, May 1960.

Sukegawa, N., "On the Formation of Alternate Bars in Straight Alluvial Channels," *Trans. Jpn. Soc. Civ. Eng.,* **2**, pp. 257–261, 1970.

Sukegawa, N., "Criterion for Alternate Bar Formation," Memoirs of the School of Science and Engineering, No. 36, Waseda University, Japan, 1972, pp. 77-84.

Sukegawa, N., "Hydraulic Investigations on the Meandering Phenomena in Straight Alluvial Channels," Bureau of Resources, Department of Science and Technology, Japan, 1971 (in Japanese).

Tanner, W. F., "Helical Flow, a Possible Cause of Meandering," *Journal of Geophysical Research*, **65**, pp. 993–995, 1960.

Thakur, T. P. and Scheidegger, A. E., "A Test of the Statistical Theory of Meander Formation," *Water Resour. Res.*, **4**(9), pp. 317–329, 1968.

Thorne, C. R., "Processes and Mechanisms of River Bank Erosion," in *Gravel-Bed Rivers*, R. D. Hey, J. C. Bathurst, and C. R. Thorne, eds., John Wiley & Sons, Chichester, England, pp. 227–272, 1982.

von Schelling, H., "Most Frequent Particle Paths in a Plane," *Trans. Am. Geophys. Union*, **32**, pp. 222–226, 1951.

White, W. R., Bettess, R., and Paris, E., "Analytical Approach to River Regime," *J. Hydraul. Div.*, ASCE, **108**(HY10), pp. 1179–1193, October 1982.

Winkley, B. R., "Man-Made Cutoffs on the Lower Mississippi River: Conception, Construction, and River Response," Report 300-2, Potamology Investigations, U.S. Army Engineer District, Vicksburg, Mississippi, March 1977, 209 pp.

Yang, C. T., "On River Meanders," *J. Hydrol.*, **13**, pp. 231–253, 1971.

Yang, C. T., "Minimum Unit Stream Power and Fluvial Hydraulics," *J. Hydraul. Div.*, ASCE, **107**(HY7), pp. 919–934, July 1976.

Yang, C. T. and Song, C. C. S., "Theory of Minimum Rate of Energy Dissipation," *J. Hydraul. Div. ASCE*, **105**(HY7), pp. 769–784, July 1979.

Yang, C. T. and Song, C. C. S., "Theory of Minimum Energy and Energy Dissipation Rate," *Encyclopedia of Fluid Mechanics*, Gulf Publishing Co., Houston, pp. 353–399, 1986.

# PART IV

## MODELING OF RIVER CHANNEL CHANGES

# 13

# MATHEMATICAL MODEL FOR ERODIBLE CHANNELS

Alluvial rivers are self-regulatory in that they adjust their characteristics in response to any change in the environment. These environmental changes may occur naturally, as in the case of climatic variation or changes in vegetative cover, or may be a result of such human activities as river training, damming, diversion, sand and gravel mining, channelization, bank protection, and bridge and highway construction. Such changes distort the natural quasi-equilibrium of a river; in the process of restoring the equilibrium, the river will adjust to the new conditions by changing its slope, roughness, bed-material size, cross-sectional shape, or meandering pattern. Within the existing constraints, any one or a combination of these characteristics may adjust as the river seeks to maintain the balance between its ability to transport and the load provided.

River channel behavior often needs to be studied for its natural state and responses to the aforementioned human activities. Studies of river hydraulics, sediment transport, and river channel changes may be through physical modeling or mathematical modeling, or both. Physical modeling has been relied upon traditionally to obtain the essential design information. It nevertheless often involves large expenditure and is time consuming in model construction and experimentation. What limits the accuracy of physical modeling is the scale distortion which is almost unavoidable whenever it involves sedimentation.

Mathematical modeling of erodible channels has been advanced with the progress in the physics of fluvial processes and computer techniques. Since the actual size of a river is employed in mathematical modeling, there is no scale distortion. The applicability and accuracy of a model depend on the physical foundation and numerical techniques employed.

There are many mathematical models for water and sediment routing. Several models were evaluated by the National Academy of Sciences (1983), including HEC2SR (HEC-2 with sediment routing) by Simons, Li, and Associates, Inc. (1980), HEC-6 by the U. S. Army Corps of Engineers (1977), FLUVIAL-11 by Chang and Hill (1976) and Chang (1982), and so on. Other models have been developed by Krishnappan (1981), Bettess and White (1979), and Holley and Karim (1986), among others. A review of the existing models was made by Dawdy and Vanoni (1986).

The analyses of rivers presented in Part III pertain to regime rivers and their long-term adjustments in equilibrium. The hydraulic geometry, flow, and sediment transport processes exhibited in the process of adjustments are outside the previous scope but will be included in the present discussion. This chapter will also address the more rapid process-response or the transient behavior of alluvial rivers. The discussion is on the unsteady flow and sediment transport in river channels with a changing boundary and certain physical constraints. The physical foundation and numerical techniques for the transient process-response of the FLUVIAL model will be described and illustrated by examples. Applications of the FLUVIAL model will be presented in Chapter 14.

## 13.1 PHYSICAL FOUNDATION OF FLUVIAL PROCESS-RESPONSE

Mathematical modeling of river channel changes requires adequate and sufficient physical relationships for the fluvial processes. Although the processes are governed by the principles of continuity, flow resistance, sediment transport, and bank stability, such relations are insufficient to explain the time and spatial variations of channel geometry in an alluvial river. Generally, width adjustment occurs concurrently with changes in river bed profile, slope, channel pattern, roughness, and so on. These changes are closely interrelated; they are delicately adjusted to establish or to maintain the dynamic state of equilibrium. While any factor imposed upon the river is usually absorbed by a combination of the above responses. The extent of each type of response is inversely related to the resistance to change. For example, in response to a deficit in sediment supply, the slope of a river is generally reduced more through meandering development than through degradation because the latter is usually inhibited by the coarsening of the bed material. Also, there tends to be more adjustment in width in erodible bank materials than in erosion-resistant bank materials.

Dynamic equilibrium is the condition toward which each river channel evolves. The transient behavior of an alluvial river undergoing changes must reflect its constant adjustment toward dynamic equilibrium, although the true equilibrium may never be attained. For a short river reach of uniform discharge, the conditions for dynamic equilibrium given in Sec. 9.2 are (1) equal sediment load along the channel and (2) uniformity in power expenditure $\gamma QS$. If the energy gradient is approximated by the water-surface slope, then a uniform energy gradient is equivalent to a linear (straight-line) water-surface profile along the channel. A river channel undergoing changes usually does not have a linear water-surface profile or uniform sediment load and may have significant nonuniformities. However, river channel adjustments are such that the nonuniformities in water-surface profile and sediment load are effectively reduced. The rate of adjustment is limited by the rate of sediment movement and is subject to the rigid constraints such as grade-control structures, bank protection, abutments, bedrock, and so on.

The energy gradient usually varies significantly along an alluvial channel reach. From the results of a hydraulic computation, such as the HEC-2 output, the energy gradient exhibits nonuniformity even if it is for a fairly uniform channel. This spatial variation is much more pronounced in disturbed rivers. A mathematical modeler realizes that a river channel will change in order to attain streamwise uniformity in sediment load. It is equally important to perceive that it will also adjust toward equal energy gradient along the channel. Because sediment discharge is a direct function of $\gamma QS$, channel adjustment in the direction of equal power expenditure also favors the uniformity in sediment discharge. The sediment discharge in the reach will match the inflow rate when the equilibrium is reached.

## 13.2 CHANNEL WIDTH ADJUSTMENTS DURING SCOUR AND FILL

A stream channel's adjustment in the direction of equal power expenditure, or linear water-surface profile, provides the physical basis for the modeling of channel width changes. However, this adjustment does not necessarily mean movement toward uniformity in channel width. For one thing, the power expenditure is also affected by channel roughness and channel-bed elevation, in addition to the width. But, more importantly, the adjustment toward uniformity in power expenditure is frequently accomplished by significant streamwise variation in width. Such spatial width variation generally occurs concurrently with streambed scour or fill, to be illustrated in the following by several examples.

The transient behavior can be more clearly demonstrated by more dramatic river channel changes in the short term. Field examples of this nature are selected herein to illustrate how the significant spatial variation in width is related to river channel's adjustment toward uniform power expenditure.

Figure 13.1 shows a short reach of the San Diego River at Lakeside, California on February 25, 1981 during the initial stage of a storm. The estimated discharge of 600 cfs persisted for several subsequent days. Prior to the storm, this sandy streambed was graded, because of sand mining, to a wavy profile. During the initial period of flow, the water-surface profile was not straight because its gradient was steeper over higher streambed areas than over lower areas. Gradually, these higher streambed areas were scoured while lower places were filled. Small widths formed with streambed scour, whereas large widths developed with fill as shown in Fig. 13.1. The width development in this case was rapid in occurrence in the sandy material, its significant streamwise variation is depicted by the natural streamlines visible on the water surface. This pattern of significant spatial variation in width actually represents the stream's adjustment toward equal power gradient, as explained in the following.

The streambed area undergoing scour had a steeper energy gradient (or water-surface slope) than its adjacent areas. Formation of a narrower and deeper channel was effective to reduce the energy gradient due to decreased boundary resistance and lowered streambed elevation. In addition, the cross section developed a somewhat circular shape, which conserved power as a result of being

**Figure 13.1** Streamwise variation in width during stream channel adjustment toward establishing a straight water-surface profile, San Diego River at Lakeside, California.

closer to the best hydraulic section. On the other hand, the streambed area undergoing fill had a lower streambed elevation and a flatter energy gradient. Channel widening at this area was effective to steepen its energy gradient due to the increasing boundary resistance and rising streambed elevation. In summary, these adjustments in channel width effectively reduced the spatial variation in power expenditure or nonuniformity in water-surface profile. Because sediment discharge is a direct function of stream power $\gamma QS$, channel adjustment in the direction of equal power expenditure also favors the equilibrium, or uniformity, in sediment discharge.

The significant spatial width variation shown in Fig. 13.1 was temporary. The small width lasted while streambed scour continued, and the large width persisted with sustained fill. At a later stage when scour and fill ceased, the energy gradient or water-surface slope associated with the small width became flatter than that for the large width. The new profile of energy gradient or water surface became a reversal of the initial profile. Then, the small width started to grow wider while the large width began to slide back into the channel, resulting in a more uniform width along the channel.

The above example illustrates that a regime relationship for channel width may not be used in simulating transient river channel changes. Under the regime

**Figure 13.2** Santa Cruz River near Tucson, Arizona after the October 1983 flood. The change in channel width was greater than the concomitant scour and fill in the bed (photo by P. Kresan, GeoPhoto Pub. Co.).

relationship, the width is a function of the discharge (see Sec. 11.3); but under transient changes, the channel can have very different widths even though the discharge is essentially uniform along the channel.

The characteristic changes in channel width during channel-bed scour and fill were also observed by Andrews (1982) on the East Fork River in Wyoming. This river was in its natural state, undisturbed by human activities. River channel changes were induced by the variation in discharge.

In 1906, the Associated Press filed a well-written report that seemed to end one of the world's most spectacular stories. In the story, the AP reported that the Colorado River flooded; the water moved from the All American Canal to the New River and poured down to the Salton Sea. The sea rose 7 in. per day. The water became a cascade and its force cut back the banks. Soon the bank was receding faster and faster, moving upstream into the valley at a pace of 4000 ft a day and widening the New River channel to a gorge of more than 1000 feet. This example also illustrates the dramatic widening of river channel associated with a rising bed elevation.

A field study by the U. S. Bureau of Reclamation (1963) upstream of Milburn Diversion Dam on the Middle Loup River, central Nebraska also exemplifies the aggradation of a channel and associated channel widening.

> The construction of Milburn Diversion Dam was completed in May, 1956... By May 1957, two months after the reservoir was impounded for the first irrigation season, the channel had aggraded an average of 1.6 feet, with a rise in the channel thalweg elevation of 2.2 feet; and by October of the same year, the total rise in the streambed averaged 2.2 feet. The cross section obtained in December 1957 shows a continued rise in the thalweg elevation. During the same period, the width of the channel had increased by 70 feet, from 475 feet in 1951 to 545 feet in 1957. One-third of this increase occurred during the June, 1956–December, 1957 period.

For these two case histories, both alluvial rivers entered reservoirs with a rising base level. The transient changes are characterized by rising channel bed and increasing channel width. Although measurements of the discharge and other parameters are not available, it is possible to describe, at least in trend, the nature of power transformation in the river channel. At the river mouth, the base level was controlled in the reservoir. The rising base level first caused a lower velocity and energy gradient in the river channel near the mouth in relation to its upstream reach. In response to this change, channel adjustments through widening and aggradation near the mouth provided greater flow resistance and power expenditure at this location partly due to the increased boundary resistance. This process resulted in a more uniform power expenditure per unit channel length along the river.

A lowering base level, on the other hand, would result in a higher energy gradient in the river channel near the mouth. The higher energy gradient could be reduced through the development of a narrower and deeper channel at this location. This process would also result in a more uniform power expenditure along the channel. Such morphological features for deltas are also applicable to alluvial fans and hill slopes.

## 13.3 ANALYTICAL BASIS OF THE FLUVIAL MODEL

The FLUVIAL model (different versions) has been developed for water and sediment routing in rivers while simulating river channel changes, as documented in a series of publications by Chang and Hill (1976, 1977) and Chang (1982, 1984, 1985). River channel changes simulated by the model include channel-bed scour and fill (or aggradation and degradation), width variation, and changes caused by curvature effect. Because changes in channel width and channel-bed profile are closely interrelated, modeling of erodible channels must include both changes. In fact, width changes are usually greater than the concomitant scour and fill in the bed, particularly in ephemeral streams (see Figs. 13.1 and 13.2). The analytical background of the FLUVIAL model is described in this chapter; its applications are discussed in the next chapter.

The FLUVIAL model has the following five major components: (1) water routing, (2) sediment routing, (3) changes in channel width, (4) changes in channel-bed profile, and (5) changes in geometry due to curvature effect. These interrelated components are described in the following sections.

This model employs a space–time domain in which the space domain is represented by the discrete cross sections along the channel; the time domain is represented by discrete time increments. Temporal and spatial variations in flow, sediment transport, and channel geometry are computed following an iterative procedure. Water routing, which is coupled with the changing curvature, is assumed to be uncoupled from the sediment processes because sediment movement and changes in channel geometry are slow in comparison to the flow hydraulics. A flow chart showing the major steps of the computation is given in Fig. 13.3.

## 13.4 WATER ROUTING

Water routing provides temporal and spatial variations of the stage, discharge, energy gradient, and other hydraulic parameters in the channel. The water routing component has the following three major features: (1) numerical solution of the continuity and momentum equations for longitudinal flow, (2) evaluation of flow

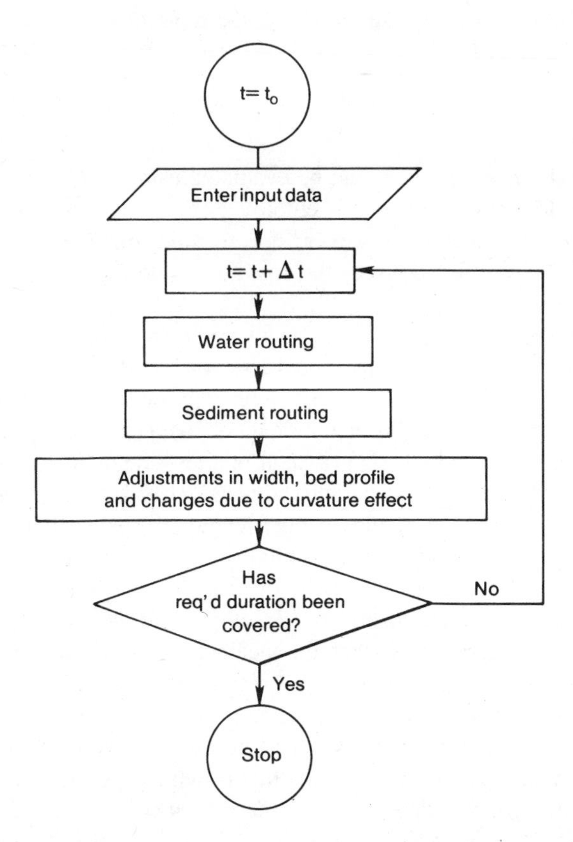


**Figure 13.3** Flow chart showing the major steps of computation for the mathematical model.

resistance due to longitudinal and transverse flows, and (3) upstream and downstream boundary conditions. The continuity and momentum equations in the longitudinal direction, derived in Sec. 3.9, are

$$\frac{\partial A}{\partial t} + \frac{\partial Q}{\partial s} - q = 0 \tag{13.1}$$

$$\frac{1}{A}\frac{\partial Q}{\partial t} + g\frac{\partial Z}{\partial s} + \frac{1}{A}\frac{\partial}{\partial s}\left(\frac{Q^2}{A}\right) + gS - \frac{Q}{A^2}q = 0 \tag{13.2}$$

where $Q$ is the discharge, $A$ is the cross-sectional area of flow, $t$ is the time, $s$ is the curvilinear coordinate along discharge centerline measured from the upstream entrance, $q$ is the lateral inflow rate per unit length, $Z$ is the stage or water-surface elevation, and $S$ is the energy gradient. The upstream boundary condition for water routing is the inflow hydrographs; the downstream condition is the stage–discharge relation or the base-level variation. Techniques for numerical solution of Eqs. 13.1 and 13.2 are described in Sec. 3.9.

In a curved channel, the total energy gradient $S$ in Eq. 13.2 is partitioned into the longitudinal energy gradient $S'$ and the transverse energy gradient $S''$ due to secondary currents, that is,

$$S = S' + S'' \tag{13.3}$$

The longitudinal energy gradient can be evaluated using any valid flow-resistance relationship. If Manning's formula is employed, the roughness coefficient $n$ must be selected by the modeler. However, if a formula for alluvial bed roughness (e.g., Brownlie's formula) is used, the roughness coefficient is predicted by the formula.

The transverse energy gradient is evaluated using Eq. 8.74. Because of the streamwise changing curvature, the transverse energy loss varies with the growth and decay of secondary currents. Analytical relationships pertaining to curved channels are often based on the mean channel radius $r_c$. Under the streamwise changing curvature, the application of such relationships is limited to fully developed transverse flow for which the curvature is defined. Streamwise variation of transverse flow, over much of the channel length, is characterized by its growth and decay. In order to describe this spatial variation, the mean flow curvature, defined as the flow curvature along the discharge centerline, is employed. It is assumed that analytical relationships for developed transverse flows are applicable for developing transverse flows when the mean channel curvature $r_c$ is replaced by the mean flow curvature $r_f$. Upon entering a bend, the mean flow curvature increases with the growth in transverse circulation. In a bend, the transverse flow becomes fully developed if the flow curvature approaches the channel curvature. In the case of exiting from a bend to the downstream tangent, in which the channel curvature is zero, the flow curvature decreases with the decay of transverse circulation.

The reason that the flow curvature lags behind the channel curvature during circulation growth and decay is attributed to the internal turbulent shear that the flow has to overcome in transforming from parallel flow into the spiral pattern and vice versa. From the dynamic equation for the transverse velocity (Eq. 8.79), an equation governing the streamwise variation in transverse surface velocity $\tilde{v}$ was derived as detailed in See. 8.7. The change in $\tilde{v}$ over the distance $\Delta s$ given by Eq. 8.87 is written in finite-difference form as

$$\tilde{v}_{i+1} = \left[\tilde{v}_i + \overline{F}_1 \frac{\overline{U}}{\overline{r}_c} \exp[\overline{F}_2 \Delta s] \Delta s\right] \exp[-\overline{F}_2 \Delta s] \tag{13.4}$$

where $\tilde{v}$ is the transverse surface velocity along discharge centerline, $U$ is the average velocity of a cross section, $i$ and $i + 1$ are $s$-coordinate indices counted from upstream toward downstream, $F_1$ and $F_2$ are functions of $D$, $f$, and $\kappa$ defined in Eq. 8.88, and the overbar denotes averaging over the incremental distance between $i$ and $i + 1$. Equation 13.4 provides the spatial variation in $\tilde{v}$, from which the mean flow curvature may be obtained using the transverse velocity profile. Similar transverse velocity profiles are given in Sec. 8.2. From the velocity profile given by Eq. 8.22, the mean flow curvature $r_f$ is related to the transverse surface velocity as follows:

$$r_f = \frac{D}{\kappa} \frac{U}{\tilde{v}} \left[\frac{10}{3} - \frac{1}{\kappa} \frac{5}{9} \left(\frac{f}{2}\right)^{1/2}\right] \tag{13.5}$$

At each time step, the mean flow curvature at each cross section is obtained using Eqs. 13.4 and 13.5. Accuracy of computation for the finite-difference equation (Eq. 13.4) is maintained if the step size $\Delta s < 2D$. For this reason, the distance between two adjacent cross sections is divided into smaller increments if necessary. Flow parameters for these increments are interpolated from values known at adjacent cross sections.

If the temporal terms in Eqs. 13.1 and 13.2 are ignored, water routing may be simplified by computing water-surface profiles at successive time steps. This option is available in the model. Computation of the water-surface profile at each time step is based on the standard-step method (see Sec. 3.8) using techniques similar to the HEC-2 computer model (U.S. Army Corps of Engineer, 1982). For many cases, spatial variation in discharge due to channel storage is small and this technique produces results that are very similar to those of the unsteady routing.

## 13.5 SEDIMENT ROUTING

The sediment routing component for the FLUVIAL model has the following major features: (1) computation of sediment transport capacity using a suitable formula for the physical conditions, (2) determination of actual sediment discharge by making corrections for availability, sorting, and diffusion, (3) upstream conditions for sediment inflow, and (4) numerical solution of the continuity equation for sediment. These features are evaluated at each time step, and the results so obtained are used in determining the changes in channel configuration.

### Determination of Sediment Discharge

To treat the time-dependent and nonequilibrium sediment transport, the bed material at each section is divided into several, say five, size fractions; the size for each fraction is represented by its geometric mean. For each size fraction, sediment transport capacity is first computed using a sediment-transport formula. At present, five options for sediment-transport equations, reflecting different physical conditions, are included in the model. The actual sediment rate is then obtained by considering sediment material of all size fractions already in the flow, as well as the exchange of sediment load with the bed using the method by Borah et al. (1982) described in Sec. 7.9. If the stream carries a load in excess of its capacity, it will deposit the excess material on the bed. In the case of erosion, any size fraction available for entrainment at the bed surface will be removed by the flow and added to the sediment already in transport. During sediment removal, the exchange between the flow and the bed is assumed to take place in the active layer at the surface. Thickness of the active layer is based on the relation defined by Borah et al. given by Eq. 7.109. This thickness is a function of the material size and composition, but also reflects the flow condition. During degradation, several of these layers may be scoured away, resulting in the coarsening of the bed material and formation of an armor coat. However, new active layers may be deposited on the bed in the process of aggradation. Materials eroded from the channel banks, excluding that portion in the wash-load size range, are included in the accounting. Bed armoring develops if bed shear stress is too low to transport any available size.

The nonequilibrium sediment transport is also affected by diffusion, particularly for finer sediments. Because of diffusion, the deposition or entrainment of sediment is a gradual process and it takes certain travel time or distance to reach the transport capacity for a flow condition. Therefore, the actual sediment discharge at a section depends not only on the transport capacity at the section but also on the supply from upstream and its gradual adjustment toward the flow condition of this section. In the model, the sediment discharge is corrected for the diffusion effects on deposition and entrainment using the method by Zhang et al. (1983) described in Sec. 7.8. The procedures for computing sediment transport rate, sediment sorting, and diffusion are applied to the longitudinal and transverse directions. They are also coupled with bed-profile evolution, which is described later in this section.

Sediment discharge may be limited by availability, as exemplified by the flow over a grade-control structure or bed rock. The very high transport capacity at such a section, associated with the high velocity, is limited by the supply rate from upstream; that is, the sediment discharge at such a section is under upstream control.

### Upstream Boundary Conditions for Sediment Inflow

The rate of sediment inflow into the study reach is provided by the upstream boundary condition for sediment. If this rate is known, it may be included as a part of the input and used in the simulation. Unfortunately, sediment rating data are rarely very reliable or simply not available. For such cases, it is assumed that the river channel remains unchanged above the study reach, and sediment inflow rate is computed at the upstream section at each time step, the same way they are computed at other cross sections. For this reason, the study reach should extend far enough upstream so that the channel beyond may be considered basically stable. Factors that may induce river channel changes must be included in the study reach.

### Numerical Solution of Continuity Equation for Sediment

Changes in cross-sectional area, due to longitudinal and transverse imbalances in sediment discharge, are obtained based on numerical solution of continuity equations for sediment in the respective directions. First, the continuity equation for sediment in the longitudinal direction (Eq. 7.94) is

$$(1 - \lambda)\frac{\partial A_b}{\partial t} + \frac{\partial Q_s}{\partial s} - q_s = 0 \tag{13.6}$$

where $\lambda$ is the porosity of bed material, $A_b$ is the cross-sectional area of channel within some arbitrary frame, $Q_s$ is the bed-material discharge, and $q_s$ is the lateral inflow rate of sediment per unit length. According to this equation, the time change of cross-sectional area $\partial A_b/\partial t$ is related to the longitudinal gradient in sediment discharge $\partial Q_s/\partial s$ and lateral sediment inflow $q_s$. In the absence of $q_s$, longitudinal imbalance in $Q_s$ is absorbed by channel adjustments toward establishing uniformity in $Q_s$.

The change in cross-sectional area $\Delta A_b$ for each section at each time step is obtained through numerical solution of Eq. 13.6. This area change will be applied to the bed and banks following correction techniques for channel width and channel-bed profile.

From Eq. 13.6, the correction in cross-sectional area of channel for a time increment can be written as

$$\Delta A_b = -\frac{\Delta t}{1 - \lambda}\left(\frac{\partial Q_s}{\partial s} - q_s\right) \tag{13.7}$$

At a section $i$, the lateral sediment inflow may be written as

$$q_{s_i} = \tfrac{1}{2}(q_{s_i}^j + q_{s_i}^{j+1}) \tag{13.8}$$

where superscripts $j$ and $j + 1$ are the times at $t$ and $t + \Delta t$, respectively. Chang and Hill (1976) suggested an upstream difference in $s$ and a centered difference in $t$ for the partial derivative $\partial Q_s/\partial s$ in Eq. 12.6, that is,

$$\left(\frac{\partial Q_s}{\partial s}\right)_i = \frac{2}{\Delta s_i + \Delta s_{i-1}}\left(\frac{Q^j_{s_i} + Q^{j+1}_{s_i}}{2} - \frac{Q^j_{s_{i-1}} + Q^{j+1}_{s_{i-1}}}{2}\right) \tag{13.9}$$

where $i$ is counted from upstream to downstream, $\Delta s_i$ is the distance between sections $i$ and $i + 1$, and $\Delta s_{i-1}$ is the distance between $i - 1$ and $i$. With this upstream difference for $\partial Q_s/\partial s$, the change in cross-sectional area at a section $i$ depends on sediment rates at this section and its upstream section $i - 1$; it is independent of the sediment rate at the downstream section. In other words, it is under upstream control. Contrary to this, the stage at a section in subcritical flow depends on the downstream stage and is independent of the upstream stage; that is, the stage is under downstream control in a subcritical flow.

## 13.6 SIMULATION OF CHANGES IN CHANNEL WIDTH

The change in cross-sectional area $\Delta A_b$ obtained in sediment routing represents the correction for a time increment $\Delta t$ that needs to be applied to the bed and banks. With $\Delta A_b$ being the total correction, it is possible for both the bed and banks to have deposition or erosion; it is also possible to have deposition along the banks but erosion in the bed and vice versa. The direction of width adjustment is determined following the stream power approach, and the rate of change is based on bank erodibility and sediment transport described as follows.

### Direction of Width Adjustment

For a time step, width corrections at all cross sections are such that the spatial distribution of stream power along the reach moves toward uniformity; these corrections are subject to the physical constraint of rigid banks and limited by the amount of sediment removal or deposition along the banks within the time step. A river channel undergoing changes usually has nonuniform spatial distribution in power expenditure $\gamma QS$. Usually the spatial variation in $Q$ is small, but that in $S$ is pronounced. An adjustment in width reflects the river's adjustment in flow resistance, that is, in power expenditure. A reduction in width at a cross section is usually associated with a decrease in energy gradient for the section, whereas an increase in width is accompanied by an increase in energy gradient. To determine the direction of width change at a section $i$, the energy gradient at this section, $S_i$, is compared with the weighted average of its adjacent sections, $\overline{S}_i$. Here

$$\overline{S}_i = \frac{S_{i+1}\,\Delta s_{i-1} + S_{i-1}\,\Delta s_i}{\Delta s_i + \Delta s_{i-1}} \tag{13.10}$$

If the energy gradient $S_i$ is greater than $\overline{S}_i$, channel width at this section is reduced so as to decrease the energy gradient. On the other hand, if $S_i$ is lower, channel width is increased in order to raise the energy gradient. These changes are subject to the rate of width adjustment and physical constraints.

Width changes in alluvial rivers are characterized by widening during channel-bed aggradation (or fill) and reduction in width at the time of degradation (or scour). Such river channel changes represent the river's adjustments in resistance to seek equal power expenditure along its course. A degrading reach usually has a higher channel-bed elevation and energy gradient than do its adjacent sections. Formation of a narrower and deeper channel at the degrading reach decreases its energy gradient as a result of reduced boundary resistance. On the other hand, an aggrading reach is usually lower in channel-bed elevation and energy gradient. Widening at the aggrading reach increases its energy gradient as a result of increasing boundary resistance. These adjustments in channel width reduce the spatial variation in energy gradient and total power expenditure of the channel.

## Rate of Width Adjustment

For a time increment, the amount of width change depends on the sediment rate, bank configuration, and bank erodibility. The slope of an erodible bank is limited by the angle of repose of the material. The rate of width change depends on the rate at which sediment material is removed or deposited along the banks. For the same sediment rate, width adjustment at a tall bank is not as rapid as that at a low bank. The rates of width adjustment for cases of width increase and decrease are somewhat different, as described in the following paragraphs.

An increase in width at a channel section depends on sediment removal along the banks. The maximum rate of widening occurs when sediment inflow from the upstream section does not reach the banks of this section while bank material at this section is being removed. River banks have different degrees of resistance to erosion; therefore, the rate of sediment removal along a bank needs to be modified by a coefficient. For this purpose, the bank erodibility factor is introduced as an index for the erosion of bank material, and the four bank types reflecting the variation in erodibility are classified as follows:

1. Nonerodible banks.
2. Erosion-resistant banks, characterized by highly cohesive material or substantial vegetation, or both.
3. Moderately erodible banks having medium bank cohesion.
4. Easily erodible banks with noncohesive material.

Values of the bank erodibility factor vary from 0 for the first type to 1 for the last type of banks. The values of 0.2 and 0.5 have been empirically determined for the second and third types, respectively, based on test and calibration of the model using field data from rivers in the western United States. However, the bank erodibility factor should still be calibrated whenever data on width changes are available.

A decrease in channel width is accomplished by sediment deposition along the banks or by a decrease in stage, or both. For practical reasons, deposition does not exceed the stage in the model. The maximum amount of width reduction at a section occurs when sediment inflow from the upstream section is spread out at this section and the sediment removal from the bank areas at this section is zero.

Within the limit of width adjustment, changes in width are made at all cross sections in the study reach toward establishing uniformity in power expenditure.

## 13.7 SIMULATION OF CHANGES IN CHANNEL-BED PROFILE

After the banks are adjusted, the remaining correction for $\Delta A_b$ is applied to the bed. Distributions of erosion and deposition, or scour and fill, at a cross section are usually not uniform. Generally speaking, deposition tends to start from the low point and is more uniformly distributed because it tends to build up the channel bed in nearly horizontal layers. This process of deposition is often accompanied by channel widening. On the other hand, channel-bed erosion tends to be more confined with greater erosion in the thalweg. This process is usually associated with a reduction in width as the banks slip back into the channel. Such characteristic channel adjustments are effective in reducing the streamwise variation in stream power as the river seeks to establish a new equilibrium. In the model, the allocation of scour and fill across a section for a time step is assumed to be a power function of the effective tractive force $\tau_0 - \tau_c$, that is,

$$\Delta z = \frac{(\tau_0 - \tau_c)^m}{\sum_B (\tau_0 - \tau_c)^m \Delta y} \Delta A_b \tag{13.11}$$

where $\Delta z$ is the local correction in channel-bed elevation, $\tau_0$ (given by $\gamma DS$) is the local tractive force, $\tau_c$ is the critical tractive force, $m$ is an exponent, $y$ is the horizontal coordinate, and $B$ is the channel width. The value of $\tau_c$ is zero in the case of fill.

The $m$ value in Eq. 13.11 is generally between 0 and 1; it affects the pattern of scour–fill allocation. For the schematic cross section shown in Fig. 13.4, a small value of $m$, say 0.1, would mean a fairly uniform distribution of $\Delta z$ across the section; a larger value, say 1, will give a less uniform distribution of $\Delta z$ and the local change will vary with the local tractive force or will vary roughly with the depth. The value of $m$ is determined at each time step such that the correction in channel-bed profile will result in the most rapid movement toward uniformity in power expenditure, or linear water-surface profile, along the channel.

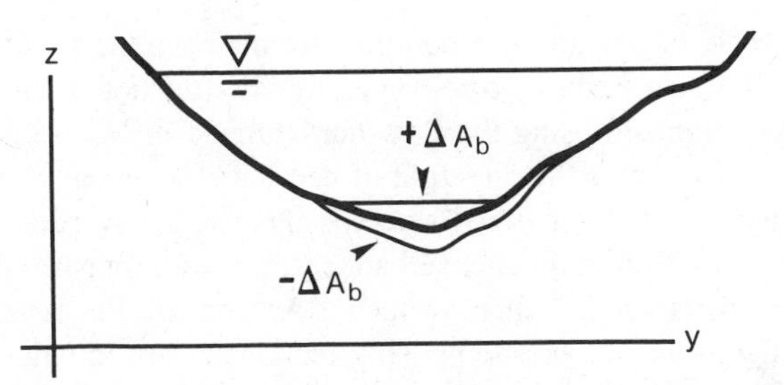


**Figure 13.4** Corrections of bed profile for aggradation and degradation. They are made in such ways that water-surface profile or power expenditure moves toward uniformity.

Equation 13.11 may only be used in the absence of channel curvature. The change in bed area at a cross section in a curved reach is

$$\Delta A_b = \frac{1}{r_f} \int r dz\, dr \tag{13.12}$$

where $r_f$ is the radius of curvature at the discharge centerline or thalweg. Because of the curvature, adjacent cross sections are not parallel and the spacing $\Delta s$ between them varies across the width. Therefore, the distribution of $\Delta z$ given in Eq. 13.11 needs to be weighted according to the $r$-coordinate with respect to the thalweg radius $r_f/r$, that is,

$$\Delta z = \frac{(\tau_0 - \tau_c)^m / r}{\sum_B (\tau_0 - \tau_c)^m / r \Delta r} \Delta A_b \tag{13.13}$$

## 13.8 SIMULATION OF CHANGES DUE TO CURVATURE EFFECT

Simulation of curvature-induced scour and deposition is based on the flow curvature for which the streamwise variation is given by Eqs. 13.4 and 13.5. The major features of transverse sediment transport and changes in bed topography are described as follows.

Sediment transport, in the presence of transverse flow, has a component in that direction. Sediment movement in the transverse direction contributes to the adjustment of transverse bed profile. In an unsteady flow, the transverse bed profile varies with time and is constantly adjusted toward equilibrium through scour and deposition. The transverse bed load per unit channel length $q_b{}'$ can be related to the streamwise transport $q_b$. Such a relationship is given by Eq. 8.64, which can be written in parametric form as follows:

$$\frac{q_b'}{q_b} = F\left(\tan \delta, \frac{\partial z}{\partial r}\right) \tag{13.14}$$

where $\delta$ is the angle of deviation of bottom currents from the streamwise direction (see Fig. 8.6). The near-bed transverse velocity is a function of the curvature (see Sec. 8.2) and is computed using the flow curvature.

Equation 13.14 relates the direction of bed-load movement to the direction of near-bed velocity and transverse bed slope $\partial z/\partial r$. As transverse velocity starts to move sediment away from the concave bank, it creates a transverse bed slope that counters the transverse sediment movement. An equilibrium is reached, that is, $q_b' = 0$, when the effects of these opposing tendencies are in balance. Transverse bed-profile evolution is related to the variation in bed-material load. Since bed-material load is usually concentrated near the bed, it is assumed to follow the direction given by Eq. 13.14. Ikeda and Nishmura (1986) developed a method for estimating transport and diffusion of fine sediments in the transverse direction by vertical integration of suspended load over the depth. Their model for predicting the transverse bed slope is also substantiated with experimental data.

Changes in channel-bed elevation at a point due to transverse sediment movement are computed using the transverse continuity equation for sediment:

$$\frac{\partial z}{\partial t} + \frac{1}{1-\lambda}\frac{1}{r}\frac{\partial}{\partial r}(rq_s') = 0 \tag{13.15}$$

Written in finite difference form with a forward difference for $q_s'$, this equation becomes

$$\Delta z_k = \frac{\Delta t}{1-\lambda}\frac{2}{r_k}\frac{r_{k+1}q'_{s_{k+1}} - r_k q'_{s_k}}{r_{k+1} - r_{k-1}} \tag{13.16}$$

where $k$ is the radial (transverse) coordinate index measured from the center of radius. Equation 13.16 provides the changes in channel-bed elevation for a time step due to transverse sediment movement. These transverse changes, as well as the longitudinal changes, are applied to the stream bed at each time step. Bed-profile evolution is simulated by repeated iteration along successive time steps.

## 13.9 TEST AND CALIBRATION OF MATHEMATICAL MODEL

The accuracy of a mathematical model depends on the physical foundation, numerical techniques, and physical relations for momentum, flow resistance, and sediment transport. Test and calibration are important steps to be taken for the more effective use of a model. Because of the difference in sensitivity of simulated results to each relation or empirical coefficient, more attention needs to be paid to those that generate sensitive results. Major items that require calibration include the roughness coefficient, sediment transport equation, bank erodibility factor, bed erodibility factor, and so on.

To determine the sensitivity of flow, the sediment transport, and the channel changes caused by the variation of each variable, different values of the variable need to be used in simulation runs, and the results so obtained are compared. Generally speaking, the rate of channel changes is more sensitive to the sediment rate computed from a sediment equation, but the equilibrium channel configuration is less sensitive. For example, the constriction scour at a bridge crossing, or the equilibrium local scour at a bridge pier, is found to be more or less independent of the sediment equation, or sediment size, since both inflow and outflow rates of sediment are affected by about the same proportion. It may also be stated that the rate of widening is sensitive to the bank erodibility factor but that the equilibrium width is not nearly as sensitive.

Field data are generally required for test and calibration of a model. Generally, the channel configuration before and after the changes, a flow record, and a list of sediment characteristics are required. Data sets with more complete information are also more useful. Several data sets that are useful for test and calibration are described in Chapter 14.

# 14

# COMPUTER-AIDED STUDY OF ALLUVIAL RIVERS

Because the transient river behavior is governed by so many convoluted physical relationships, river studies are facilitated with the aid of an appropriate mathematical model. Types of problems related to river channel changes are presented as computer-aided studies in this chapter. The primary objective of this approach is to elucidate the physical nature of a problem that may not otherwise be quantified based on traditional methods. The computer-aided approach is illustrated by case studies for the following types of problems: (1) general scour at bridge crossings, (2) simulation of gradual breach morphology, (3) stream channel changes induced by sand and gravel mining, (4) tidal responses of river and delta system, (5) water and sediment routing through a curved channel, (6) fluvial design of river bank protection, and (7) stream gaging of fluvial sediment.

These case studies will also exemplify (1) the types of problems that require an engineering solution and (2) the respective nature of their fluvial process-response. For some of these studies, measurements are available for comparison with model simulations.

## 14.1 GENERAL SCOUR AT BRIDGE CROSSINGS

River bed scour at bridge crossings may be considered as consisting of *local scour* and *general scour*. Local scour that occurs around bridge piers and abutments is caused by the local obstruction to flow. Formulas for predicting local scour are given in Secs. 5.9 and 5.10. General scour refers to the change in river channel configuration provoked by sediment imbalance, due to natural or human-made causes, between the supply and transport capacity of the river channel. The bridge structure is one such human-made cause if it interferes with the flow pattern. General scour at bridge crossings is related to the flow and sediment-transport processes of the adjacent river as a system; therefore, evaluation of such scour requires modeling of the river channel for water and sediment routing.

Bridges are often constructed with the span shorter than the channel width, particularly across broad flood plains in semiarid regions. Therefore, river flow is often constricted at the bridge crossing, resulting in higher velocities and channel-bed scour. Flow constriction at the bridge crossing varies with the discharge or stage. During high flow, the river channel is wide and thus the constriction effect is more pronounced; therefore, greater general scour at the bridge crossing can usually be expected. During low flow, the width of the stream flow may be less than the bridge opening and thus flow constriction no longer exists. Therefore, general scour caused by the constriction of high flow may be refilled to the pre-flood level during the subsequent low flow period. Since it is generally difficult to observe the river bed through the muddy flood water, such major scours may occur but may not be noticed. However, for the sake of bridge safety, such scour development should be evaluated in bridge design and restoration.

General scour develops when more sediment is removed from an area than is supplied from upstream. In addition to the constriction effect described above, general scour is also induced by the deficit in sediment supply and channel-curvature effect. In fact, the scour development may be related to a combination of such factors. Figure 14.1 shows the general scour at a bridge crossing in an ephemeral stream induced primarily by sand mining in the adjacent stream channel. The engineer in the picture points at the original stream-bed elevation prior to the flood event.

General scour is accompanied by other changes, including that in bed-material composition due to hydraulic sorting. These changes provide the mechanisms with which the river seeks to establish the dynamic equilibrium in sediment transport. Net scour at a bridge crossing ceases when sediment equilibrium, that is, uniformity in sediment discharge, is established or when bed armoring forms to prevent further scour.

Simulation of general scour at a bridge crossing using the FLUVIAL model is illustrated by two case studies described in the following. In the case of constriction scour, the rate of scour development is sensitive to the sediment-transport equation used. However, the quasi-equilibrium scour depth, which occurs when the sediment discharge becomes uniform along the channel, is not sensitive to the sediment equation or to the sediment size. Along the same vein, the local scour depth given by empirical equations in Sec. 5.9 is independent of the sediment size.

### Santa Margarita River Study

General scour at a bridge crossing caused primarily by the effect of flow constriction is illustrated by the Santa Margarita River study. The study reach is near the Basilone Road Bridge (see Fig. 14.2) at Camp Pendleton, California. The width of flow in this ephemeral river varies significantly with the discharge. It is about 1000 ft wide during the 100-yr flood, but the low flow channel is less than 100 ft in width. The Basilone Road Bridge has long approach embankments and a span of 200 ft. Except at very high flood stages, when flood waters overtop the approach embankments, flood flows are confined to the small bridge opening in the broad flood plain.

**Figure 14.1** General scour at a bridge crossing, induced by sand mining.

**Figure 14.2** Basilone Road Bridge on Santa Margarita River after a flood.

A major flood with a magnitude approximately that of a 50-yr flood occurred in the winter of 1978. The picture in Fig. 14.2 was taken soon after this flood; it shows no sign of significant channel-bed scour at the bridge crossing. However, there is strong evidence of severe scour at this location during the flood. For example, very high velocities of muddy flow through the constriction were observed. Bridge abutments were undermined; eroded bridge abutments were repaired and reinforced after the flood. During inspection excavation of the bridge footings, which reached about 10 ft below the river bed, a broken reinforced concrete pile was found; its 40-ft lower section was never found. This could only be explained by the fact that this pile section was washed away during the flood.

A simulation study of the Santa Margarita River near the bridge crossing was made for the 50-yr flood shown in Fig. 14.3. Simulated results given in Fig. 14.4 consist of the profiles of the water surface, channel bed, and mean velocity at different time intervals. At the peak discharge ($t$ = 30 hr), maximum channel-bed scour at the bridge crossing is predicted to reach a minimum elevation of 57.3 ft, which means a scour depth of 15.4 ft from the original bed level. At this discharge, the adjacent flood plain has a width that is about five times that of the bridge opening. However, the width of flood flow decreases with discharge during the falling limb of the hydrograph. As the flow-constriction effect becomes gradually less, so does the channel-bed scour at the bridge crossing. Restoration of the channel bed, more or less to its preflood level, at the end of the flood is predicted.

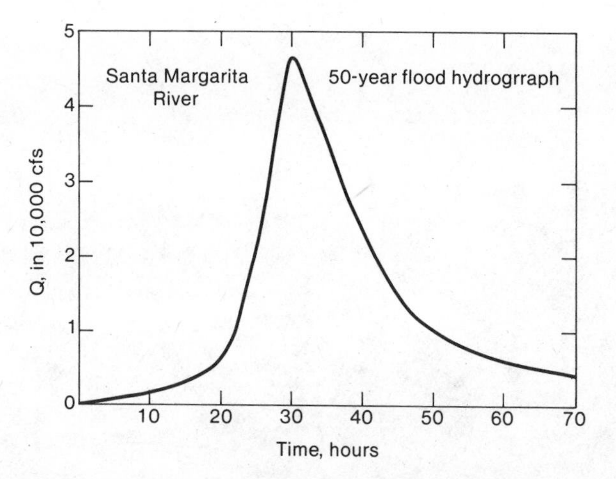


**Figure 14.3** The 50-yr flood hydrograph for Santa Margarita River.

At the peak flood, the river reach has an uneven width, primarily because of the small bridge opening. Although the river flow has established a more or less uniform sediment discharge along the reach through the constriction, the energy gradient is not constant, as indicated by the uneven water-surface profile near the bridge shown in Fig. 14.4. The physical constraint in width at the bridge crossing prevents the formation of a linear water-surface profile. This example demon-

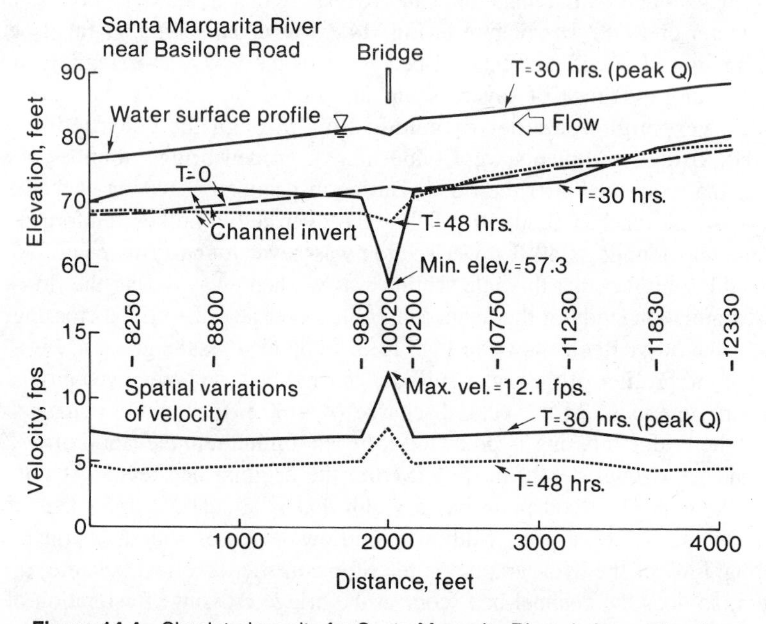


**Figure 14.4** Simulated results for Santa Margarita River during a 50-yr flood.

strates that continuity in sediment transport (uniform sediment discharge) does not necessarily mean equal energy gradient or constancy in power expenditure along the channel.

## San Dieguito River Study

The San Dieguito River at Rancho Santa Fe, California went through significant changes in a 2-mile reach (see Fig. 14.5) during the 1978 and 1980 floods. The bridge on Via de Santa Fe Road (see Fig. 14.6) was damaged as a consequence of channel-bed scour and high velocities. Documentation of river channel changes and flood hydrographs were made by the County of San Diego (1979, 1980), providing a valuable set of field data for river studies. The study reach has a wide and flat natural configuration; it is about 4 miles from the ocean and about 5 miles below Lake Hodges Dam. Along this reach, the slope and bed-material size decrease significantly in the downstream direction; the bed material varies from coarse sand ($d_{50}$ = 0.85 mm) at the upstream end to fine sand ($d_{50}$ = 0.24 mm) downstream.

The natural channel configuration was altered before these flood events by human activities, including sand mining and construction of the Via de Santa Fe Road and Bridge (section 51). As a result of sand mining, several large borrow pits, both upstream and downstream of the bridge, with a depth as large as 25 ft were created. The natural wide channel was encroached on by the approach embankment on each side of the bridge. The river channel had erodible bed and banks; the banks, however, were constrained by the hills at the south bank of section 51 and along the north banks of sections 60–63 and by bank protection at the north banks of sections 51 and 58.

Two floods passed through the river in March 1978 (peak flow = 4400 cfs) and in February 1980 (peak flow = 22,000 cfs) when Lake Hodges spilled. Hydrographs of these floods are shown in Fig. 14.7. Before these events, Lake Hodges had not spilled for 26 yrs. Significant changes in the river channel were observed after the March 1978 flood. Channel-bed scour occurred near borrow pits and notably at the bridge crossing where measurements were made (see Figs. 14.8 and 14.9). Deposition was observed in the borrow pits. Because of limited flood discharge and duration, these borrow pits were only partly refilled. Major changes in the river channel occurred during the greater February 1980 flood. These changes included channel-bed scour and fill, width variation, and lateral migration of the channel.

The mathematical model FLUVIAL-12 was used to simulate river channel changes in the San Dieguito River during the 1978 and 1980 floods (Chang, 1984). Graf's equation for bed-material load was used in computing the sediment discharge. Channel roughness in terms of Manning's $n$ was selected to be 0.035 in consideration of channel irregularity and scattered vegatation; it was estimated to be 0.04 at the bridge crossing. The combined duration of 140 hr for these two floods was computed using 2000 time steps.

Simulated results are shown in Figs. 14.8–14.10. Changes in the longitudinal channel-bed profile shown in Fig. 14.8 are characterized by fill in the borrow

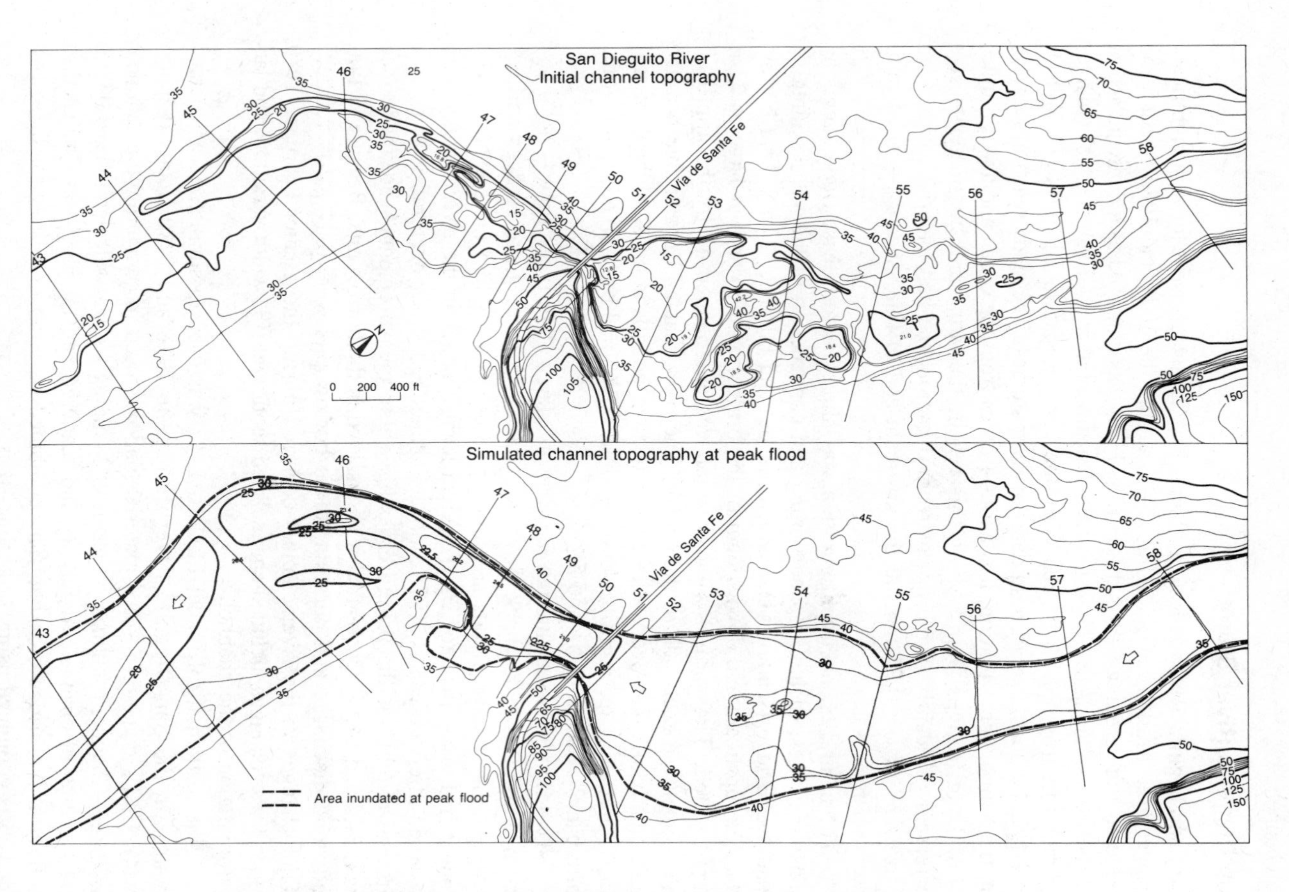


**Figure 14.5** Topographies and cross-section locations.

**Figure 14.6** Aerial view looking south of damaged bridge on San Dieguito River on Feb. 21, 1980.

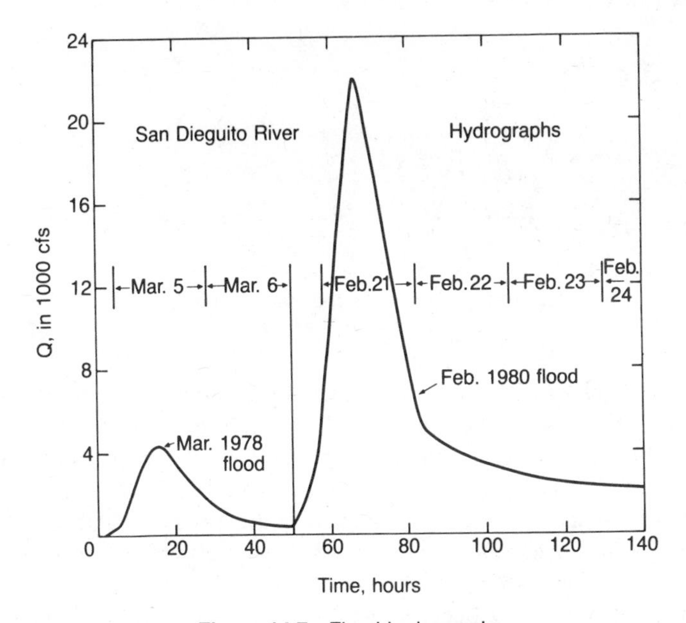


**Figure 14.7** Flood hydrographs.

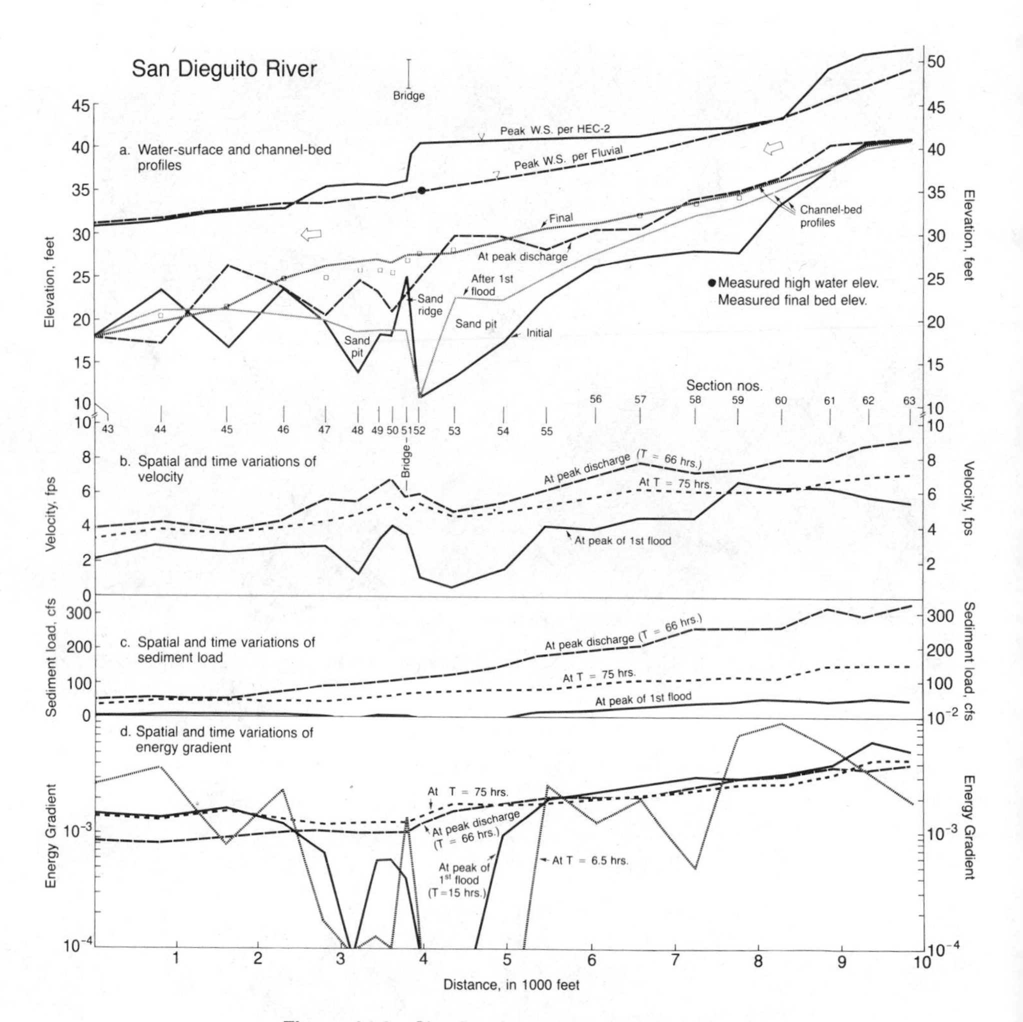


**Figure 14.8** Simulated and measured profiles.

pits, scour at higher grounds, and the gradual formation of a basically smooth channel-bed profile at the end of the flood. In that process, considerable longitudinal variation in channel-bed elevation through the downstream portion of the river reach is predicted at peak flood, as shown in Figs. 14.5 and 14.8. The higher channel-bed elevations at sections 45, 46, and 48 are associated with large channel widths, and the lower elevations at sections 47 and 50 are due to their smaller widths.

Changes in width that occur concurrently with variations in channel-bed elevation are simulated. Width changes are characterized by the gradual widening of the initially narrow sections, notably sections 47, 49–51, and 57–59 and reductions in width of initially wide sections, notably sections 53 and 54. Simulated

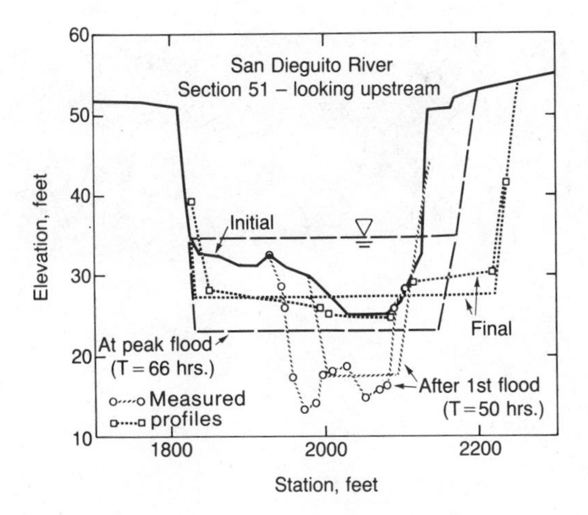


**Figure 14.9** Simulated and measured cross-sectional profiles at a bridge crossing.

channel width at peak flood shown in Fig. 14.5 is highly uneven in its spatial variation along the river, but this variation is gradually reduced during the flood. Widening of a section is by bank erosion; reduction in width is usually through sand-bar formation along the bank, as illustrated in Figs. 14.10 and 14.11.

The interrelated changes in channel width and channel-bed elevation are illustrated by the simulated time variation of the cross-sectional profile at the bridge crossing (section 51) given in Fig. 14.9. This section is initially on a sand ridge with borrow pits on both sides (see Figs. 14.5 and 14.8). Gully erosion through this sand ridge during the first flood is simulated, followed by gradual widening and lessening of the gully depth during the second flood. The maximum scour depth is predicted to occur in the initial narrow gully. While the scour depth

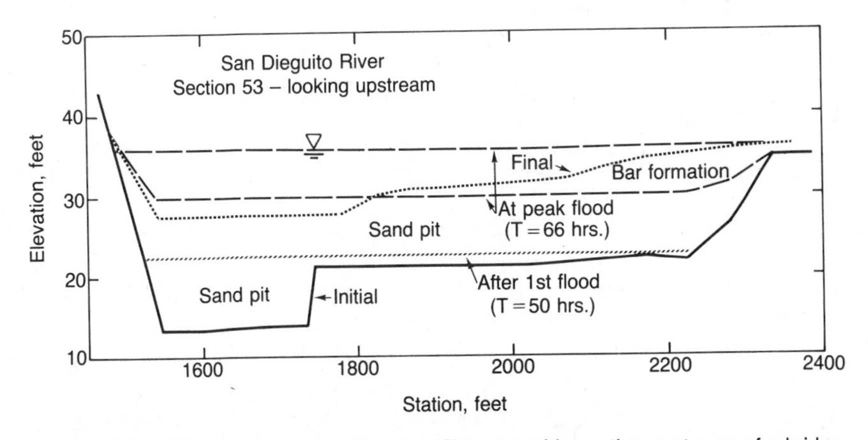


**Figure 14.10** Simulated cross-sectional profiles at a wide section upstream of a bridge.

**Figure 14.11** View looking south of a sand-bar formation near section 53.

is measured by the change in channel-bed elevation, its development is also closely related to the variation in channel width. These interrelated changes in channel width and bed elevation reflect dynamic adjustments toward uniformities in sediment transport and power expenditure. At this bridge crossing, the maximum change of about 20 ft in bed elevation is small in comparison to the change of about 80 ft in width. This illustrates, for ephemeral rivers, that the width variation during a major event is generally greater in magnitude than the associated bed scour.

## 14.2 GRADUAL BREACH MORPHOLOGY

There are approximately 50,000 dams in the United States; the majority of them are earth embankments. Earth dam failures under the erosive action of water are typically gradual. The breach of a small earth dam is shown in Fig. 14.12. In simulating the gradual breach of an earth dam or levee, the breach morphology and its transient changes are perhaps the least understood. The breach morphology that governs the outflow discharge from a reservoir is also caused by the flow. The convoluted relationship between breach morphology and flow condition lends itself to computer simulation. The accuracy of any dam breach analysis hinges

**Figure 14.12** Breach of a small earth dam.

heavily upon the prediction of the breach morphology and its transient responses to flow.

## Transient Changes in a Breach

The transient breach morphology observed in the San Diego River during the 1878 floods was documented. The natural channel configuration was altered, prior to the flood, primarily by sand mining, as shown in Fig. 14.13. Two large borrow pits created on each side of the bridge left a sand ridge between them. The downstream borrow pit, with an average width of 350 ft at the bottom, was between section 72 and 79. The upstream pit, shaped like an inverted cone, was between section 80 and 82.

A small flood cut a breach through the sand ridge in February 1978, as shown in Fig. 14.14; it was subsequently backfilled in order to protect the bridge footings. The sequence of events during the March 1978 flood can be separated into the following four phases:

*Phase 1.* At the beginning of the flood, the sand ridge acted as a dam, ponding water and sediment in the upstream channel and borrow pit.

*Phase 2.* After overtopping, the sand ridge was eroded in the shape of a deep narrow gully similar to that shown in Fig. 14.14. The backwater effect of the sand ridge continued to pond water in the upstream borrow pit. The plan view of the channel was highly nonuniform because of the narrow breach and the resulting backwater effect.

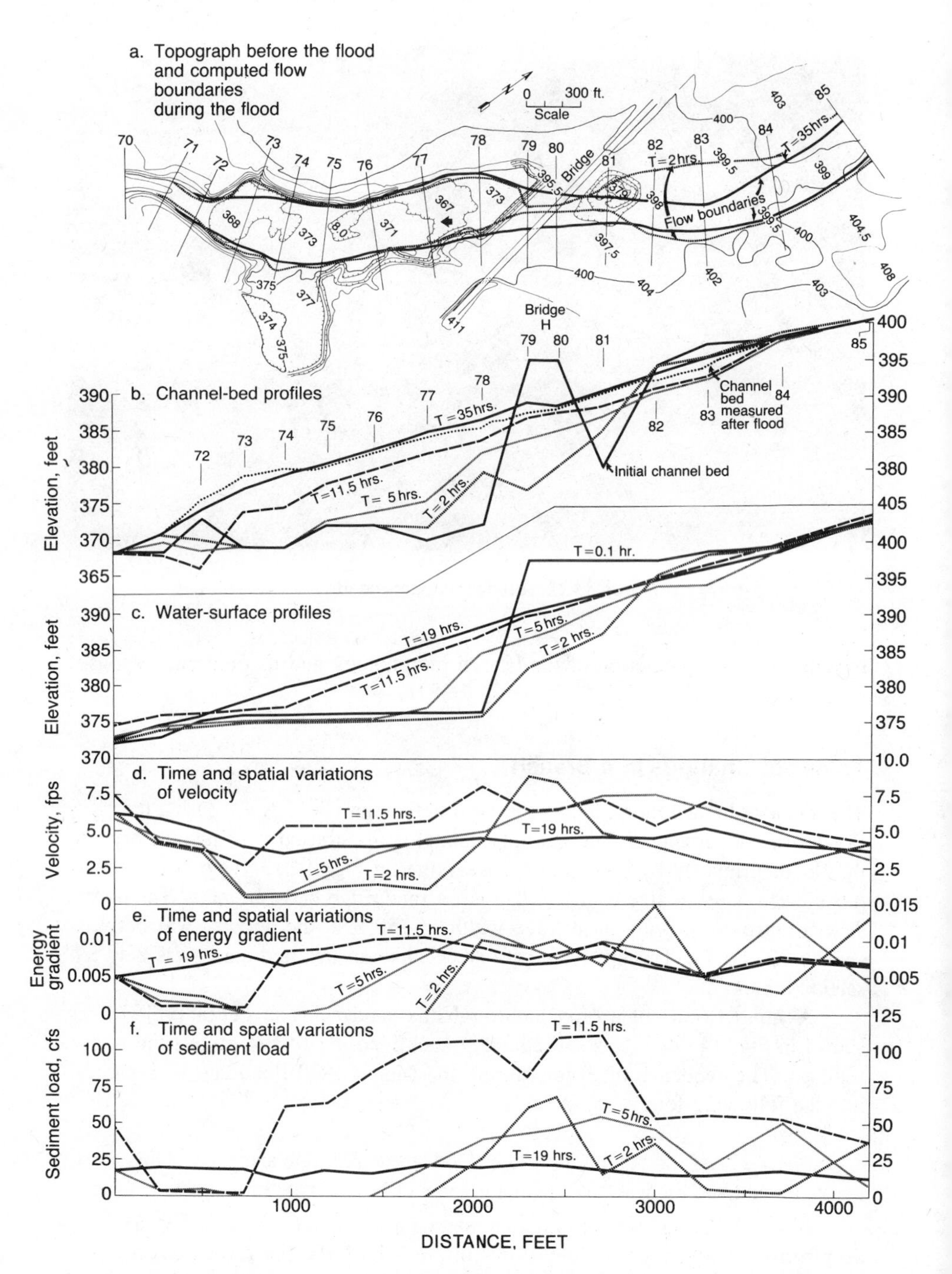


**Figure 14.13** Topography of San Diego River: Simulated results and measurements.

**Figure 14.14** Breach through a sand ridge at a bridge crossing.The breach was caused by a small flood in February 1978 and was backfilled before the March 1978 flood.

*Phase 3*. As the flood flow continued through the breach, significant changes took place. The initial narrow and deep breach through the ridge developed gradually into a wider and shallower configuration.

*Phase 4*. The sand ridge was totally removed when the flood flow receded. With the breach as wide as the channel, the aerial view of the channel becomes quite uniform, as shown in Fig. 14.15.

## Simulation and Results

The mathematical model FLUVIAL-12 was employed to simulate the changes in river configuration, including the breach morphology for the San Diego River during the March 1978 flood (Chang, 1982). The hydrograph measured by the County of San Diego (1978) is shown in Fig. 14.16. Roughness in terms of Manning's $n$ was found to be 0.03 in the channel, based on several measurements; it was estimated to be 0.035 at the bridge crossing. The bed material consisted of sand with a median diameter of 0.8 mm; its size showed no significant change before and after the flood. Graf's equation for sediment transport was used in the model. The 35 hr flood was computed using 1200 time steps.

**Figure 14.15** Aerial view of a river when a flood was receding.

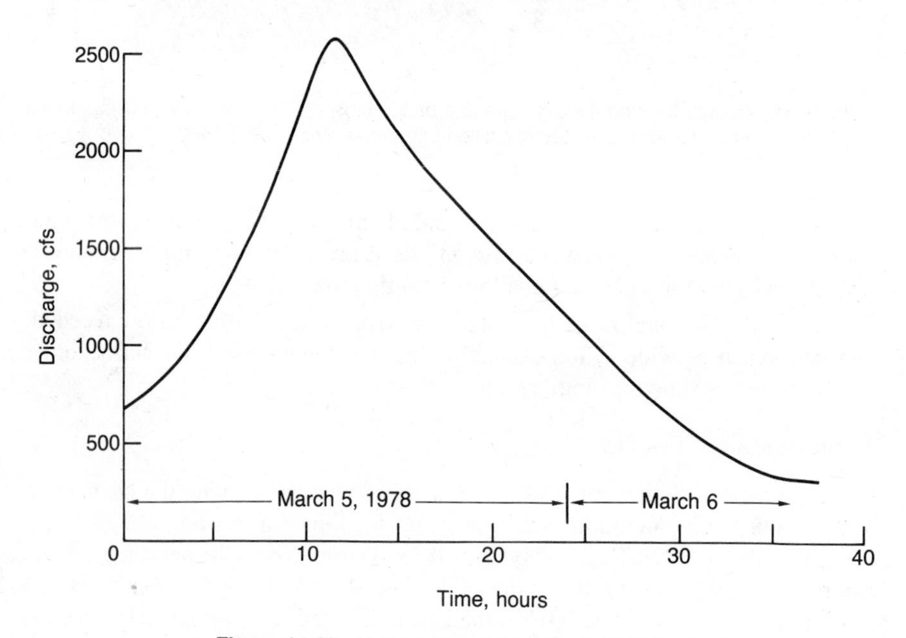


**Figure 14.16** Hydrograph of the March 1978 flood.

Simulated results using the mathematical model are presented in Figs. 14.13 and 14.17. Time and spatial variations of the width, bed elevation, water-surface profile, velocity, energy gradient, and sediment load along the reach are shown in Fig. 14.13. Time variation of the breach (section 79) is shown in Fig. 14.17. Breach configuration measured at the end of the flood is also plotted for comparison with the simulated results.

Changes of the breach morphology shown in these figures are characterized by the initial gully formation through the sand ridge and its gradual widening. The maximum depth of channel-bed scour is predicted to occur in the initial gully, followed by gradual widening and a decrease in scour depth.

Changes in the width of flow shown in Fig. 14.13a are depicted by an initial significant width variation along the reach, followed by the gradual formation of a more uniform channel. Since the initial breach has a small width, it also causes a higher stage and a wider channel in its upstream sections. This backwater effect is gradually relieved with the enlargement of the breach.

Changes in channel-bed profile depicted in Fig. 14.13b consist of (1) scour at higher grounds and (2) fill in the borrow pits. The initial uneven channel-bed profile is gradually smoothed by the flow. Time and spatial variations of velocity, energy gradient, and sediment load are shown in Figs. 14.13d, 14.13e, and 14.13f, respectively; they follow the same trend in that their spatial variations are gradually reduced with time. These changes are rapid during the initial period and then they slow down gradually. With sufficient flow duration, the stream will eventually establish a new equilibrium with a basically uniform flow.

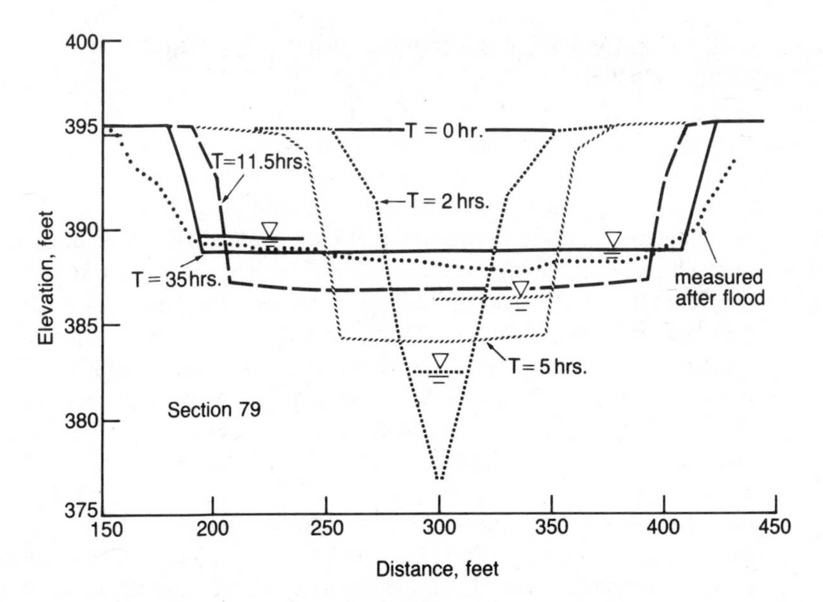


**Figure 14.17** Simulated time variation of a breach profile.

### Breach Morphology in Relation to Power Expenditure

Transient changes in breach morphology are accompanied by adjustments in flow resistance and, hence, in power expenditure. The $\gamma QS$ product represents the rate of energy expenditure per unit channel length. Simulated adjustments in breach morphology reflect the gradual development toward uniform power expenditure, or equal $\gamma QS$, along the channel, as illustrated by the sequential breach profiles in Fig. 14.17.

During the initial stage of development, the flow is over the sand ridge, where the energy gradient is much greater than those of the adjacent sections. This pronounced spatial variation in energy gradient is reduced by (1) gully formation through the sand ridge and (2) deposition in adjacent sections. The gully, which is small in width and has a low channel-bed elevation, provides the least possible flow resistance and, hence, the lowest possible energy gradient at this section; it also reduces the backwater effect on the upstream section and thereby increases its energy gradient. This development is therefore effective in reducing the spatial variation in power expenditure. At subsequent time intervals, the energy gradient through the breach becomes less than those of adjacent sections. Under this situation, changes of the breach profile are characterized by fill in the bed and erosion of the banks to increase the width. These changes are associated with a gain in boundary resistance and, hence, in energy gradient at this section, favoring the establishment of equal power expenditure along the reach. In summary, the pattern of morphological changes for the breach, characterized by the formation of a narrow channel during channel-bed scour and widening during fill, is in the direction of uniform power expenditure.

## 14.3 STREAM CHANNEL CHANGES INDUCED BY SAND AND GRAVEL MINING

Sand and gravel are important mineral resources used as construction materials. The supply of such materials is vital to any regional economic growth. Sand and gravel are usually mined from a stream bed; a borrow pit so created in the Salt River near Phoenix, Arizona is shown in Fig. 14.18. Simulation of stream channel changes induced by sand mining is illustrated by the case studies given in Secs. 14.1 and 14.2; a case study for gravel mining is presented in this section. Gravel is usually found in relatively steep streams, which are generally supplied by smaller watersheds. Sand and gravel mining significantly distort the natural equilibrium of stream channels, thereby inducing stream channel changes. While alluvial streams are a supplier of sand and gravel, special attention must be paid to instream mining because of the impacts on stream morphology.

As storm flow enters a stream channel with gravel excavation, the initial water-surface profile can be quite nonuniform and the sediment discharge also varies greatly along the channel. These nonuniformities will gradually be reduced through changes in width, channel-bed elevation, and bed-material composition, each of which contributes toward uniformities in sediment discharge and power

**Figure 14.18** Gravel pit in Salt River, Arizona.

expenditure. The magnitude of each change must be such that the movement toward uniform power expenditure is expedited subject to the physical constraints and physical conditions governing the flow and sediment transport processes.

## San Juan Creek and Its Changes

Sand and gravel were mined from San Juan Creek downstream of the Bell Canyon confluence in Orange County, California (see Figs. 14.19 and 14.20). The gravel pit at its finished stage in 1977 had an average depth of about 45 ft, a length of 1700 ft, and a maximum width of 800 ft. The supply watershed from San Juan Creek has an area of 55.6 square miles; and its two tributaries, Bell Canyon and Verdugo Canyon, covers 17.2 and 4.0 square miles, respectively.

Hydrographs of the major storms of 1978, originally presented by Vanoni et al (1980), are shown in Fig. 14.21. The March 4th storm was estimated to have a return period in excess of 100 yrs. As a result of these major storms, significant changes in channel configuration occurred. The channels of San Juan Creek, Bell Canyon, and Verdugo Canyon upstream of the gravel pit went through severe headward erosion, and deltas formed concurrently at the stream mouths as the eroded materials were deposited in the pit. Such a development at the mouth of Verdugo Canyon is shown in Fig. 14.20. Measured stream-bed profiles at different time intervals for San Juan Creek are shown in Fig. 14.22b. The stream channel development was characterized by headward erosion, downstream erosion,

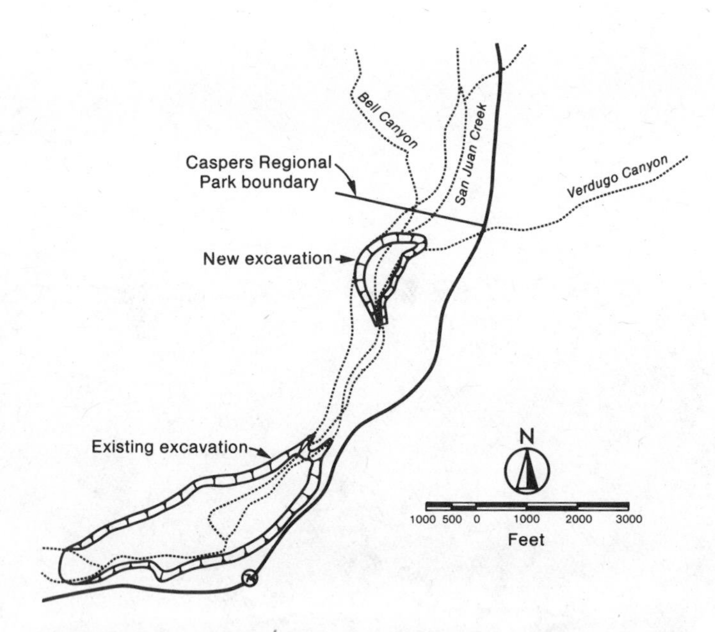


**Figure 14.19** Vicinity map of a mining operation on San Juan Creek.

**Figure 14.20** Mouth of Verdugo Canyon at a gravel pit on January 17, 1978 showing headward erosion and delta formation.

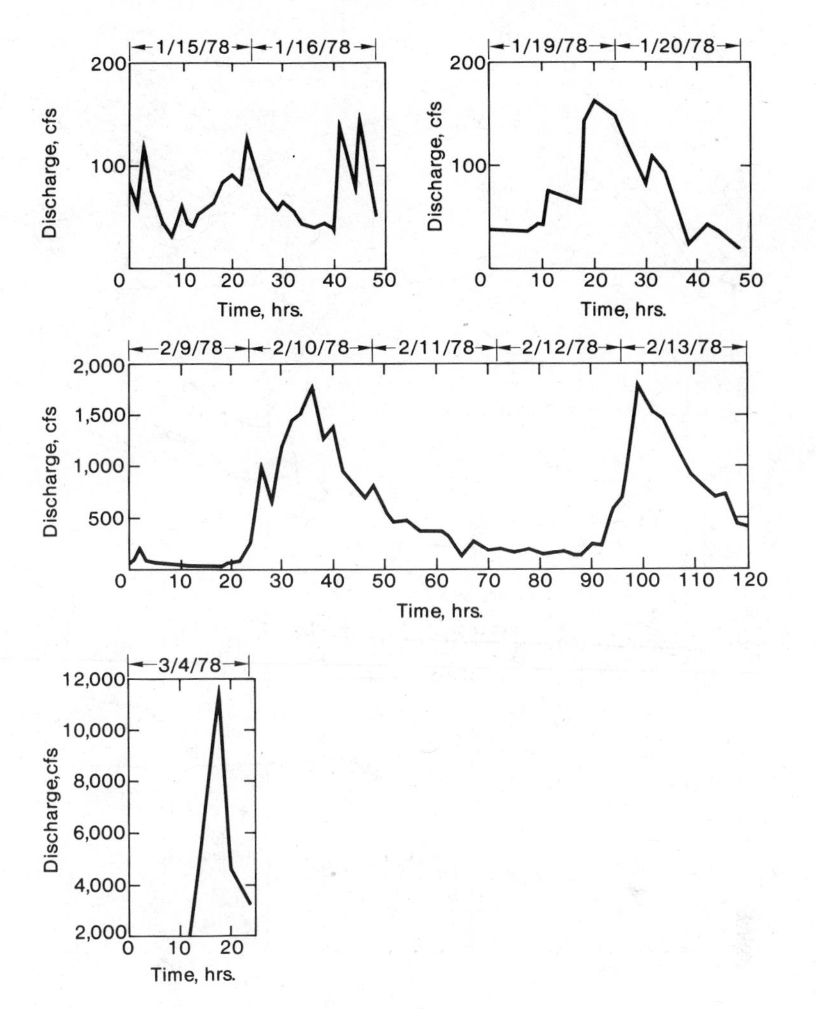


**Figure 14.21** Hydrographs of the major 1978 storms in San Juan Creek downstream of Verdugo Canyon confluence (Vanoni et al., 1982).

deltaic deposition, and gradual refill of the pit. Gravel transport through the gravel pit was reestablished after the refill.

These vertical, or elevation, changes were accompanied by even greater lateral changes in channel width. During the initial stage of severe headward erosion, a relatively narrow channel developed upstream of the pit, as shown in Figs. 14.22a and 14.23 by the survey of February 6, 1978. This narrow channel widened immediately as it flowed into the gravel pit, resulting in significant spatial variation in channel width. During subsequent storms, this spatial width variation was gradually reduced, principally because of the progressive widening development upstream of the pit. This widening, as illustrated in Fig. 14.23 for section 12, was accompanied by a gradual rise in stream-bed elevation.

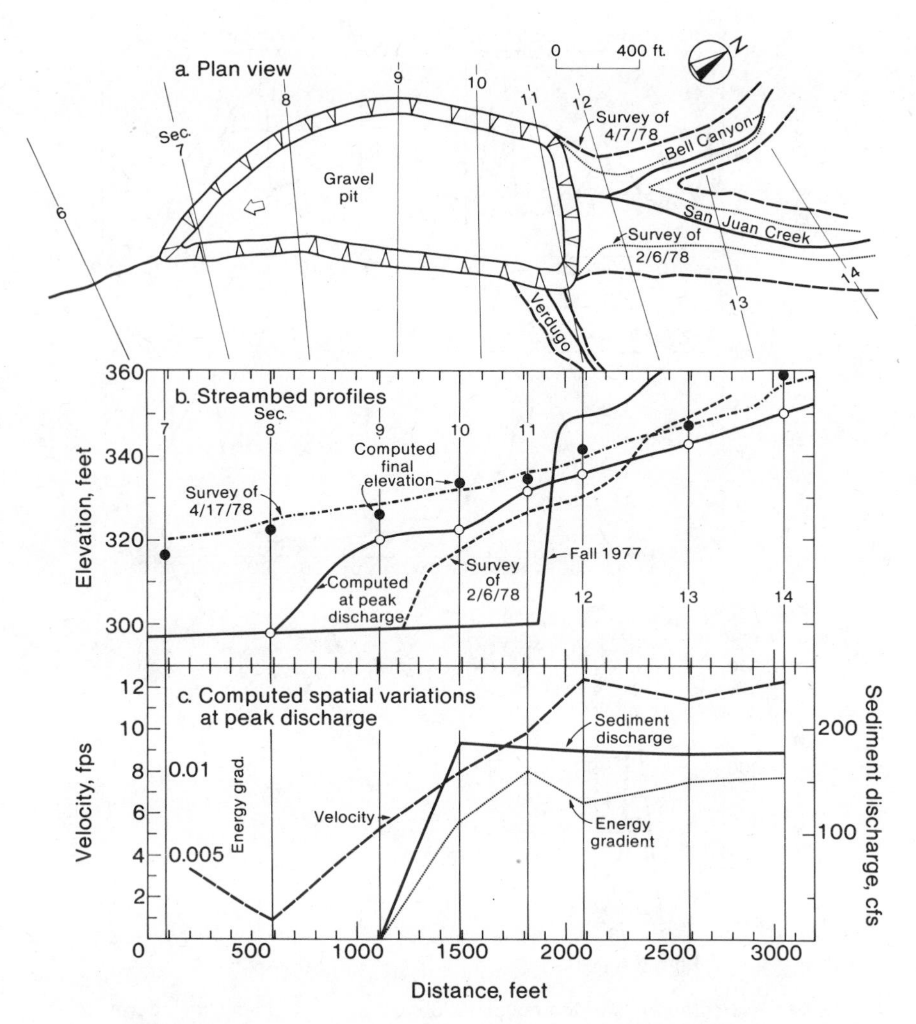


**Figure 14.22** Plan and profiles of San Juan Creek near a gravel pit.

## Simulation and Results

The FLUVIAL-12 model was used to simulate stream channel changes near the gravel pit. Cross sections were selected along the stream reach covering a total length of two miles. Stream channel changes outside this reach were not considered. Sediment discharge was computed using Graf's formula. Sediment inflow into the reach was based on the computed sediment discharge at the upstream section using the same formula.

Simulated results shown in Figs. 14.22b, 14.22c, and 14.23 are used to describe the fluvial processes near the gravel pit. The simulated longitudinal variations of velocity, sediment discharge, and energy gradient shown in Fig. 14.22c are during the peak discharge of 11,600 cfs on March 4, 1978. At this point in

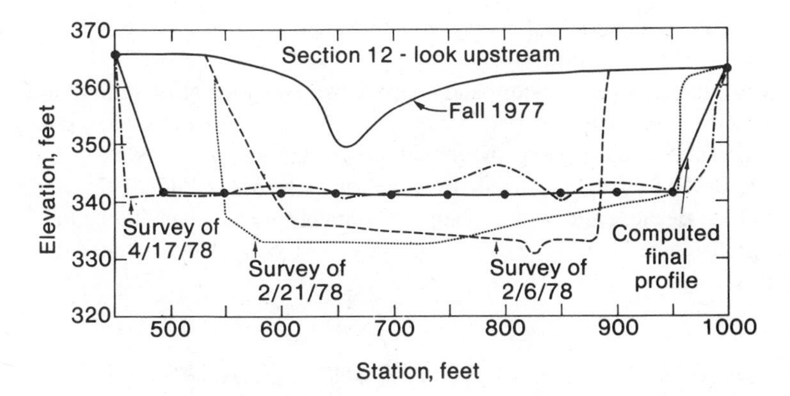


**Figure 14.23** Cross-sectional changes at an upstream entrance of a gravel pit.

time, the gravel pit was still being refilled. This process of refill can be seen from the nonuniformity in sediment discharge, which was high in the upstream refilled portion of the gravel pit and practically zero in the remaining pit. With continued refill, a basically uniform sediment discharge was gradually established along the channel.

Stream channel changes occurred not only because of sediment imbalance, but also because of nonuniformity in power expenditure per unit channel length, $\gamma QS$. During most of the storm periods, the power expenditure was not uniform along the channel, and the water-surface profile was not a straight line. Since channel roughness was assumed constant in this case study, adjustment in power expenditure was through changes in channel-bed elevation and channel width. During the initial headward erosion, a steep energy gradient existed at section 12 above the pit and a flat energy gradient was present at section 11 in the pit. The formation of a gully at section 12 was the most efficient way to reduce the energy gradient or power expenditure at this section. The opposite occurred at section 11, where deposition was accompanied by widening; the power expenditure was thus effectively raised because of the increasing wetted perimeter and rising bed elevation. The nonuniformity in water-surface profile between sections 11 and 12 was also reduced through such fluvial processes.

It is interesting to note that, during the initial stage, stream channel adjustments toward uniform power expenditure and sediment discharge were actually accomplished by developing significant spatial variation in width on the delta. But at a later point in time, such as at the peak discharge, stream channel adjustments were toward reducing the spatial variation in width by channel widening upstream of the pit. This sequential development may be explained by the computed spatial variations in sediment discharge and energy gradient shown in Fig. 14.22c. With reference to section 12, it had a sediment discharge similar to those at the adjacent sections, but it had a lower energy gradient because of the smaller width. In the process of establishing a uniform energy gradient, this section gradually widened so as to increase the energy gradient, and the spatial variation in width was thus reduced. The simulated final cross-sectional shape compares favorably with measurement, as shown in Fig. 14.23.

Now, the changes in width and channel-bed elevation at section 12 are compared based on the measurements shown in Fig. 14.23. From the initial configuration to the final configuration, the maximum change in channel-bed elevation was about 30 ft. The change in width from February 6 to April 17, 1978 alone was over 200 ft. Because the width change was much greater than the bed-elevation change, it becomes clear that width changes must be included in modeling such processes.

## 14.4 TIDAL RESPONSES OF RIVER AND DELTA SYSTEM

As a river enters a reservoir, lake, or ocean, a delta is usually formed by sediment deposition. Delta formation at a river mouth, as well as alluvial fan formation at foothills, has been of interest not only to engineers but also to geologists. During the formation, the delta stream (Fig. 14.24) manifests different patterns of flow, either as a single concentrated flow or braided in distributary channels. Such morphological patterns are also seen on alluvial fans, as depicted in Fig. 14.25. It is

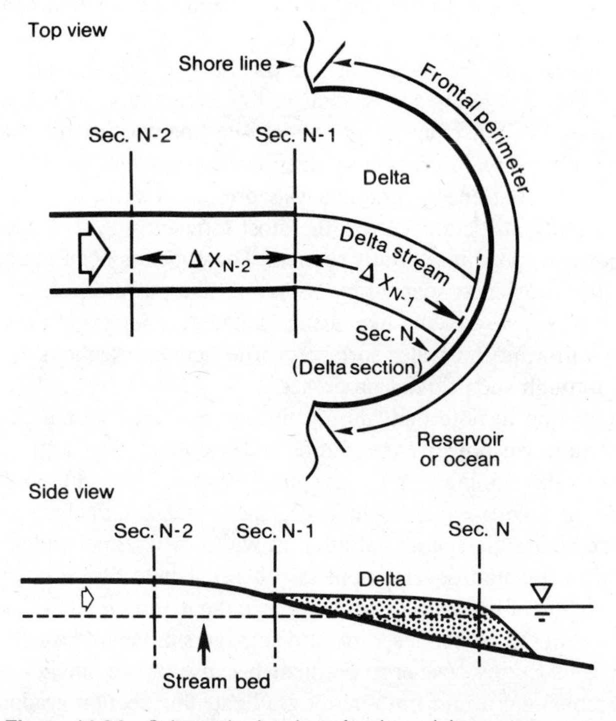


**Figure 14.24** Schematic drawing of a river–delta system.

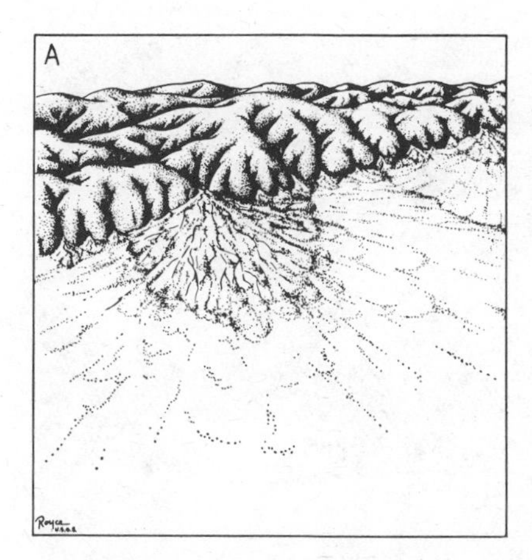


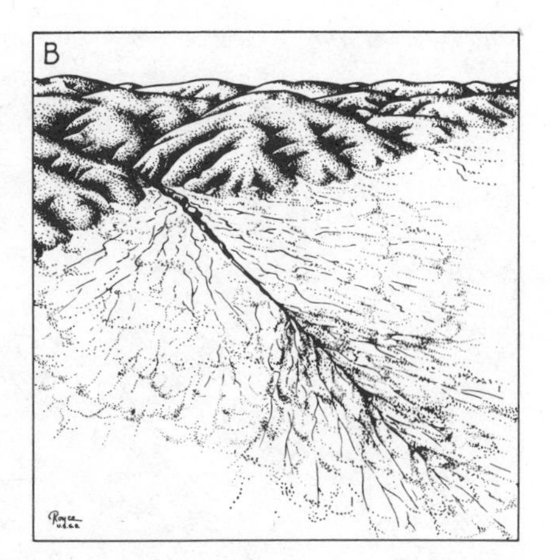


**Figure 14.25** Two types of alluvial fans: (a) area of deposition near mountain front, (b) area of deposition near fan tip due to upstream channel entrenchment (Bull, 1968).

necessary to determine the variation of the flow pattern during a delta formation, since it affects other hydraulic parameters (such as flow velocity, energy gradient, and sediment load) of the river as well as the rate of delta growth. Modeling of delta formation provides a means for evaluating the sand influx into the ocean or a reservoir.

## Morphological Changes of River–Delta System

Changes in a river channel are in direct response to changes in discharge, sediment load, tectonic events, or base level (water-surface elevation in a downstream reservoir or lake). These changes affect the pattern of deposition and thus affect the delta or fan shape. With the freedom of change, river channel adjustments occur primarily through a change in channel pattern, which is accompanied by a large change in channel width and a moderate change in channel length. Distributary channels have a much larger overall width than a comparable single channel, with the overall width varying in direct relation to the number of distributary channels.

Delta formations were studied in a laboratory flume, as shown in Fig. 14.26 (Chang, 1967). The variation of the stream pattern on the delta was observed to depend on the aggradation and degradation (fill and scour) of the delta stream. Aggradation of the delta stream was induced by a rise of the base level, a decrease in water discharge, or an increase in sediment inflow; degradation was caused by the opposite of these factors. Based on laboratory observations, an aggrading delta stream tends to widen and become braided into branch channels; during degradation, however, the branches tend to merge into a single stream.

It is important to note that braiding or widening of the delta stream, from the observations, was not necessarily related to water overflowing the delta stream

**Figure 14.26** Laboratory study of delta formation at Colorado State University.

banks. For example, with an increase in water discharge (cause degradation), the water surface was above the stream banks on the delta; instead of overflowing to the sides, the delta stream actually assumed a single concentrated flow pattern. On the other hand, with a decrease in water discharge (cause aggradation), the water surface was below the stream banks; braiding of the delta stream was also observed.

With the change in base level, water discharge, or sediment discharge and delta growth, the delta stream undergoes constant aggradation or degradation. During aggradation, widening of the delta stream contributes to the decrease in flow velocity and thus facilitates the deposition of sediment. On the other hand, a stream tries to seek a lower bed elevation during degradation, converging of the delta stream concentrates the erosion power to a narrow area, representing the most effective way to reach the lower elevation. The following mathematical simulation will show that these characteristic adjustments are toward minimization of the power expenditure for the stream reach.

### Simulation of Delta-Stream Responses

To describe the geometry of the river and delta stream, cross sections along the channel are selected as shown in Fig. 14.24. The cross section at the downstream end of the delta (delta section) is designated as section N, that at the beginning of the delta is designated as section N − 1, others belonging to the upstream river are designated as N − 2, N − 3, . . . and 1. The delta stream in noncohesive sand has almost total freedom in its pattern adjustment. For practical purposes, the ratio of width to cross-sectionally averaged depth may be taken as an indication of braiding tendencies. That is, a large width–depth ratio is indicative of multiple stream branches or even sheet flow on the delta, whereas a small ratio (say less than 40) signifies a single delta stream.

The total stream power of the river–delta-stream system is given by

$$P = \int_L \gamma Q S \, dx \tag{14.1}$$

where $P$ is the total power expenditure of the channel reach, $L$ is the reach length, $\gamma$ is the specific weight of the water–sediment mixture, $Q$ is the discharge, and $S$ is the energy gradient. In the simulation, the values of $Q$ and $S$ in Eq. 14.1 are computed at each time step. The variation of width at section N affects the flow as well as the power expenditure of the river and delta stream. Usually, only that portion of the river adjacent to the delta is affected. For this reason, the value of $P$ in Eq. 14.1 needs to be evaluated only for the stream reach affected by the variation at section N.

Equation 14.1 can be written in the following finite-difference form

$$P = \sum_{i=1}^{N-1} \frac{1}{2} \gamma (Q_i S_i + Q_{i+1} S_{i+1}) \Delta x_i \tag{14.2}$$

The total stream power given by this equation is a function of $B$, the stream width at section N. Given the channel configuration at time step $j$, the width at a new time step $j + 1$ is determined based on the condition that $P^{j+1}$ is a minimum. For this purpose, $P^{j+1}$ for the stream is established as a function of the width $B^{j+1}$ applying the following steps:

1. Assume a value for $B^{j+1}$, the width at section N for time step $j + 1$.
2. Compute the depth $D^{j+1}$ at the section based on the following analysis. At section N, the new cross-sectional area of flow, $A^{j+1}$, is the original area,

$A^j$, plus the area increase due to base-level variation, minus the bed area increase due to sediment movement, that is,

$$A^{j+1} = A^j + \frac{B^j + B^{j+1}}{2}\Delta Z - \Delta A_b \tag{14.3}$$

where $\Delta Z$ is the change in base level. Substituting Eqs. 13.7 and 13.9 for $\Delta A_b$ into Eq. 14.3 yields

$$B^{j+1}D^{j+1} = A^j + \frac{B^j + B^{j+1}}{2}\Delta Z - \frac{\Delta t}{1-\lambda}\left[\frac{(Q_s)^j_{N-1} + (Q_s)^{j+1}_{N-1}}{(\Delta x_{N-1} + \Delta x_N)} - \frac{(Q_s)^j_N + (Q_s)^{j+1}_N}{(\Delta x_{N-1} + \Delta x_N)}\right] \tag{14.4}$$

where sediment discharges $(Q_s)_{N-1}$ and $(Q_s)_N$ depend on $B^{j+1}$, $D^{j+1}$, and $Q_N$. However, as long as the variation is small for a time step, the values of $(Q_s)^{j+1}_{N-1}$ and $(Q_s)^{j+1}_N$ may be approximated by their respective counterparts at time step $j$. For an assumed width $B^{j+1}$, Eq. 14.4 contains $D^{j+1}$ as the only explicit unknown that can be readily computed.

3. Compute the flow condition in the stream using the techniques for water routing described in Sec. 13.4.
4. Compute the stream power according to Eq. 14.2 for the width assumed.

With the stream power $P^{j+1}$ established as a function of the width $B^{j+1}$, the value of $B^{j+1}$ at the minimum $P^{j+1}$ may now be obtained. An iterative procedure is employed to obtain the solution; the computation is carried to an accuracy of 1 ft for the width.

The width of the delta stream at section N is subject to the constraint of the frontal perimeter defined in Fig. 14.24. The frontal perimeter is smaller for the delta formed in a narrow canyon.

## Case Study

The entrance channel for San Elijo Lagoon on the southern California coast is shown in Fig. 14.27. A simulation study was made for the stream–delta system by Chang and Hill (1977). The channel was blocked by beach sand but was opened by excavation at 3:35 p.m., February 4, 1975 in order to drain storm water stored in the lagoon. Configurations of the delta formed at the channel mouth are shown in Figs. 14.28 and 14.29. The delta stream was very wide and fan-shaped at 3:50 p.m., February 4, 1975, as shown in Fig. 14.28. In sharp contrast to this channel pattern, the delta stream was narrower and concentrated at 11:50 a.m., February 5, 1975, as shown in Fig. 14.29.

Figure 14.27 Entrance channel for San Elijo Lagoon.

**Figure 14.28** Initial delta with a fan-shaped flow at 3:50 p.m., Feb. 4, 1975.

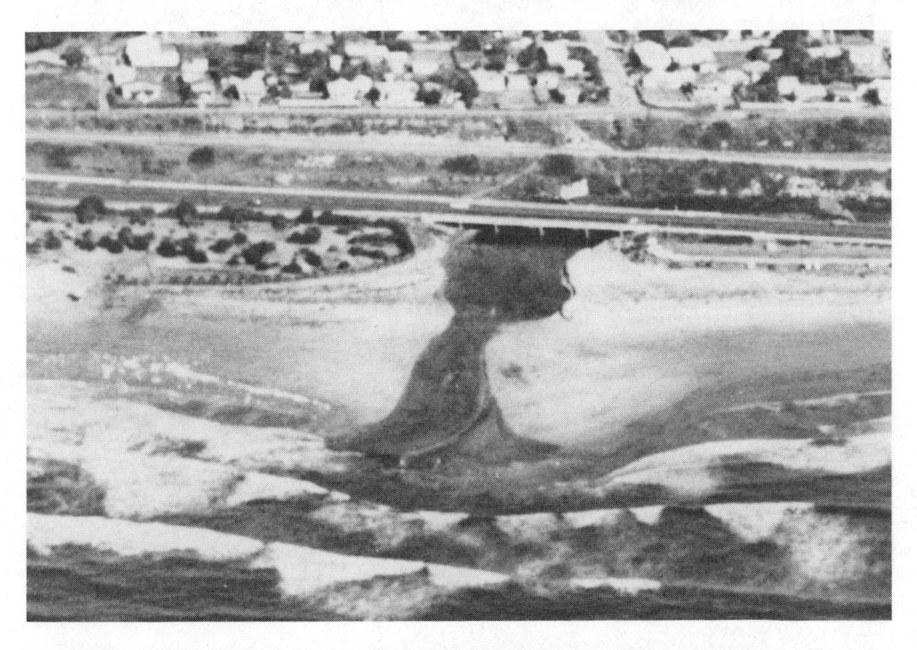

**Figure 14.29** Delta with a concentrated stream flow at 11:50 a.m., Feb. 5, 1975.

The bed material in the channel was primarily sand, with a median diameter of 0.25 mm. The DuBoys formula was used to compute the sediment discharge, which was primarily bed load. The Manning's coefficient of 0.025 was used for channel roughness. The time increment $\Delta t$ employed in the computation was in the range of 20–400 sec.

The study is summarized in Fig. 14.30, which contains tidal variation of the ocean, simulated results, and certain experimental data. The simulated hydrograph, time variations of mean velocity, width, and width–depth ratio at section N are described below.

During the time interval 9.5–16.5 hr, the stream width at section N was constrained by the frontal perimeter defined in Fig. 14.24. Outside this time interval, this width was unconstrained; its variation, shown in Fig. 14.30c, can be correlated with other changes. Figure 14.30 shows that a rising tide, which induces fill in the delta stream, is generally associated with a decreasing velocity and an increasing width at the delta section. Conversely, a lowering tide, which causes scour of the stream bed, is concomitant to an increasing velocity and a decreasing width. Since these characteristic changes were obtained by minimizing the total stream power, it therefore follows that these pattern changes are in the direction of best fluvial efficiency subject to the given constraints.

Variations of the stream power $P$ with the width $B$ of the delta section (section N) computed at two time steps are shown in Fig. 14.31. At the time of 21.3 hr, the delta section is unconstrained by the frontal perimeter; the value of $P$ has a minimum at $B = 98$ ft. At the time of 10.3 hr, the width is constrained by the frontal perimeter, which has a value of 693 ft.

Simulated and measured results for the channel are also compared. The initial and final water-surface profiles in the channel are shown in Fig. 14.32. These results indicate that the channel bed degraded during the process of lagoon flushing.

## 14.5 WATER AND SEDIMENT ROUTING THROUGH A CURVED CHANNEL

The analytical background of the FLUVIAL model pertaining to water and sediment routing through curved channels is given in Chapter 13. This model incorporates the major effects of transverse flow, inherent in curved channels, on the flow and sediment processes. In the simulation of stream-bed profile changes, the effects of transverse flow are tied in with the scour and fill development. Changes in bed topography induced by the curvature are characterized by the transverse bed slope. Greater depths and higher velocities near concave banks produced by secondary currents, in addition to scour and fill, are necessary considerations in the design of bank protection.

The FLUVIAL-12 model was applied to simulate flow and river bed changes in the San Lorenzo River (see Figs. 14.33 and 14.34) (Chang, 1985). This river drains into the Santa Cruz Harbor on the Pacific coast through the City of Santa

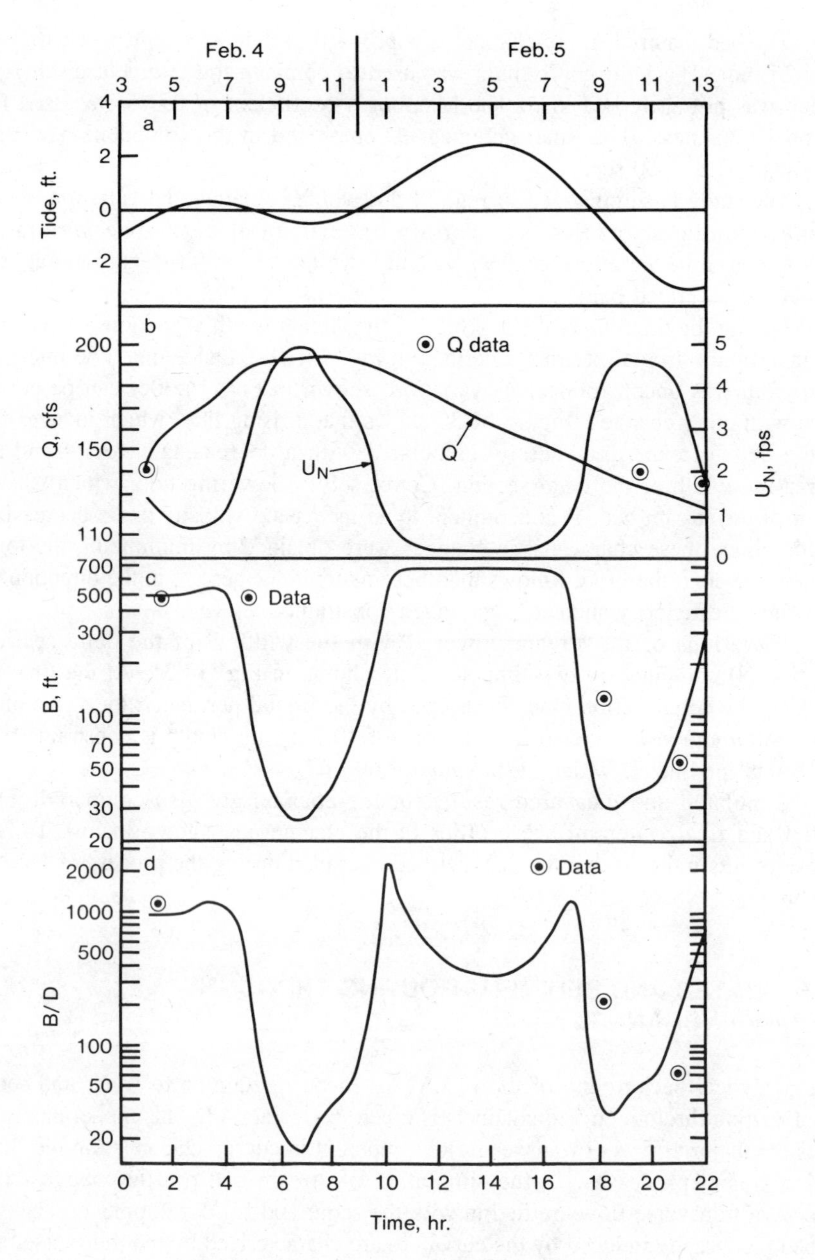


**Figure 14.30** Summary of a simulation study.

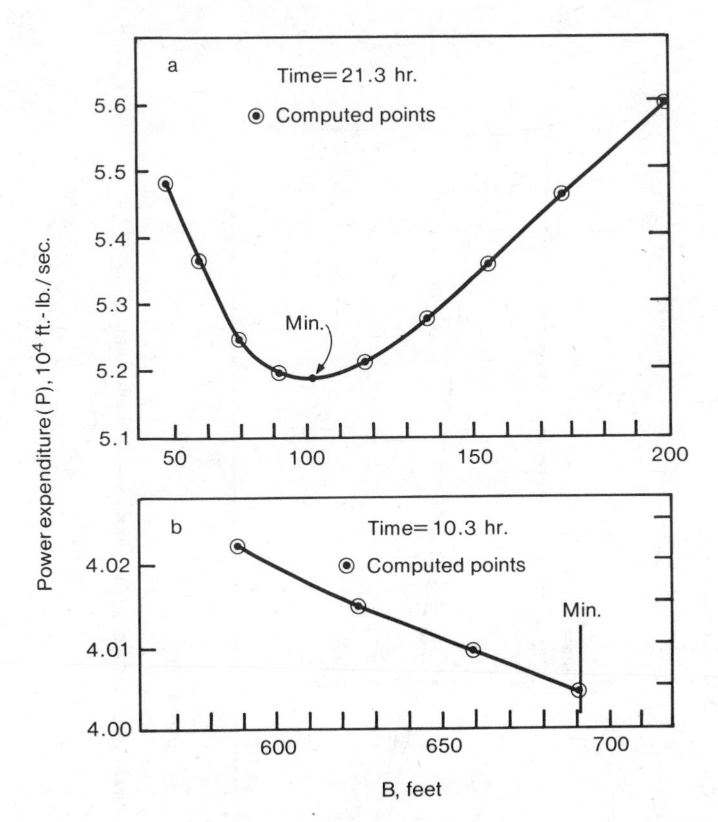


**Figure 14.31** Relationship between total stream power with delta section width at two time steps.

Cruz, California. A field study, sponsored by the San Francisco District (1980) of the U.S. Army Corps of Engineers, was made during the February 1980 flood for the purpose of analyzing river bed scour-fill processes during the flood. This river reach, which has several channel bends near the mouth, is protected by riprap on its bank slopes.

The lower two-mile reach of the San Lorenzo River was simulated using the mathematical model for the February 1980 flood. Measured hydrographs for this flood event and tidal variation at the harbor are shown in Fig. 14.35. The former provided the upstream boundary conation; the latter provided the downstream condition of the simulation. The 90-hr flood duration was computed using 800 time steps. Initial bed materials in the river varied from very coarse sand (median diameter = 1.05 mm) at the upstream end to coarse sand (median diameter = 0.64 mm) at the mouth. Sediment load was computed using the Engelund–Hansen formula.

Selected results obtained in this simulation are shown in Figs. 14.36–14.38 for the curved reach, where significant river bed changes are predicted. These results are compared with measurements made at the gaging stations G2–G6 on

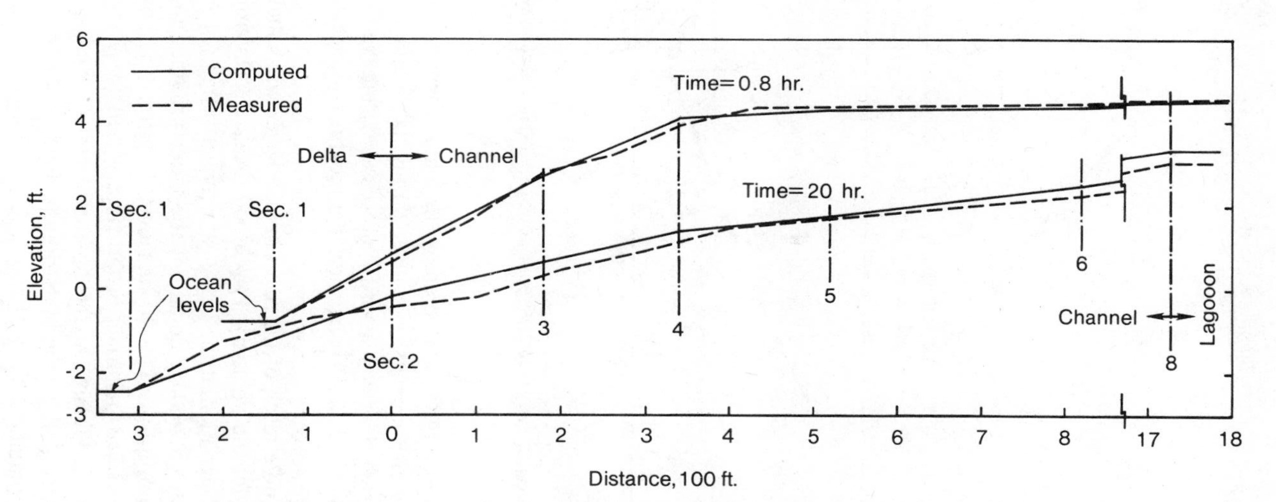


**Figure 14.32** Simulated and measured water-surface profiles in an entrance channel.

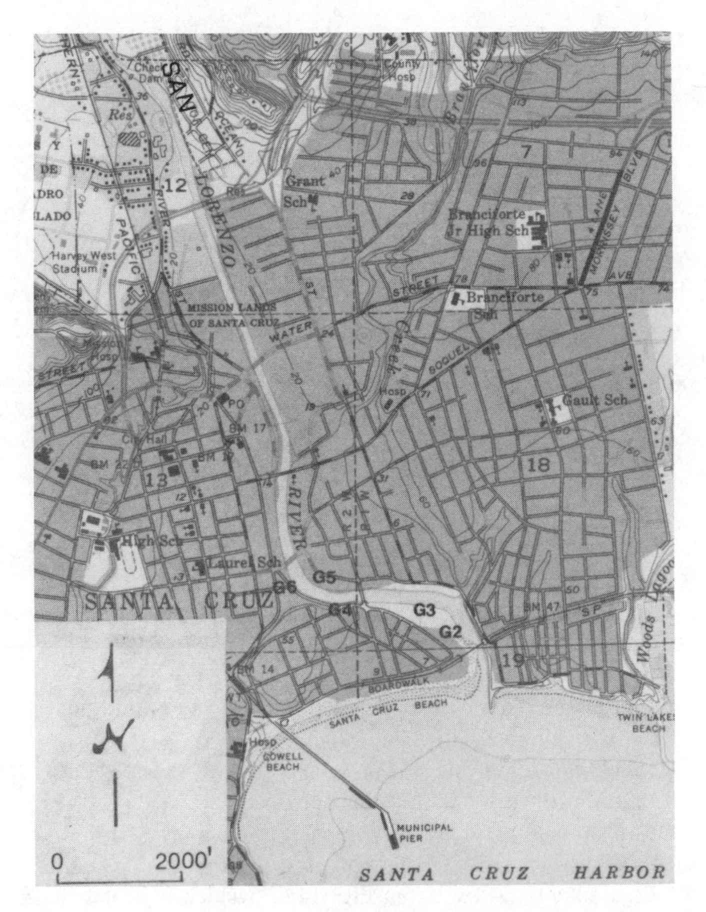


**Figure 14.33** San Lorenzo River in Santa Cruz, California with gaging stations G2–G6.

February 19, 1980, around the time of 78.5 hr on the hydrograph. Simulated water-surface profiles at the time of the peak flood and at the time of measurement (time = 78.5 hr) are shown in Fig. 14.36b together with the simulated profile of minimum river bed elevation at the time of measurement.

The simulated river bed profile is closely related to the streamwise variation in the secondary flow, as shown in Fig. 14.36c. The mean flow curvature (defined in Eq. 13.5) increases as the flow enters a bend; it decreases after leaving the bend exit. Upon entering a bend, the rate of increase in flow curvature is more rapid initially and then it slows down gradually. In a long bend, the flow curvature will approach the channel curvature, and thus the transverse circulation becomes fully developed. Such a condition is predicted at sections 10, 11, 19, and 22. The simulated transverse flow is not fully developed in other shorter bends, where the flow curvature remains less than the channel curvature. Upon leaving a bend, the flow curvature decreases following an exponential decay curve, and it persists for a

**Figure 14.34** San Lorenzo River: Looking upstream from G2.

considerable distance downstream, consistent with the experimental findings reported in Secs. 8.7 and 8.8.

The simulated results in Fig. 14.36 indicate that river bed scour is related to the flow curvature, with the maximum scour occurring at the bend exit. These results are consistent with previous experimental findings by Rozovskii (1957) and Yen (1970). Simulated cross-sectional profiles at the five gaging stations are compared with measurements, as shown in Fig. 14.37, which indicates that the maximum scour depth is attributed to the development of transverse bed profile. Simulated bed topography through a bend is illustrated by the results from section 9 to section 14, as shown in Fig. 14.38. It depicts increasing transverse bed slope and scour near the concave bank in the downstream direction, consistent with the growth in flow curvature. Maximum scour is reached at the bend exit, followed by a gradual decrease in transverse bed slope and scour depth with the decline in flow curvature.

While the scour depth is found to be generally in direct relation to the flow curvature, river bed evolution is also affected by scour and fill, which are provoked by the longitudinal imbalance in sediment load. Spatial variation in sediment load at the peak discharge, shown in Fig. 14.36e, has an increasing trend in the downstream direction associated with a low tide in the harbor. This indicates a general trend of river bed scour induced by the low tide, as more sediment is removed from the reach than the amount supplied. At the time of 78.5 hr, on the other hand, the sediment load has a decreasing trend in the downstream direction associated with a high tide. River bed fill is predicted at this time as some sediment is stored in the reach. These fluvial processes exemplify that modeling of

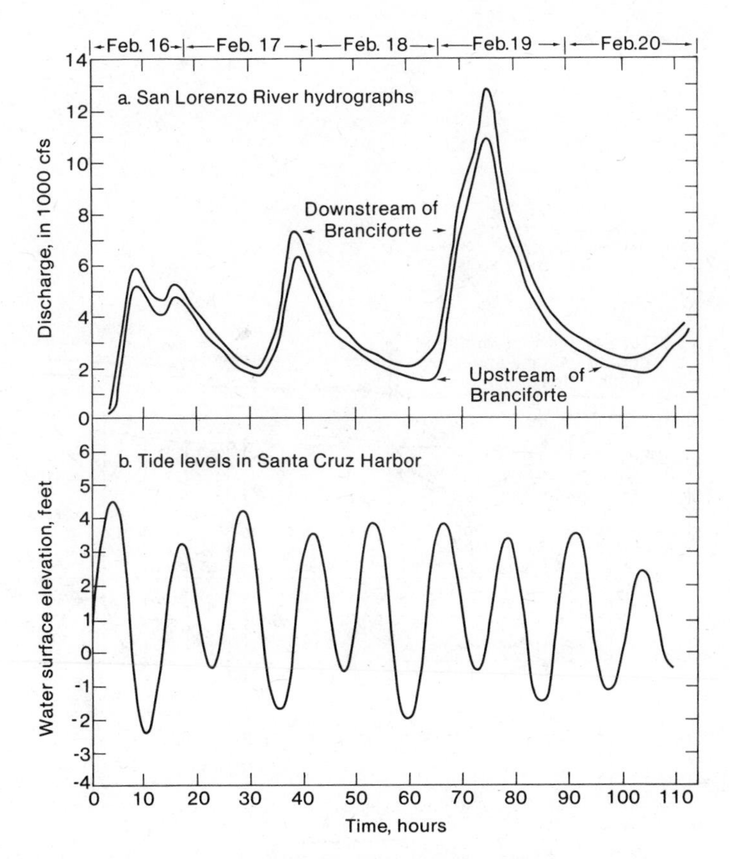


**Figure 14.35** Hydrographs and tidal variation during the 1980 flood.

river bed changes induced by transverse circulation must also be tied in with the scour and fill development.

## 14.6 FLUVIAL DESIGN OF RIVER BANK PROTECTION

Natural rivers through urban regions usually need to be stabilized in order to prevent channel migration. A economical and environmentally desirable practice is to protect channel banks while maintaining an alluvial bed. Types of bank protection are described in Sec. 15.1. In general, bank protection should contain the design flood, it must be strong enough to withstand the design flow, and its toe depth should extend beyond the potential scour to prevent undermining.

While the alluvial bed is subject to scour and fill that are induced by the imbalance in longitudinal sediment discharge, such channel-bed development may also be caused by transverse sediment movement due to channel curvature. Despite bank protection, the channel still has certain freedom in width adjustment

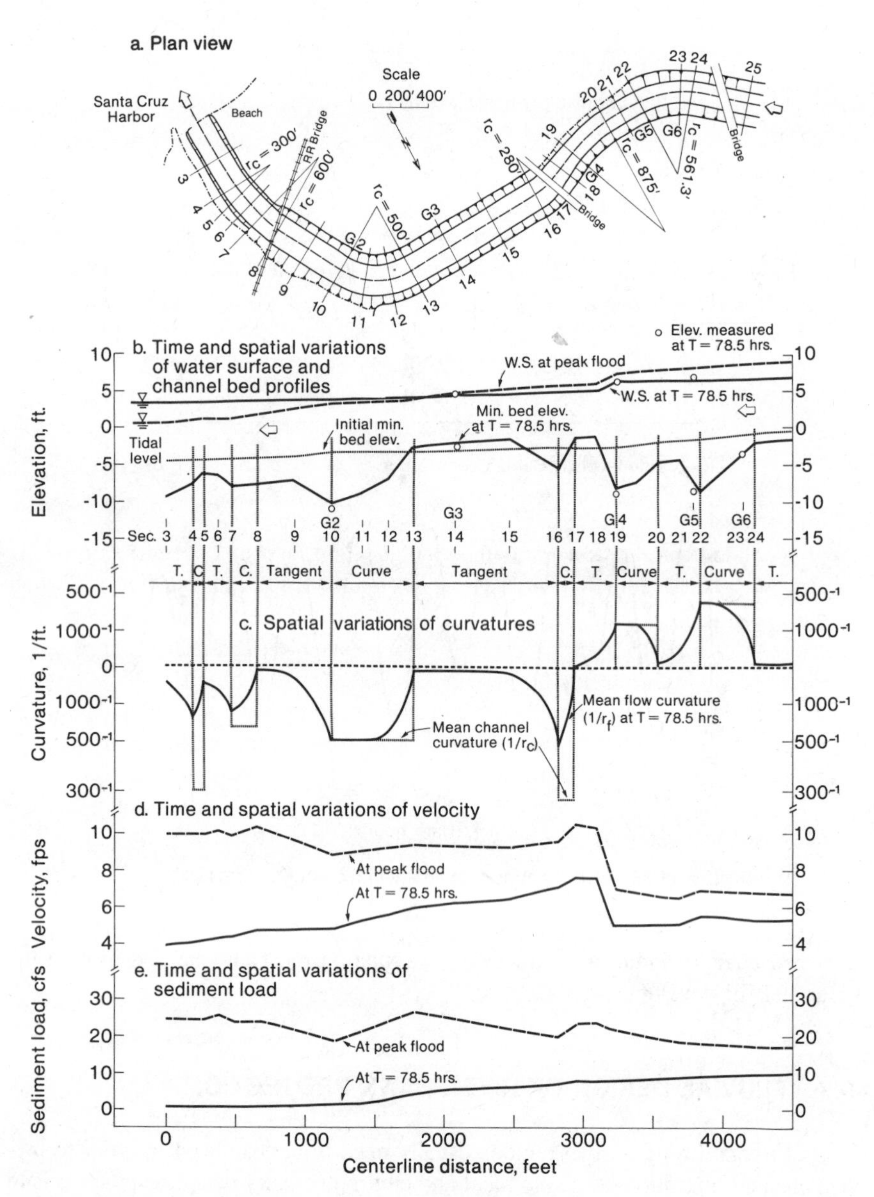


**Figure 14.36** Simulated results and measured elevations for the curved reach of the San Lorenzo River.

within the bank constraints, particularly if the width between rigid banks varies along the channel. Therefore, curvature effects, scour and fill, and width changes need to be considered in the simulation study.

The transverse bed slope in curved channels is related to the spiral motion or secondary currents. Because of the streamwise variation in spiral motion, uneven

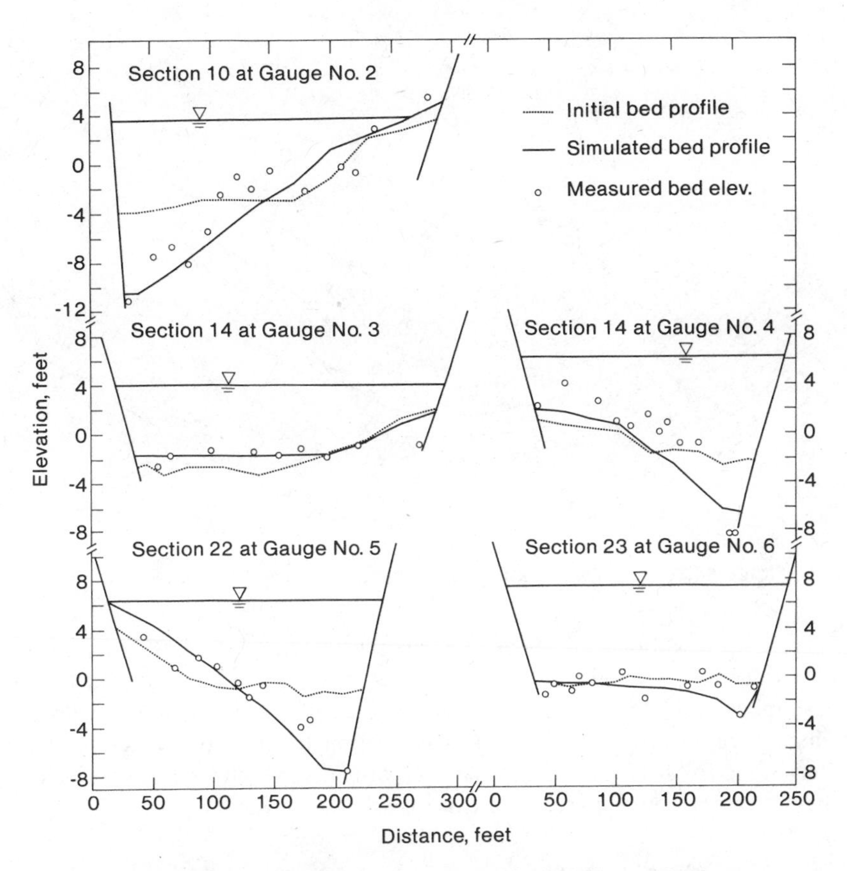


**Figure 14.37** Simulated and measured bed profiles at 78.5 hr.

bed topography is usually produced, characterized by a lower bed elevation near the concave bank. The intensity of spiral motion is directly related to the discharge. Therefore, the nonuniformity in bed topography is more pronounced at high flow and becomes partially eliminated during the subsequent low flow. This explains why an observer of the post-flood channel may fail to recognize the uneven bed scour under the muddy water at high flow. If the bank protection for a river is designed based on the simulated pattern of channel-bed scour, variable toe elevations for the banks should be used to provide an effective protection.

Computer-based design of bank protection was made by Chang et al. (1985) for the Rillito River, which flows through the City of Tucson, as shown in Fig. 14.39. This river has a total drainage area of 935 square miles at the confluence with the Santa Cruz River. The Rillito River underwent significant changes during the October 1983 flood, which reached 100-yr magnitude at several places in the county. A picture of the river near the Dodge Boulevard Bridge taken soon after the flood is shown in Fig. 14.40. Lateral migration and channel widening resulted in the failure of one bridge abutment and endangered other adjacent properties.

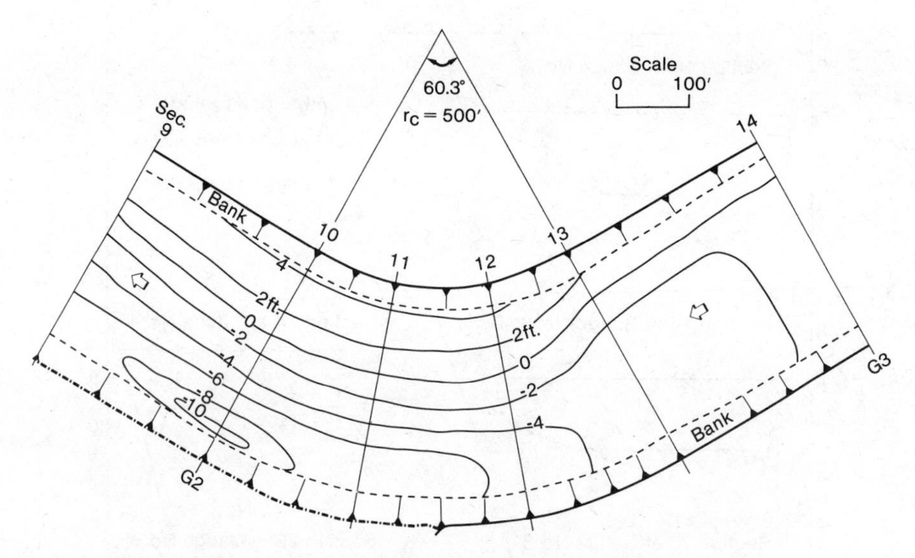


**Figure 14.38** Simulated bed topography through a river bend at 78.5 hr.

The river reach for which bank protection was considered started from the La Cholla Bridge at 2.75 miles upstream from the confluence to the Craycroft Bridge at 11.94 miles upstream. The study was made to determine the appropriate designs for channel geometry, slope, and bank protection for this river reach. Certain portions of the river channel already had soil–cement bank protection, either along one bank or along both banks. These existing structures would be utilized as much as feasible.

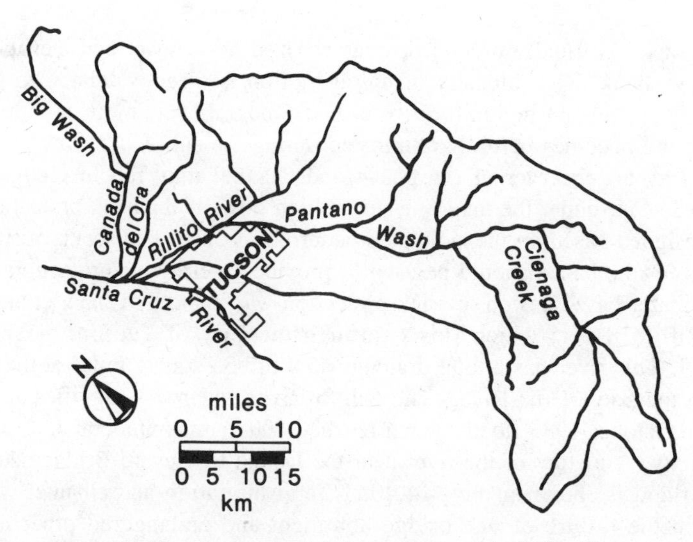


**Figure 14.39** Location and drainage basin of the Rillito River.

**Figure 14.40** Bank erosion caused by the October 1983 flood near Dodge Boulevard.

## Selection of Design Configuration for Channel

The 100-yr flood hydrograph for the Rillito River, as shown in Fig. 14.41, has a peak discharge of 34,000 cfs. For this design flood, the final design configuration for the channel was arrived at from a preliminary assumed configuration defined by cross sections. For the initial assumed configuration, river channel changes were evaluated using the FLUVIAL-12 model. The simulated results served as a feedback to be used as a guide in revising the design. The final configuration was selected based on the major considerations of flood control, flow velocity, river channel changes, material balance, and economy.

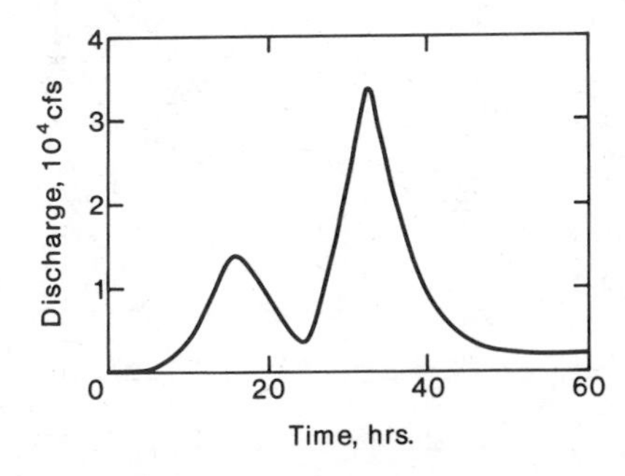


**Figure 14.41** The 100-yr flood hydrograph for the Rillito River.

Except for certain existing bank protection, the design channel is trapezoidal in cross section, with a bottom width of 320 ft and 1-on-1 side slopes. At the peak discharge of 34,000 cfs, the average depth of flow is about 8.5 ft. Longitudinal alignment of the channel design consists of straight reaches and simple curves as exemplified in Fig. 14.42, which also depicts the longitudinal profile of design channel bed. Because of the slight decrease in sediment size from upstream to downstream, the channel slope is designed to decrease gently downstream.

## Simulated Results for Design Configuration

In the model simulation, Manning's coefficient of 0.03 was used for the channel, and Yang's unit stream power equation for sediment transport was employed. Selected results are shown in Figs. 14.42 and 14.43. The water-surface profile at the peak discharge based on the FLUVIAL-12 computation is shown in

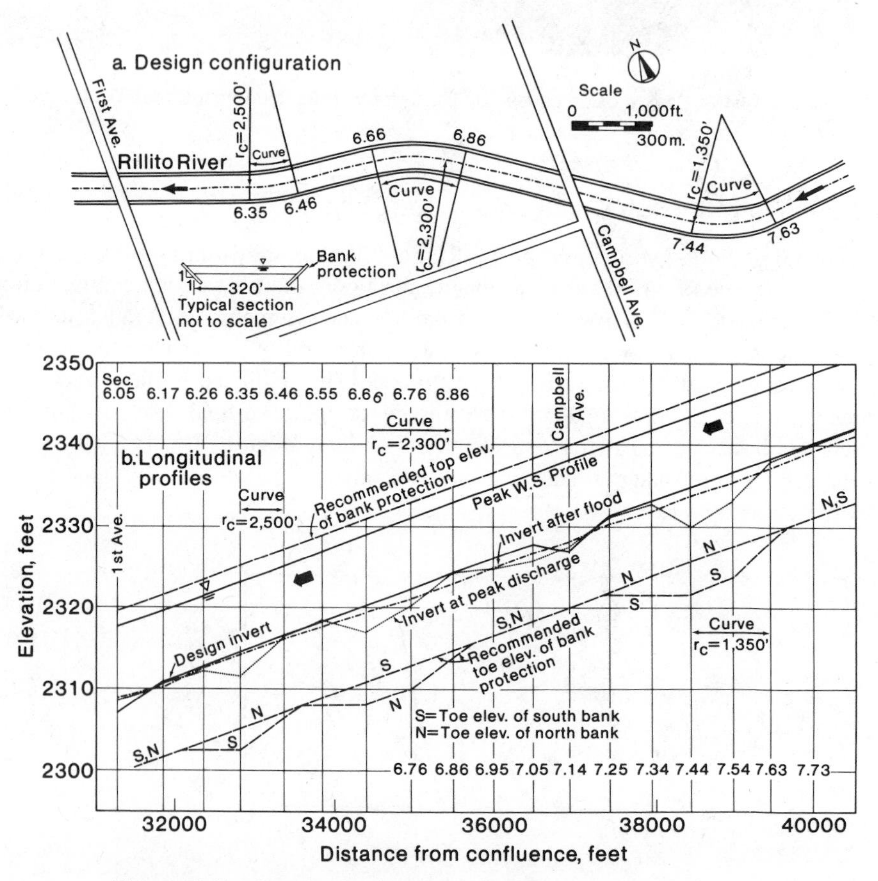


**Figure 14.42** Sample of a design configuration and simulated results.

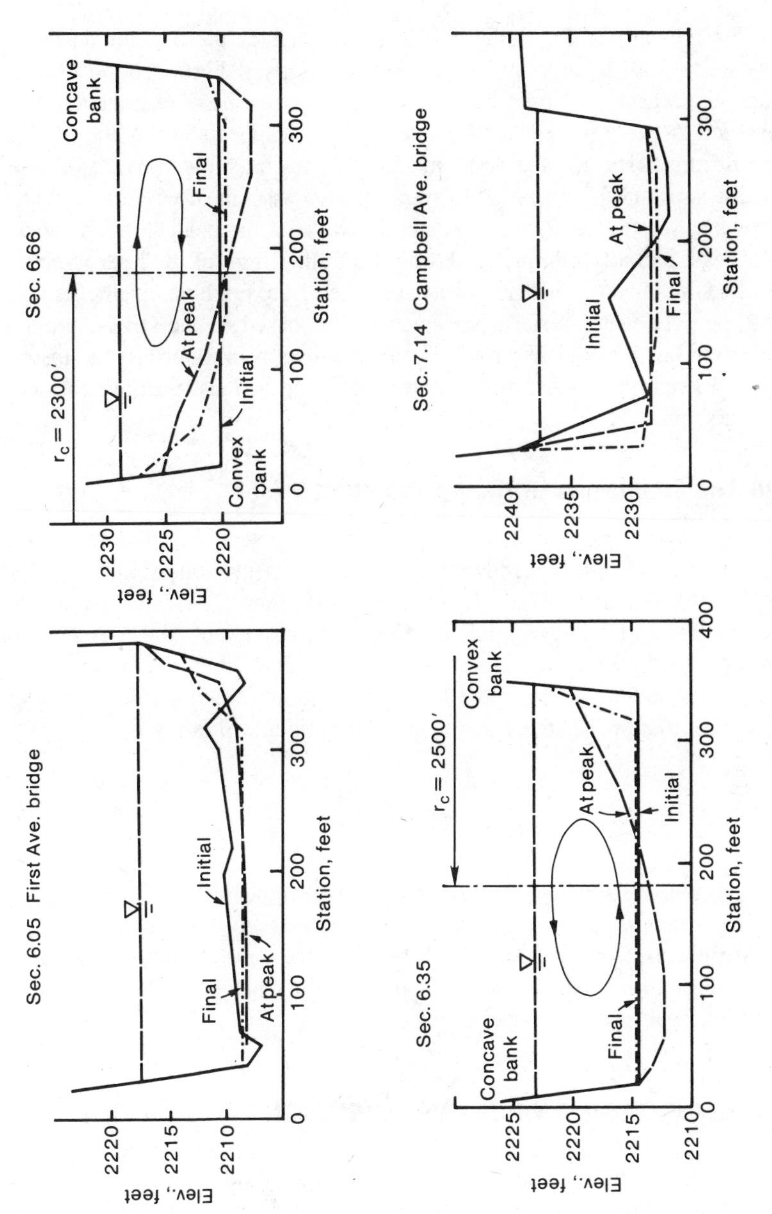


**Figure 14.43** Samples of cross-sectional changes.

Fig. 14.42, together with the respective channel-bed profiles at the peak discharge and at the end of flooding. The channel-bed profile at the end of flooding does not reflect the effect of extended low flow which usually develops an incised small channel.

Cross-sectional profiles shown in Fig. 14.43 include those of the design (the initial), at the peak discharge and at the end of flooding. For a section in a river bend, such as section 6.35 or 6.66, the channel-bed profile is characterized by a transverse bed slope at the peak discharge, which disappears at the end of flooding. The nonuniform channel-bed profile at the peak discharge, shown in Fig. 14.42b, is primarily attributed to spiral motion associated with the curvature, which grows as the flow enters a bend and decays after the bend exit. This type of scour is at least partially eliminated during the falling limb of the hydrograph.

The FLUVIAL profile for the water surface of the erodible channel is nearly a straight line. The water-surface profile computed based on the HEC-2 program, which is a fixed-bed model, is less uniform; it shows backwater effects upstream of bridges. In reality, such bridge obstructions are at least partially offset by channel-bed scour.

### Top and Toe Elevations of Bank Protection

The top and toe elevations of bank protection were selected on the basis of the simulated results and other considerations described in the following paragraphs. The top elevations given in Fig. 14.42 are 2 ft above the peak water-surface profile based on the FLUVIAL model, plus the superelevation. The toe elevations shown in Fig. 14.42 were determined based on the computed maximum channel-bed scour plus one half of the wave height for antidunes and a safety margin of about 6 ft. From Eq. 6.12, the wave height for antidunes $h$ is given by

$$h = 2\pi(0.14)\frac{U^2}{2g} \tag{14.5}$$

For the maximum velocity of 13 ft/sec, the wave height is about 2.3 ft.

An important feature of this design is the variable toe elevation used to account for the difference in scour depth between the concave and convex banks, as shown in Fig. 14.43. Needless to say, this design scheme provides more effective protection against channel-bed scour.

## 14.7 STREAM GAGING OF FLUVIAL SEDIMENT

Natural rivers respond to changes in discharge by adjustments in channel geometry. The dynamic fluvial processes present a challenge with regard to obtaining continuous records of water discharge and sediment load basically because of the difficulties of accounting for the effects of river channel changes. By obtaining samples of the water–sediment mixture, measurements of sediment load are made at discrete time steps. The continuous record of sediment load may be computed at short time steps by using a suitable formula. As an expedient, the average relation of the concentration or bed-material load to the water discharge, sometimes with adjustments for other influences, has often been used to provide a sediment-transport curve for estimation of bed-material loads for various time intervals. However, all the influences cannot be well accounted for by adjustments to a sediment-transport curve. For example, sediment load is affected by channel storage (or depletion) of sediment, which occurs concurrently with river channel changes. Because of the storage effect, sediment load is not a function of water discharge alone but is also related to the pattern of discharge variation. The "hysteresis" phenomenon widely reported in fluvial applications refers to situations where a certain property, such as sediment discharge or flow depth, has different values for a given discharge during rising and falling stages.

The feasibility of applying the FLUVIAL model as an aid for obtaining continuous records of water discharge and sediment load was investigated (Chang et al., 1987). In the first phase of the study, the focus was on providing a continuous record of bed-material load that reflects river channel responses during a changing discharge. The measured water discharge was employed in this phase, but the study can be extended such that the continuous discharge record is obtained by the model based on stage measurements at a gaging station.

In order to determine continuous river channel changes at a gaging station during a flow period, the adjacent channel reach must be included in the simulation. The rate of sediment inflow into the study reach is provided by the upstream boundary condition for sediment. If this rate is known, it may be included as a part of the input and used in the simulation. Unfortunately, sediment rating data are unreliable or simply not available. For such cases, it is assumed that the river channel remains unchanged above the study reach; and sediment inflow rate is computed at the upstream section at each time step, the same way it is computed at other cross sections. For this reason, the study reach should extend far enough upstream so that the channel beyond will have insignificant effect on the sediment inflow. Factors that may induce river channel changes should be included in the study reach.

### Little Arkansas River Study

The FLUVIAL model was applied to the Little Arkansas River at Valley Center, Kansas (Fig. 14.44). Fourteen cross sections were used in the study reach near a stream gaging station. Simulation of river flow and sediment transport in a changing river boundary was made for the period from May 1 to May 9, 1961, for which detailed stream flow and sediment records are available. This river reach has a sand bed with a median diameter of 0.66 mm for the bed material. The measured water discharge as shown in Fig. 14.47 varied between 700 and 3380 cfs. Bed-material loads in the sampled zone (see Fig. 7.26) were measured at several times during the period.

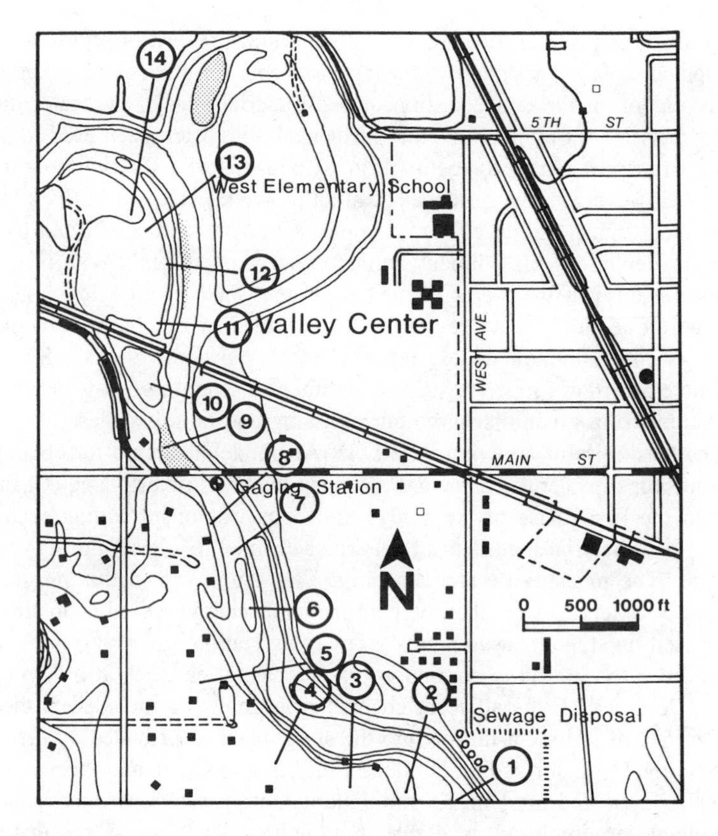


**Figure 14.44** Map of the Little Arkansas River near Valley Center, Kansas.

In the study, flow resistance was computed using the Brownlie formula for alluvial resistance (see Sec. 6.5), sediment discharge was computed using the Yang formula and the Ackers–White formula (see Sec. 7.4). Because the resistance formula is used in the model, it is not necessary to assume the roughness coefficient, a practice required in many models for water-surface computations.

In the simulation, the initial channel geometry defined at the cross sections, bed-material size distribution, and discharge hydrograph were used as input parameters. Complete simulation for the 9-day duration was made using 320 time steps, and it consumed 180 CPU seconds on the VAX 780 computer.

Results of the simulation are shown in Figs. 14.45–14.48. Cross-sectional changes during this period are shown by samples in Fig. 14.45. Although changes are relatively small in this river reach, such changes contribute significantly to sediment storage and, hence, variation in sediment load. If the peak discharge is applied to the initial channel geometry, the computed energy gradient and sediment load, as depicted by the dashed lines, are highly nonuniform (see Fig. 14.46), indicating that the channel is not in dynamic equilibrium. The dashed

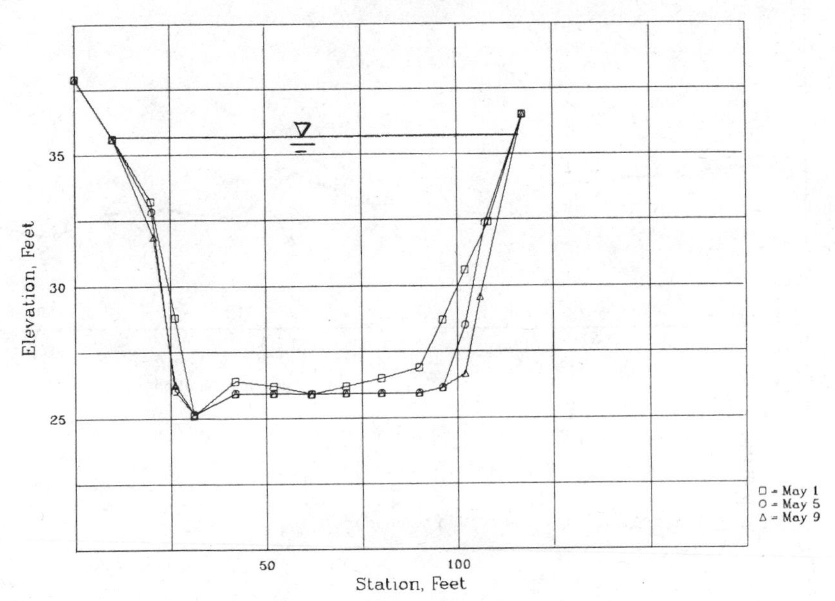


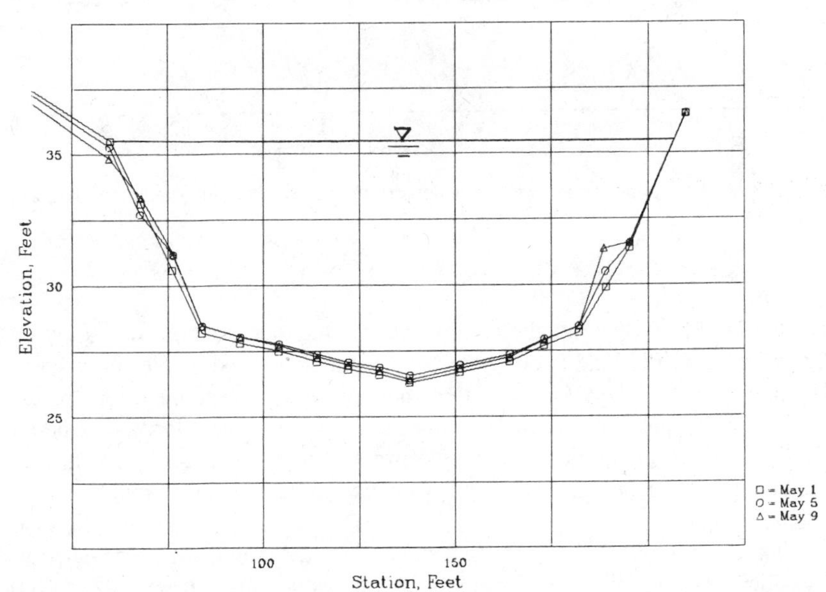


**Figure 14.45** Cross-sectional profiles at different times with maximum water-surface level.

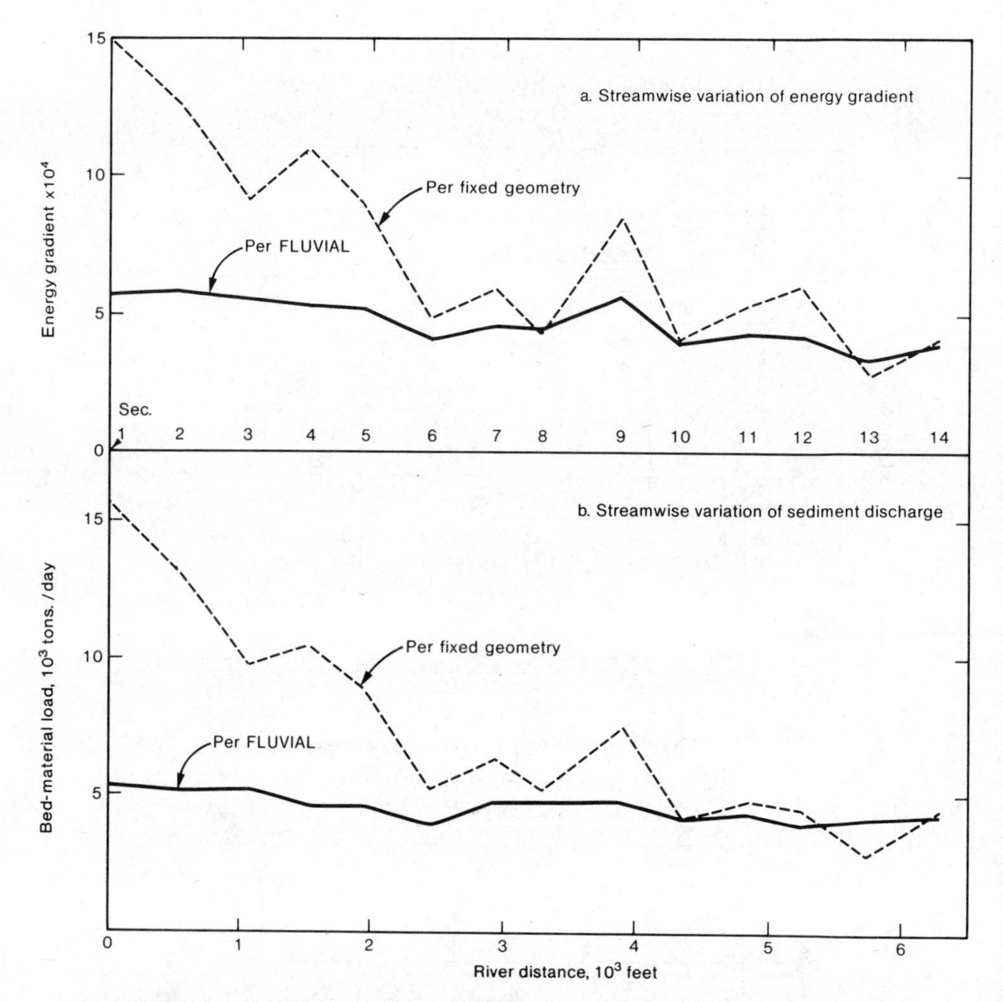


**Figure 14.46** Simulated streamwise variations of energy gradient and sediment discharge at peak flood.

lines in Fig. 14.46 illustrate the significant longitudinal variations of energy gradient and bed-material load at one time point computed based on the surveyed channel geometry a few days earlier. Such a nonuniformity in sediment load precludes meaningful computation of sediment load using any sediment formula. Therefore, this study demonstrates that sediment computation in a changing stream must be based on the transient geometry that is provided in the dynamic simulation of river channel changes.

Time variation of the roughness coefficient (Manning's $n$), simulated using Brownlie's formula, is shown in Fig. 14.47; its value ranges from 0.027 to 0.033, indicating lower regime flow. With the capability of determining the roughness

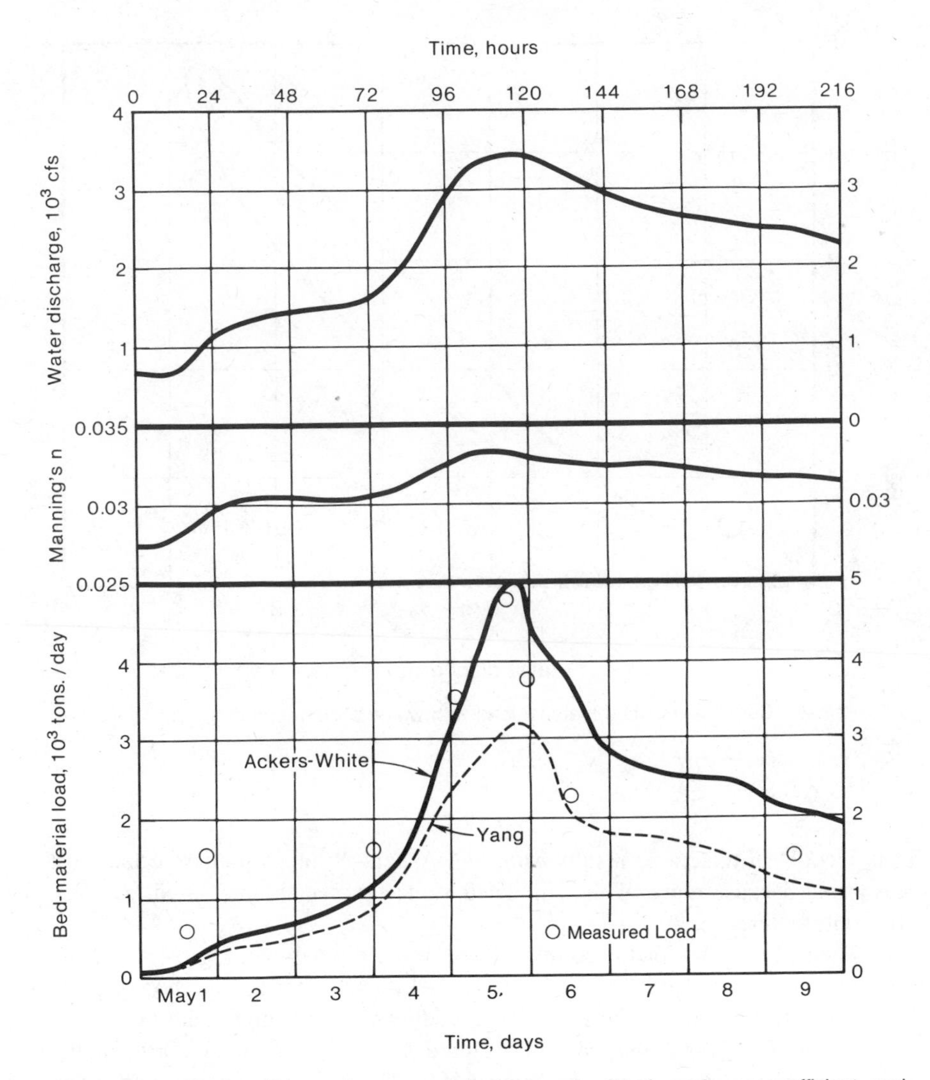


**Figure 14.47** Time variations of measured discharge, simulated roughness coefficient, and bed-material load.

coefficient, the model can be extended to make discharge determinations based on the measured stage. The discharge so determined also reflects the changing alluvial channel boundary.

Because sediment loads measured in the Little Arkansas River are only those in the sampled zone rather than being the total bed-material loads, accurate model results would exceed the measurements. Time variations of bed-material load, simulated using the Ackers–White formula and the Yang formula, are compared with measurements as shown in Fig. 14.47. This comparison serves to provide the test whereby the most appropriate formula may be selected to generate the contin-

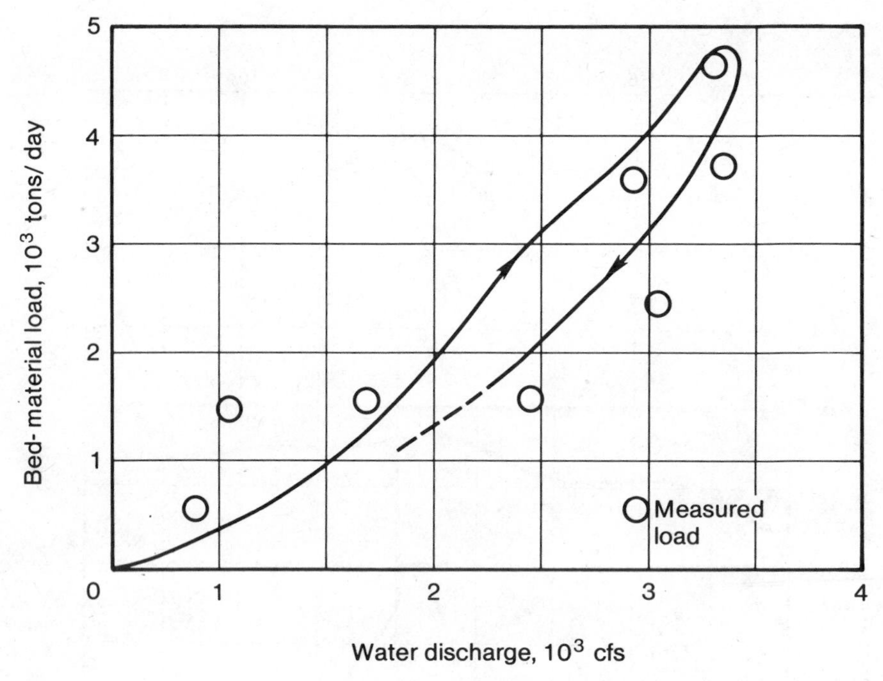


**Figure 14.48** Correlation of bed-material load with water discharge during rising and falling discharge.

uous record. In this case, results using the Ackers–White formula are the most reasonable because more of the simulated values exceed the measured (sampled zone only) values.

The bed-material load is correlated with water discharge in Fig. 14.48 during the rising and falling limbs of the flood. The "hysteresis" effect, as shown by the loop, indicates that at the same discharge, sediment load is higher during the rising stage associated with the removal of stored sediment in the stream or the initial flush of mobile sediment from the land surface. This effect also demonstrates that sediment load is not uniquely correlated with the water discharge, as the rating curve suggests. It is therefore necessary to consider the rise and fall of the hydrograph in stream gaging of fluvial sediment.

In summary, this case study demonstrates the feasibility for stream gaging of fluvial sediment using a movable bed computer model as an aid. The most suitable sediment formula can be selected on the basis of comparison with measurements at discrete points. The simulated results provide the continuous record of sediment load reflecting channel storage of sediment. Because of river channel responses to discharge variation, discharge and sediment calculations need to be made using the changing channel geometry. This study demonstrates that large errors in sediment load computation are unavoidable if a rigid boundary is assumed.

## REFERENCES FOR PART IV

Andrews, E. D., "Bank Stability and Channel Width Adjustment, East Fork River, Wyoming," *Water Resour. Res.*, **18**(4), pp. 1184–1192, August 1982.

Bettess, R. and White, W. R. 1979, "A One Dimensional Morphological River Model," Report No. IT194, Hydraulics Research Station, Wallingford, England, 1979.

Borah, D. K., Alonso, C. V., and Prasad, S. N., "Routing Graded Sediments in Streams: Formations," *J. Hydraul. Div. ASCE*, **102**(HY12), pp. 1486–1503, December 1982.

Bull W. B., "Alluvial Fans," *J. Geol. Ed.*, **16**, pp. 101–106, 1968.

Chang, H. H., "Hydraulics of Rivers and Deltas," Ph.D. Thesis, Department of Civil Engineering, Colorado State University, 1967.

Chang, H. H. and Hill, J. C., "Computer Modeling of Erodible Flood Channels and Deltas," *J. Hydraul. Div. ASCE*, **102**(HY10), pp. 1461–1477, October 1976.

Chang, H. H. and Hill, J. C., "Minimum Stream Power for Rivers and Deltas," *J. Hydraul. Div. ASCE*, **103**(HY12), pp. 1375–1389, December 1977.

Chang, H. H., "Mathematical Model for Erodible Channels," *J. Hydraul. Div. ASCE*, **108**(HY5), pp. 678–689, May 1982.

Chang, H. H., "Modeling of River Channel Changes," *J. Hydraul. Eng. ASCE*, **110**(2), pp. 157–172, February 1984.

Chang, H. H., "Water and Sediment Routing through Curved Channels," *J. Hydraul. Eng. ASCE*, **111**(4), pp. 644–658, April 1985.

Chang, H. H., Osmolski, Z., and Smutzer, D., "Computer-Based Design of River Bank Protection," Proceedings of the Hydraulics Division Conference, ASCE, Orlando, Florida, August 13–16, 1985, pp. 426–431.

Chang, H. H., Jennings, M. E., and Jordan, P. R., "Use of Calibrated Model for Continuous Record of Fluvial Sediment Load," *USGS Water Supply Paper*, 1987.

County of San Diego, "Lakeside Storm Report, March 5–6, 1978," Department of Sanitation and Flood Control, San Diego, California, 1978.

County of San Diego, "Flood Plain Changes During Major Floods," Department of Sanitation and Flood Control, San Diego, California, 1979.

County of San Diego, "Storm Report, February 1980," Flood Control District, San Diego, California, 1980, 26 pp.

Dawdy, D. R. and Vanoni, V. A., "Modeling Alluvial Channels," *Water Resour. Res.*, **22**(9), pp. 71S–81S, August 1986.

Holley, F. M. Jr. and Karim, M. F., "Simulation of Missouri River Bed Degradation," *J. Hydraul. Eng. ASCE*, **112**(6), pp. 497–517, June 1986.

Ikeda, S. and Nishimura, T., "Flow and Bed Profile in Meandering Sand–Silt Rivers," *J. Hydraul. Eng. ASCE*, **112**(7), pp. 562–579, July 1986.

Krishnappan, B. G., "Users Manual. Unsteady, Nonuniform, Mobile Boundary Flow Model–MOBED," Hydraulics Division, National Water Research Institute, Canada Center for Inland Waters, Burlington, Ontario, Canada, 1981.

National Academy of Sciences, "An Evaluation of Flood-Level Prediction Using Alluvial River Models," Committee on Hydrodynamic Computer Models for Flood Insurance Studies, Advisory Board on the Built Environment, National Research Council, National Academy Press, Washington, D.C., 1983.

Rozovskii, I. L., "Flow of Water in Bends of Open Channels," the Academy of Sciences of the Ukrainian SSR, 1957, translated form Russian by the Israel Program for Scientific Translations, Jerusalem, Israel, 1961 (available from Office of Technical Services, U.S. Department of Commerce, Washington, D.C., PST Catalog No. 363, OTS 60-51133).

San Francisco District, Sponsor, "San Lorenzo River, Field and Simulation Studies," prepared for San Francisco District, U. S. Army Corps of Engineers, prepared by Jones–Tillson and Associates, Water Resources Engineers, and H. Esmaili and Associates, September 1980.

Simons, Li and Associates, Inc., "Erosion, Sedimentation, and Debris Analsis of Boulder Creek, Boulder, Colorado," prepared for UPS Company, Denver, Colorado, 1980.

U. S. Army Corps of Engineers, "HEC-6, Scour and Deposition in Rivers and Reservoirs, Users Manual," Hydrologic Engineering Center, Davis, California, 1977.

U. S. Army Corps of Engineers, "HEC-2 Water Surface Profiles, Users Manual," Hydrologic Engineering Center, Davis, California, November 1982.

U. S. Bureau of Reclamation, "Aggradation and Degradation in the Vicinity of Milburn Diversion Dam," Sedimentation Section, 1963.

Vanoni, V. A., Born, R. H., and Nouri, H. M., "Erosion and Deposition at a Sand and Gravel Mining Operation in San Juan Creek, Orange County, California," *Storms, Floods, and Debris Flows in Southern California and Arizona 1978 and 1980*, Proceedings of a Symposium, September 17–18, 1980, National Academy Press, Washington, D.C., 1982, pp. 271–289.

Yen, C. L., "Bed Topography Effect on Flow in a Meander," *J. Hydraul. Div. ASCE*, **96**(HY1), pp. 57–73, January 1970.

Zhang, Q., Zhang, Z., Yue, J., Duan, Z., and Dai, M., "A Mathematical Model for the Prediction of the Sedimentation Process in Rivers," Proceedings of the 2nd International Symposium on river Sedimentation, Nanjing, China, 1983.

# PART V

## RIVER ENGINEERING

# 15

# RIVER TRAINING

River training, in the broad subject sense, covers all the engineering works in a river to regulate the river flow and sediment transport for the sake of flood control, navigation, irrigation, and channel stabilization. River training by embankments can be traced to early human history, exemplified by the works on the Nile, the Yellow River, the Euphrates and Tigris, and the River Ganga. Such practice still continues in modern times. Rivers are often trained because of urbanization; land reclamation from flood plains has resulted in the channelization of many rivers in different parts of the world.

The principle of river training is twofold: (1) The training works must be designed to withstand the design flow, and (2) the impacts (consequences) on the river should be understood and evaluated whenever feasible. Therefore, the training works must be strong enough for the design velocity; they must also extend beyond the potential scour in order to safeguard against undermining. A river, as a system, is subject to changes in response to any type of training or regulation in the system. Such responses will usually occur at places where the bed is mobile or the where the bank is unprotected. Assume a river reach is straightened and channelized; its freedom for meandering is thus constrained. In response to this change in slope due to straightening, changes in the adjacent natural reaches may occur in order to restore the equilibrium of the river system.

Any plan for river training needs to be evaluated with regard to channel responses; the major steps that should be followed in any river engineering project are illustrated in Fig. 15.1 (Ingram, 1986). The procedures suggest that channel responses are the criteria for determining the adequacy of a design scheme. Modifications in design are often necessary in arriving at the final plan. In modern river engineering, a very challenging task is the prediction of river channel responses, which can be made either by physical modeling or by mathematical means. Analytical methods, including mathematical modeling, for evaluating river channel responses are covered in the first four parts of this book. Particular attention is called to the existence of thresholds in the hydraulic geometry of rivers as described in Sec. 11.3.

Types of training works, their features, and design considerations are described in this chapter. The commonly used types of training works can be

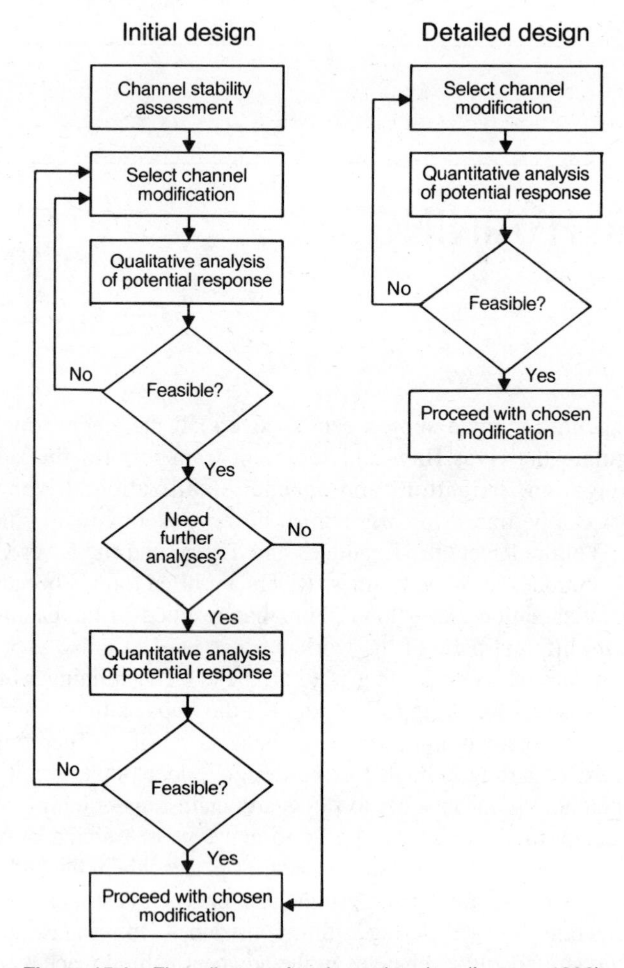


**Figure 15.1** Flow diagram for channel project (Ingram, 1986).

broadly classified as bank protection, dikes, and grade-control structures; they are described in the following sections.

## 15.1 BANK PROTECTION

Different types of bank protection are in use, including riprap, rock trench, mattress, gabion, soil cement, concrete blocks, cribs, and so on. In general, the top elevation of bank protection should stay above the design high water. Within curved reaches, the superelevation of water surface should also be considered. A freeboard of various heights is usually used for water waves; it also serves as a safety margin for the unaccounted factors, such as erratic hydrologic phenomena, changes in flood plain vegetation, unforeseen riprap settlement, channel-bed aggradation, accumulation of trash and debris, and so on.

For any structure built in erodible material, the toe elevation should extend below the expected scour by a minimum of 5 vertical feet in medium to large streams (see Fig. 15.2). The purpose of the toe is to prevent undermining but is not to support the structure above. On the concave bank of sharp river bends, scour is particularly severe and the toe protection should be deeper than that in straight reaches. Methods for evaluating potential scour due to channel curvature alone are given in Sec. 8.4, and those due to curvature and gradation changes are in Sec. 14.6. The commonly used types of bank protection—rock riprap and soil cement—are described as follows.

## Rock Riprap

Rock riprap is a popular material for bank protection because of its availability and effectiveness. Design Criteria for rock riprap are given by the Corps of Engineers (1970), the Federal Highway Administration (Searcy, 1967; Norman, 1975), and state departments of transportation. Rock riprap is capable of providing protection even with minor undermining of the toe since the loose stones will settle into the scour hole and thus extend the protection. The stability of riprap revetment depends on the major factors of stone weight, stone shape, gradation, and riprap layer thickness. In addition, the manner of placement and treatment of extremities also affect its stability. The selection of appropriate riprap for a particular project is usually based on the tractive force method given in Chapter 5.

Stones used for riprap should be hard, durable, and angular in shape. Slab-like stones, which are susceptible to hydrodynamic forces, should be avoided. Well-graded material should be used for the riprap blanket so that the interstices formed by large stones are filled by smaller ones in an interlocking fashion. Use of poorly graded material may result in a blanket with large pockets and loss of the bank material. However, poor gradations of rock used as riprap can be reme-

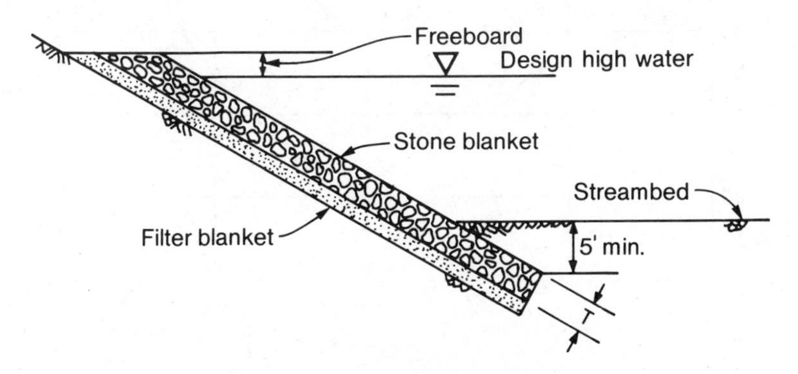


**Figure 15.2** Riprap bank protection.

died with a proper filter placed between the riprap and bank material. Criteria providing guidelines for establishing gradation limits of riprap are recommended by the Corps of Engineers (1970). The size of median-weight stone $W_{50}$ is the representative size to withstand the design shear. Suggested gradation by Simons, Li and Associates (1982) is given in Fig. 15.3. Under this gradation, the ratio of maximum size to median size $d_{50}$ is about 2, and small sizes range down to gravel. Control of the riprap gradation is made by visual inspection.

Riprap should be placed on the filter blanket (see Fig. 15.2) or directly on the prepared slope when the filter blanket is not required. The placement may be accomplished by dumping stones from trucks directly or by hand. Draglines with buckets and other power equipment are also used in the placement (see Fig. 15.4). The riprap should be so placed that there is approximate uniformity with no segregation in size. The thickness of the stone blanket should be at least equal to the maximum stone size and should be at least 12 in. for practical placement. The Corps of Engineers (1970) also requires the thickness to be at least 1.5 times the spherical diameter of the $W_{50}$ stone.

The riprap layer should normally be extended to the required toe elevation. But in rivers having a considerable depth at low water stages, placement of toe protection becomes very difficult. The Corps of Engineers (1970) has an alternative method as shown in Fig. 15.5. In this method, a thicker horizontal rock toe is provided, which will settle as degradation develops.

## Filter Layer

A filter layer is usually needed beneath the riprap cover to prevent the water from removing bank material through the voids. The filter layer may be either a granu-

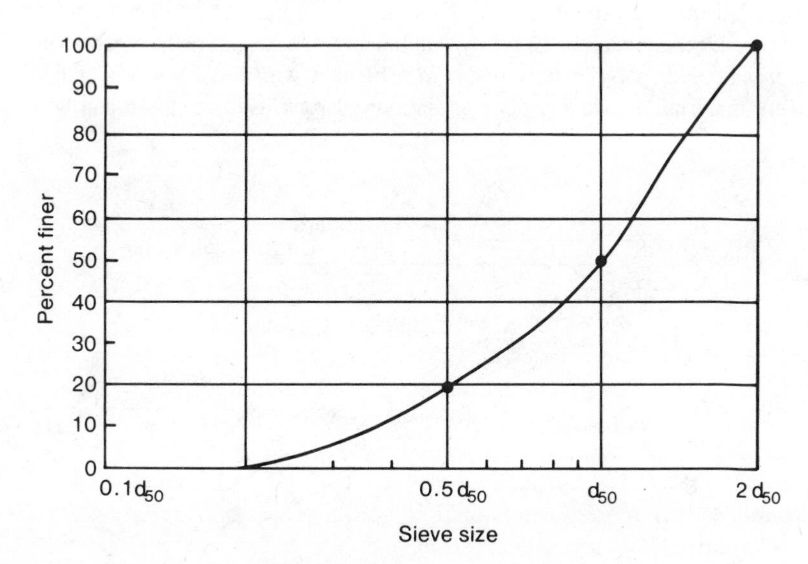


**Figure 15.3** Suggested gradation for rock riprap (after Simons, Li and Associates, 1982).

**Figure 15.4** Placement of riprap using equipment.

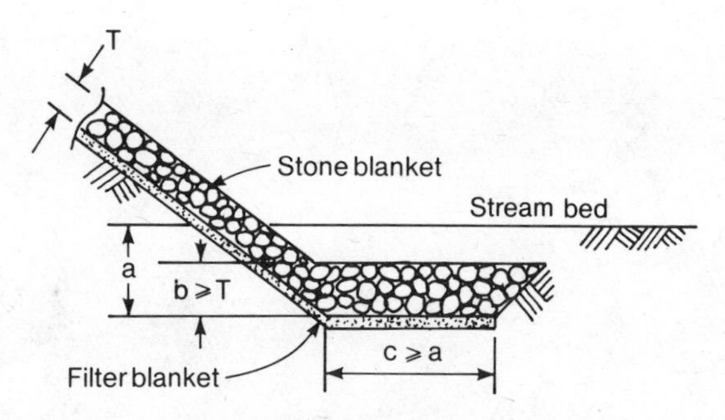


**Figure 15.5** Bank protection with thicker horizontal toe (after Corps of Engineers, 1970).

lar filter blanket or plastic filter cloth. The former provides a transition between the rock layer and the bank material, with sizes of gravel ranging from about 3/16 in. to about 3 in. Whether this transition is needed depends on the size and gradation of the rock layer in relation to the bank material. The filter is not required if the stones are sufficiently small. The suggested criteria for gradation are as follows:

$$\frac{d_{15} \text{ of filter}}{d_{85} \text{ of base}} < 5 < \frac{d_{15} \text{ of filter}}{d_{15} \text{ of base}} < 40 \tag{15.1}$$

and

$$\frac{d_{50} \text{ of filter}}{d_{50} \text{ of base}} < 40 \tag{15.2}$$

where filter refers to the overlying material, and base refers to the underlying material. These criteria are applicable to any two adjacent layers among the riprap, filter blanket, and base material.

In the case of very large stones in the rock layer, multiple filter layers with gradual size variations are required. The thickness of filter material ranges from 6 to 15 in. for a single layer or from 4 to 8 in. for individual layers of a multiple layer blanket. Design of the filter blanket is illustrated by the following example from Anderson et al. (1970).

▶ ***EXAMPLE 15.1.*** A channel will be constructed on a base material. The characteristics of the base material and riprap are given as follows:

| **Base Material** | **Riprap** |
|---|---|
| $d_{15} = 0.167$ mm | $d_{15} = 100$ mm |
| $d_{50} = 0.5$ mm | $d_{50} = 200$ mm |
| $d_{85} = 1.5$ mm | $d_{85} = 400$ mm |

Determine the necessary filter layers and their respective size distributions.

*Solution:* Application of the criteria given by Eqs. 15.1 and 15.2 yields

$$\frac{d_{15} \text{ of riprap}}{d_{85} \text{ of base}} = \frac{100}{1.5} = 66.7 > 5$$

$$\frac{d_{15} \text{ of riprap}}{d_{15} \text{ of base}} = \frac{100}{0.167} = 600 > 40$$

$$\frac{d_{50} \text{ of riprap}}{d_{50} \text{ of base}} = \frac{200}{0.5} = 400 > 40$$

These figures do not meet the recommended criteria, a filter blanket is therefore needed. The required size gradation of the filter with respect to the base material is determined from

$$\frac{d_{50} \text{ of filter}}{d_{50} \text{ of base}} < 40; \text{ hence } d_{50} \text{ of filter} < 40 \times 0.5 = 20 \text{ mm}$$

$$\frac{d_{15} \text{ of filter}}{d_{15} \text{ of base}} < 40; \text{ hence } d_{15} \text{ of filter} < 40 \times 0.0167 = 6.7 \text{ mm}$$

$$\frac{d_{15} \text{ of filter}}{d_{85} \text{ of base}} < 5; \text{ hence } d_{15} \text{ of filter} < 5 \times 1.5 = 7.5 \text{ mm}$$

$$\frac{d_{15} \text{ of filter}}{d_{15} \text{ of base}} > 5; \text{ hence } d_{15} \text{ of filter} > 5 \times 0.167 = 0.835 \text{ mm}$$

From these results, the filter material adjacent to the base material should have the following dimensions: $d_{50} < 20$ mm, and $0.83 \text{ mm} < d_{15} < 6.7$ mm.

In the next step, the required filter dimensions with respect to the riprap is determined from

$$\frac{d_{50} \text{ of riprap}}{d_{50} \text{ of filter}} < 40; \text{ hence } d_{50} \text{ of filter} > \frac{200}{40} = 5 \text{ mm}$$

$$\frac{d_{15} \text{ of riprap}}{d_{15} \text{ of filter}} < 40; \text{ hence } d_{15} \text{ of filter} > \frac{100}{40} = 2.5 \text{ mm}$$

$$\frac{d_{15} \text{ of riprap}}{d_{85} \text{ of filter}} < 5; \text{ hence } d_{85} \text{ of filter} > \frac{100}{5} = 20 \text{ mm}$$

$$\frac{d_{15} \text{ of riprap}}{d_{15} \text{ of filter}} > 5; \text{ hence } d_{15} \text{ of filter} < \frac{100}{5} = 20 \text{ mm}$$

On the basis of these numbers, the filter layer adjacent to the riprap should have the following dimensions: $5 \text{ mm} < d_{50} < 20$ mm, $2.5 \text{ mm} < d_{15} < 20$ mm, and

$d_{85} > 20$ mm. The gradations of granular filter blanket for this example problem are shown in Fig. 15.6; those within the crosshatched area are adequate for the base material as well as for the riprap.

## Plastic Filter Cloths

Since about 1967, filter cloths have been used beneath riprap or other material for bank protection as a substitute for the granular filter blanket with considerable success. Care must be exercised during placement of stones directly on the cloth to avoid damage. Stones as large as 3000 lb have been carefully placed on the fabric without causing apparent damage. The sides and toes of the filter fabric should be sealed or trenched to contain the base material.

## Soil Cement

Soil–cement blocks, as shown in Figs. 15.7–15.9, are used as bank protection as a suitable and inexpensive alternative to rock riprap. They are especially popular at places where riprap is not readily available at a reasonable cost. Experiments were carried out in India (Joglekar, 1971) to study the strength and durability of such material used for bank protection. It was determined that soil cement with the following range of gradation responded best to stabilization:

Sand content (0.05–2 mm) 60–80 %
Silt content (0.005–0.05 mm) 12–25 %
Clay content (finer than 0.005 mm) 8–15 %

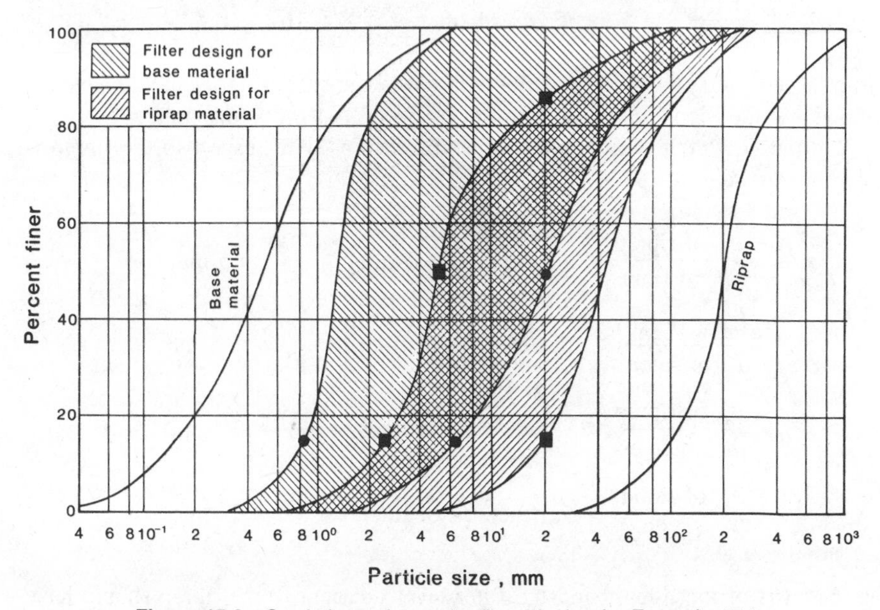


**Figure 15.6** Gradations of granular filter blanket for Example 15.1.

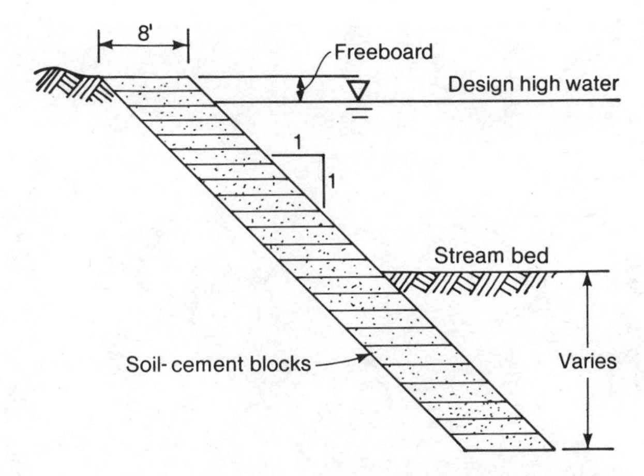


**Figure 15.7** Typical design of bank protection using soil–cement blocks.

The adequate cement content was determined to range from 9 to 10% by weight of the soil.

The soil–cement bank protection is popular in the semiarid southwestern United States, where the construction is facilitated by the usually dry river bed. Such bank protection is also desirable from the environmental viewpoint because it preserves the soil appearance. Among the experiences, the soil–cement bank protection along the Santa Cruz River in Tucson survived the October 1983 flood, which reached the 100-yr magnitude. During the peak flow, the cross-sectionally averaged velocity exceeded 15 ft/sec at many places.

## 15.2 DIKES

Dikes are training structures that extend from the bank to the river at an angle, or perpendicular, to the flow, as shown in Figs. 15.10–15.12. Most dikes are aligned in a slightly downstream direction. They are also known as groins, spurs, spur dikes, or transverse dikes and represent a very widely used type of training works. Dikes are often used to form a system covering a certain river reach. Typical layouts of the dike system on the Arkansas River are shown in Fig. 15.10.

Dikes serve one or more of the following functions: (1) training a river along a desired course, (2) creating a region of low velocity to induce siltation, (3) protecting the bank by keeping the flow away, and (4) contracting a wide river channel usually for the improvement of depth for navigation.

There are two principal types of dikes: permeable and impermeable dikes. Permeable dikes, such as timber pile dikes (Fig. 15.11) and jetty fields (Fig. 15.12), permit flow through the dikes at reduced velocities, thereby preventing bank erosion and causing deposition of suspended sediment from the flow. Experience has shown that permeable dikes are more effective than solid ones as a

**Figure 15.8** Soil–cement bank protection on Canada del Oro Wash near Tucson, Arizona.

**Figure 15.9** Aerial view of Santa Cruz River with soil–cement bank protection.

bank protection, especially in silt and sand rivers. Permeable dikes have the major advantage of being economical. They are especially useful when riprap is difficult to obtain and in deep rivers, where solid dikes are expensive. Since the flow is not severely disturbed by permeable dikes, no intensive eddies and severe scour holes will result. However, permeable dikes are not strong enough for streams with a high velocity. Submerged dikes also present a hazard for navigation.

Timber pile dikes, as exemplified by Fig. 15.11, are permeable dikes. With a number of variation in scheme, they may consist of closely spaced single, double, or multiple rows. Wire fence may be used with a pile to collect debris, thereby causing effective reduction in velocity. The base of the piles can be protected from scour with dumped rock in sufficient quantities.

Jetties, or steel jacks, consist of angles fastened together and strung with wire (see Fig. 15.13). The jetties are tied together with cables to be used in rows to form a jetty field as illustrated in Fig. 15.12. The jetty field is typically used in shallow streams to train the stream into a single narrower channel. The jetty field reduces the velocity near the bank and hence protects the bank from erosion. It is more effective in streams that have a considerable amount of debris and a high concentration of suspended load.

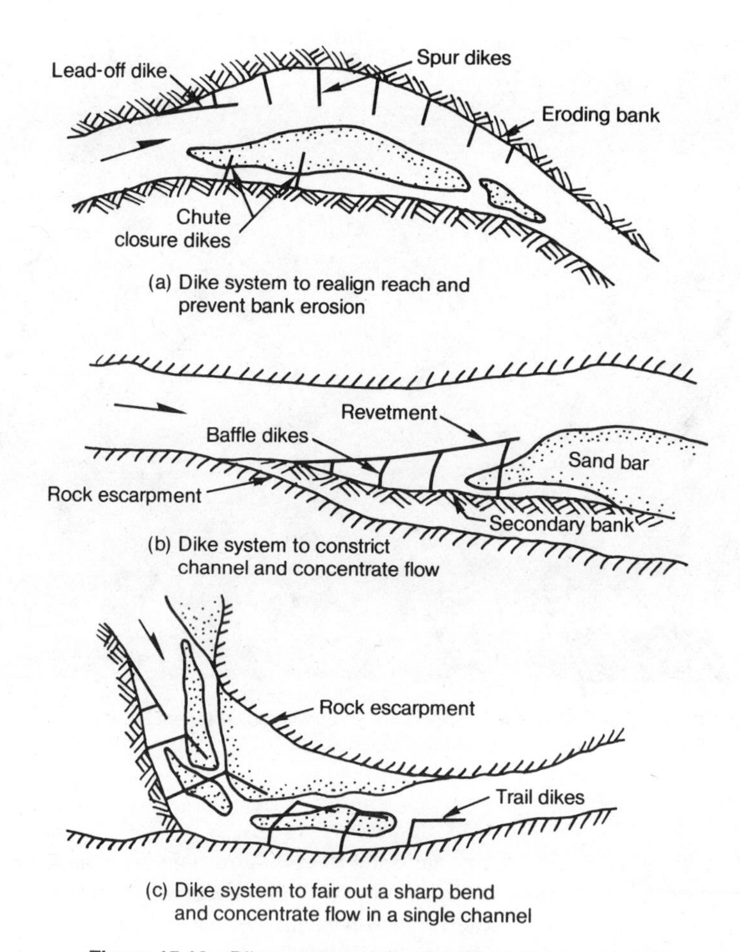


**Figure 15.10** Dike systems, Arkansas River (Petersen, 1986).

Impermeable or solid dikes are designed to attract, repel, or deflect the flow away from the bank along a desired course of flow. They are usually rock-filled or masonry structures. The rock-filled dikes are constructed with well-graded stones so that large voids are eliminated. The head and the toe of an impermeable dike usually need to be armored heavily with materials like large stones, concrete blocks, and so on. Such dikes also need to be extended sufficiently deep into the erodible bed because of the severe potential scour near the toe, around which large stones are usually dumped. In a series of repelling dikes, the uppermost dike should be especially constructed to withstand the most severe attack. Dikes may be designed to stay under water at high flow. However, the top material must be strong enough to withstand overtopping.

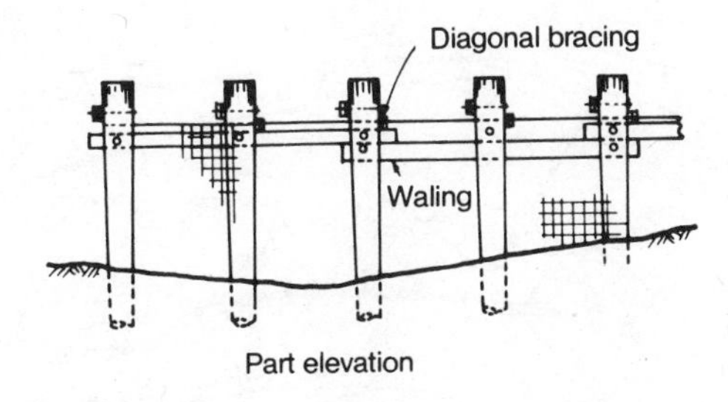


(a) Single row timber pile with wire fence

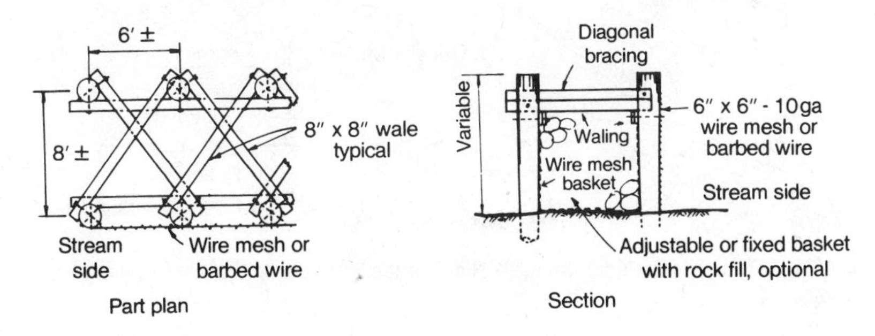


(b) Double row timber piles with rocks and wire fence

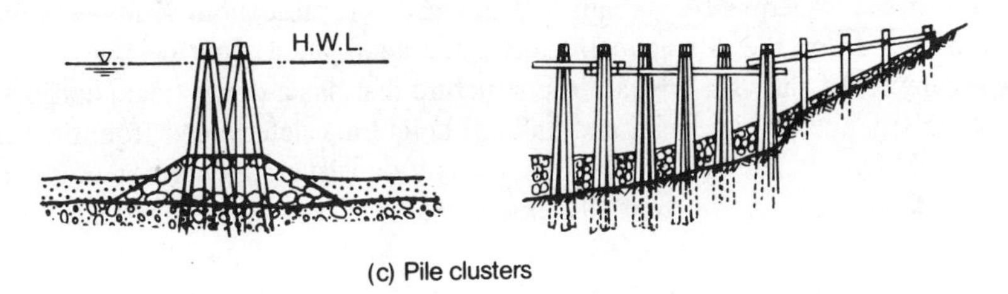


(c) Pile clusters

**Figure 15.11** Timber pile dike.

## 15.3 GRADE-CONTROL STRUCTURES

Grade-control structures—also called drop structures, stabilizers, weirs, barrages, or check dams—are generally constructed normal to the channel flow and traverse the channel bed. They are used in river channels to maintain a slope flatter than the slope of the terrain. Stabilizers refer to sediment control structures that are used primarily to stabilize the upstream channel bed where scour may endanger certain structures such as bridge foundations. The crest of a structure usually extends across the channel, and the side walls should extend into the bank and

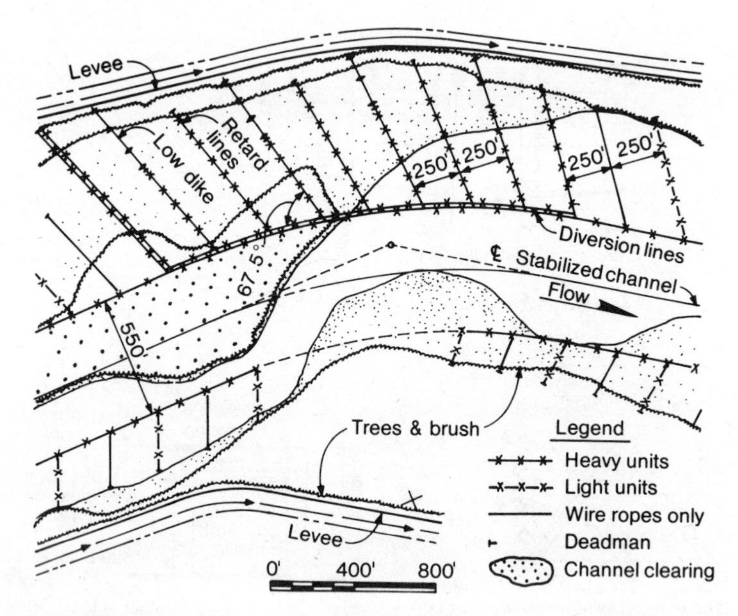


**Figure 15.12** Typical jetty field layout—Rio Grande (source: U.S. Army Engineer District, Albuquerque, NM).

have adequate bank protection to prevent flanking at high flows. Each structure should also have adequate upstream and downstream protection. Dumped stones should be placed on the downstream side to the anticipated scour depth.

Figure 15.14 shows a typical drop structure that has a downstream apron and end sill. The length of the basin and end sill height are determined from the relationships shown in the figure. A stabilizer may consist of grouted or ungrouted rock, sheet piling, or a concrete sill. Figure 15.15 shows the grouted-stone-type

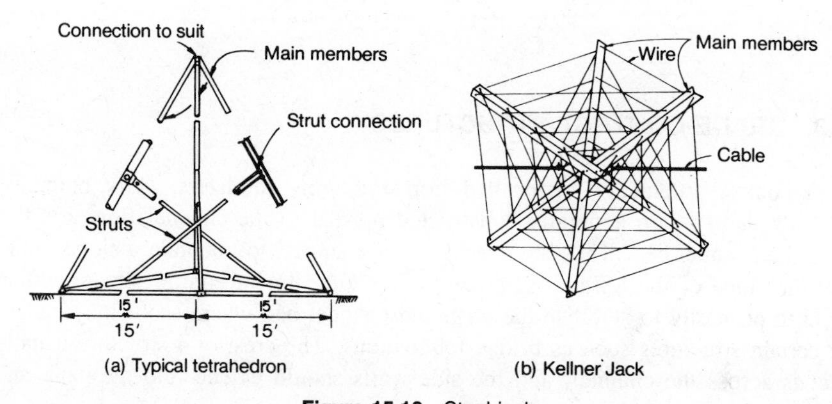


**Figure 15.13** Steel jack.

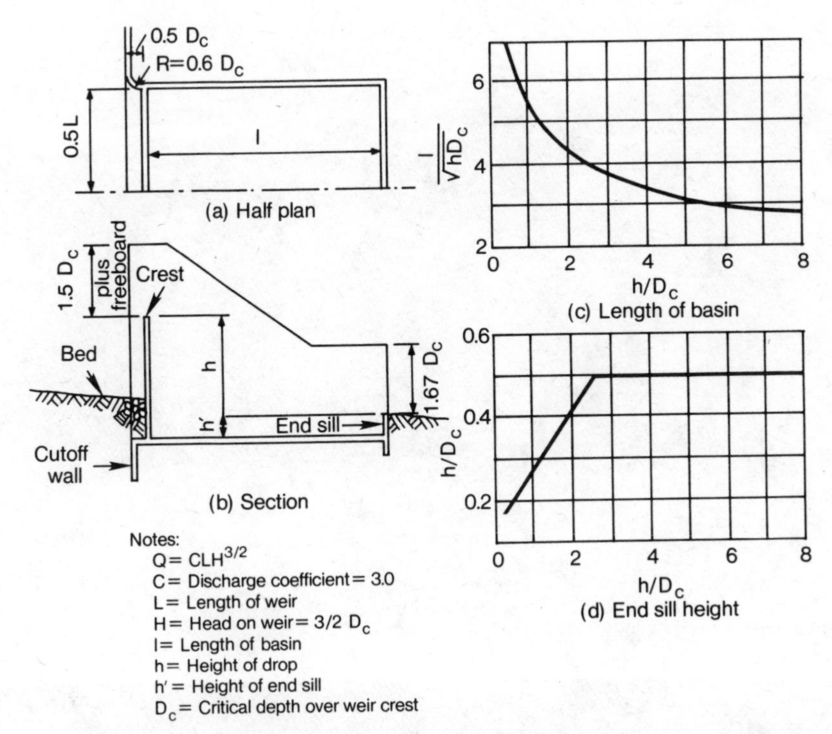


**Figure 15.14** Typical grade-control structure (after Corps of Engineers,1970).

stabilizer used by the Corps of Engineers (1970). Fig. 15.16 shows a stabilizer used to maintain the upstream channel bed, installed downstream of a bridge on the San Diego River. In this case, the structure is built with sheet piles and is protected with gabions on the downstream side.

While a grade-control structure stabilizes the upstream channel bed, it usually induces downstream changes, which are either related to the gradation change

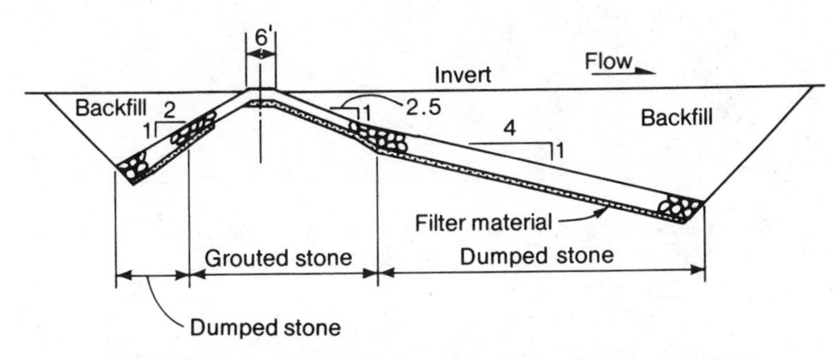


**Figure 15.15** Typical stabilizer (after Corps of Engineers, 1970).

**Figure 15.16** River-bed erosion induced by sand mining (top) stabilized by sheet-pile grade-control structure (bottom), San Diego River, San Diego, California.

in the reach or to local scour, or both. Gradation change is invoked by any imbalance in sediment transport in the river reach; it ceases when a dynamic equilibrium or regime is established. When major gradation change is expected, a series of grade-control structures may be used to limit the extent of change at each structure. Local scour is related to the local flow pattern as affected by the structure; it occurs when excess energy and high velocity are dissipated in the turbulent eddies. An important task in designing a grade-control structure is to provide protection for downstream local scour.

## REFERENCES

Anderson, A. G., Paintel, A. A., and Davenport, J. T., "Tentative Design Procedure for Riprap Lined Channels," NCHRP Report 108, Highway Research Board, National Academy of Sciences, 1970.

Corps of Engineers, "Hydraulic Design of Flood Control Channels," EM 1110-2-1601, Department of the Army, Office of the Chief of Engineers, July 1970.

Ingram, J. J., "Practical Flood-Control Channel Design," Proceedings of the 3rd International Symposium on River Sedimentation, Jackson, Mississippi, 1986, pp. 1169–1173.

Joglekar, D. V., "Manual on River Behavior Control and Training," Publication No. 60, Central Board of Irrigation and Power, New Delhi, India, September 1971, 432 pp.

Norman, J. M., "Design of Stable Channels with Flexible Lining," Hydraulic Engineering Circular No. 15, Federal Highway Administration, U.S. Department of Transportation, 1975, 136 pp.

Petersen, M. S., *River Engineering*, Prentice-Hall, Englewood Cliffs, New Jersey, 1986.

Searcy, J. K., "Use of Riprap for Bank Protection," Hydraulic Engineering Circular No. 11, Bureau of Public Roads, (Federal Highway Administration), U.S. Department of Transportation, June 1967.

Simons, Li and Associates, *Engineering Analysis of Fluvial Systems*, Fort Collins, Colorado, 1982.

# APPENDIX: SOME COMMONLY USED TABLES

**TABLE A.1 Physical Properties of Water in English Units**

| Temperature (°F) | Specific Weight $\gamma$ (lb/ft$^3$) | Density $\rho$ (slugs/ft$^3$) | Viscosity $\mu \times 10^5$ (lb · sec/ft$^2$) | Kinematic Viscosity $\nu \times 10^5$ (ft$^2$/sec) | Heat of Vaporization (Btu/lb) | Vapor Pressure $p_v$ (psia) | Vapor Pressure Head $p_v/\gamma$ (ft) | Bulk Modulus of Elasticity $E_v \times 10^{-3}$ (lb/in.$^2$) |
|---|---|---|---|---|---|---|---|---|
| 32 | 62.42 | 1.940 | 3.746 | 1.931 | 1075.5 | 0.09 | 0.20 | 293 |
| 40 | 62.43 | 1.940 | 3.229 | 1.664 | 1071.0 | 0.12 | 0.28 | 294 |
| 50 | 62.41 | 1.940 | 2.735 | 1.410 | 1065.3 | 0.18 | 0.41 | 305 |
| 60 | 62.37 | 1.938 | 2.359 | 1.217 | 1059.7 | 0.26 | 0.59 | 311 |
| 70 | 62.30 | 1.936 | 2.050 | 1.059 | 1054.0 | 0.36 | 0.84 | 320 |
| 80 | 62.22 | 1.934 | 1.799 | 0.930 | 1048.4 | 0.51 | 1.17 | 322 |
| 90 | 62.11 | 1.931 | 1.595 | 0.826 | 1042.7 | 0.70 | 1.61 | 323 |
| 100 | 62.00 | 1.927 | 1.424 | 0.739 | 1037.1 | 0.95 | 2.19 | 327 |
| 110 | 61.86 | 1.923 | 1.284 | 0.667 | 1031.4 | 1.27 | 2.95 | 331 |
| 120 | 61.71 | 1.918 | 1.168 | 0.609 | 1025.6 | 1.69 | 3.91 | 333 |
| 130 | 61.55 | 1.913 | 1.069 | 0.558 | 1019.8 | 2.22 | 5.13 | 334 |
| 140 | 61.38 | 1.908 | 0.981 | 0.514 | 1014.0 | 2.89 | 6.67 | 330 |
| 150 | 61.20 | 1.902 | 0.905 | 0.476 | 1008.1 | 3.72 | 8.58 | 328 |
| 160 | 61.00 | 1.896 | 0.838 | 0.442 | 1002.2 | 4.74 | 10.95 | 326 |
| 170 | 60.80 | 1.890 | 0.780 | 0.413 | 996.2 | 5.99 | 13.83 | 322 |
| 180 | 60.58 | 1.883 | 0.726 | 0.385 | 990.2 | 7.51 | 17.33 | 318 |
| 190 | 60.36 | 1.876 | 0.678 | 0.362 | 984.1 | 9.34 | 21.55 | 313 |
| 200 | 60.12 | 1.868 | 0.637 | 0.341 | 977.9 | 11.52 | 26.59 | 308 |
| 212 | 59.83 | 1.860 | 0.593 | 0.319 | 970.3 | 14.70 | 33.90 | 300 |

**TABLE A.2 Physical Properties of Water in SI Units**

| Temperature (°C) | Specific Weight $\gamma$ (kN/m$^3$) | Density $\rho$ (kg/m$^3$) | Viscosity $\mu \times 10^3$ (N · sec/m$^2$) | Kinematic Viscosity $\nu \times 10^6$ (m$^2$/sec) | Heat of Vaporization (J/g) | Vapor Pressure $p_v$ (kN/m$^2$, abs) | Vapor Pressure Head $p_v/\gamma$ (m) | Bulk Modulus of Elasticity $E_v \times 10^{-6}$ (kN/m$^2$) |
|---|---|---|---|---|---|---|---|---|
| 0 | 9.805 | 999.8 | 1.781 | 1.785 | 2500.3 | 0.61 | 0.06 | 2.02 |
| 5 | 9.807 | 1000.0 | 1.518 | 1.519 | 2488.6 | 0.87 | 0.09 | 2.06 |
| 10 | 9.804 | 999.7 | 1.307 | 1.306 | 2476.9 | 1.23 | 0.12 | 2.10 |
| 15 | 9.798 | 999.1 | 1.139 | 1.139 | 2465.1 | 1.70 | 0.17 | 2.15 |
| 20 | 9.789 | 998.2 | 1.002 | 1.003 | 2453.0 | 2.34 | 0.25 | 2.18 |
| 25 | 9.777 | 997.0 | 0.890 | 0.893 | 2441.3 | 3.17 | 0.33 | 2.22 |
| 30 | 9.764 | 995.7 | 0.798 | 0.800 | 2429.6 | 4.24 | 0.44 | 2.25 |
| 40 | 9.730 | 992.2 | 0.653 | 0.658 | 2405.7 | 7.38 | 0.76 | 2.28 |
| 50 | 9.689 | 988.0 | 0.547 | 0.553 | 2381.8 | 12.33 | 1.26 | 2.29 |
| 60 | 9.642 | 983.2 | 0.466 | 0.474 | 2357.6 | 19.92 | 2.03 | 2.28 |
| 70 | 9.589 | 977.8 | 0.404 | 0.413 | 2333.3 | 31.16 | 3.20 | 2.25 |
| 80 | 9.530 | 971.8 | 0.354 | 0.364 | 2308.2 | 47.34 | 4.96 | 2.20 |
| 90 | 9.466 | 965.3 | 0.315 | 0.326 | 2282.6 | 70.10 | 7.18 | 2.14 |
| 100 | 9.399 | 958.4 | 0.282 | 0.294 | 2256.7 | 101.33 | 10.33 | 2.07 |

**TABLE A.3 Conversion Factors**

| | To Convert English Unit | Multiply by | To Obtain Metric (SI) Unit |
|---|---|---|---|
| Area | in.$^2$ | 645.2 | mm$^2$ |
| | ft$^2$ | 0.0929 | m$^2$ |
| | acre | 0.4047 | hectare (ha) = $10^4$ m$^2$ |
| Density | slug/ft$^3$ | 515.4 | kg/m$^3$ |
| Energy (work or quantity of heat) | ft · lb | 1.356 | joule (J) = N · m |
| | ft · lb | $3.77 \times 10^{-7}$ | kwhr |
| | Btu = 778 ft · lb | 1055 | joule (J) = N · m |
| Flow rate | cfs | 0.0283 | m$^3$/sec = $10^3$ liter/sec |
| | mgd = 1.55 cfs | 0.0438 | m$^3$/sec = $10^3$ liter/sec |
| | 1000 gpm = 2.23 cfs | 0.0631 | m$^3$/sec = $10^3$ liter/sec |
| Force | lb | 4.448 | newton (N) |
| Kinematic viscosity | ft$^2$/sec | 0.0929 | m$^2$/sec = $10^4$ St |
| Length | in. | 25.4 | mm |
| | ft | 0.3048 | m |
| | mi | 1.609 | km |
| Mass | slug | 14.59 | kg |
| | lb (mass) | 453.6 | g (mass) |
| Power | ft · lb/sec | 1.356 | W = J/sec = N · m/sec |
| | hp = 550 ft · lb/sec | 745.7 | W |
| Pressure | psi | 6895 | N/m$^2$ = Pa |
| | psf | 47.88 | N/m$^2$ |
| Specific heat | ft · lb/(slug) (°R) | 0.1672 | N · m/(kg) (K) |
| Specific weight | lb/ft$^3$ | 157.1 | N/m$^3$ |
| Velocity | fps | 0.3048 | m/sec |
| | mph | 1.609 | km/hr |
| Viscosity | lb · sec/ft$^2$ | 47.88 | N · sec/m$^2$ = 10 P |
| Volume | ft$^3$ | 0.0283 | m$^3$ |
| | U.S. gallon = 0.1337 ft$^3$ | 3.785 | liter = $10^{-3}$ m$^3$ |
| Weight (see Force) | | | |

**TABLE A.4 Important Quantities**

| | English Unit | SI Unit |
|---|---|---|
| Acceleration of gravity | 32.2 ft/sec$^2$ | 9.81 m/sec$^2$ |
| Density of water (39.4°F, 4°C) | 1.94 slug/ft$^3$ = 1.94 lb · ft$^2$ sec$^{-4}$ | 1000 kg/m$^3$ = g/cm$^3$ or 1.0 Mg/m$^3$ |
| Specific weight of water (50°F, 15°C) | 62.4 lb/ft$^3$ | ~9810 N/m$^3$ or 9.81 kN/m$^3$ |
| Standard sea-level atmosphere | 14.7 psia | 101.32 kN/m$^2$, abs |
| | 29.92 in. Hg | 760 mm Hg |
| | 33.9 ft $H_2O$ | 10.33 m $H_2O$ |
| | | 1013.2 millibars |

# NAME INDEX

# SUBJECT INDEX